MATHEMATICAL
METHODS
IN
THEORETICAL
ECONOMICS

ECONOMIC THEORY AND MATHEMATICAL ECONOMICS

Consulting Editor: Karl Shell

UNIVERSITY OF PENNSYLVANIA
PHILADELPHIA, PENNSYLVANIA

Franklin M. Fisher and Karl Shell. **The Economic Theory of Price Indices:** ***Two Essays on the Effects of Taste, Quality, and Technological Change.***

Luis Eugenio Di Marco (Ed.). **International Economics and Development:** ***Essays in Honor of Raúl Presbisch.***

Erwin Klein. **Mathematical Methods in Theoretical Economics:** ***Topological and Vector Space Foundations of Equilibrium Analysis.***

MATHEMATICAL METHODS IN THEORETICAL ECONOMICS

TOPOLOGICAL AND VECTOR SPACE FOUNDATIONS OF EQUILIBRIUM ANALYSIS

Erwin Klein

Dalhousie University
Halifax, Nova Scotia

ACADEMIC PRESS New York and London

ACADEMIC PRESS, INC.
111 Fifth Avenue, New York, New York 10003

United Kingdom Edition published by
ACADEMIC PRESS, INC. (LONDON) LTD.
24/28 Oval Road, London NW1

LIBRARY OF CONGRESS CATALOG CARD NUMBER: 72-82654

PRINTED IN THE UNITED STATES OF AMERICA

To

Estelle

Verónica

Azar

Maria

CONTENTS

PREFACE

Topological and vector space methods have achieved a very distinguished position among the tools of theoretical economics, especially in the area of equilibrium analysis. This book aims at giving a systematic presentation of the basic principles of these two areas of mathematics. With this aim in mind, we hope the book may prove to be a useful introduction to the field for students of theoretical economics as well as a reference work for economists interested in these problems.

A year of calculus and a semester of elementary linear algebra should provide the reader with the mathematical background needed to understand this book. Some familiarity with modern algebra and differential equations is desirable though not necessary. Finally, it is assumed that the reader has already had some serious exposure to modern mathematical economics at the graduate level.

The book consists of two parts, each of which is essentially self-contained.

Part I introduces the elements of set theory and point set topology. It includes a discussion of the properties of spaces and mappings, and of the continuity of single and multivalued functions; it pays special attention to the question of the existence of fixed points. Summarizing, it presents the essentials of the mathematics of the theory of general economic equilibrium for economies with a finite number of agents and commodities. As an example of the use of the tools, a simplified version of this theory is investigated in Chapter 9.

Given the aim of this book, it has seemed appropriate to substantially limit the level of generality of the discussion. Properties of sets are never discussed for spaces more abstract than the so-called first countable spaces. Indeed, almost all results are confined to metric spaces, with the Euclidean space playing the center role.

Chapter 8, on algebraic structures, bridges the gap between the topological approach of Part I and the algebraic methods of Part II.

Part II deals with vector spaces, and vector-space homomorphisms and matrices. It is very well known that these subjects make up a great deal of the foundations of linear economic theory, in particular of the theory of multisectoral linear models. An economic example is discussed in Chapter 20, this time from the field of the theory of balanced growth.

The reader will soon discover that general equilibrium theory and linear economic theory have been the motivations for the choice of the mathematical subjects discussed. And some readers may wonder about the reasons for the exclusion of certain topics which by now form integral parts of the mathematics of the two areas of economics just mentioned—for instance, measure theory and the theory of games in respect to Part I, and linear programming in respect to Part II. The answer is twofold: On the one hand, the inclusion of such subjects, besides making the book unreasonably voluminous, might have harmed the desired unity of content; on the other hand, as originally planned by the author, those subjects should constitute part of the content of a separate book he hopes to write in the future.

Since this is a book on mathematics for theoretical economic analysis, we have tried to preserve a rather formal approach. With this in mind, the examples which often follow the definitions, theorems, and remarks have been included as aids to the understanding of the subjects discussed or simply as illustrations of the theoretical economic relevance of mathematical concepts, rather than as substitutes for rigorous reasoning.

Toward the end of each chapter, sections of relatively simple exercises and of somewhat more elaborate problems have been included. Most of the exercises are standard in the literature. Several of the problems have economic theoretical content and should serve to direct the reader beyond the content of this book. The Notes which close each chapter relate the discussion to the bibliographic References given at the end of the volume.

ACKNOWLEDGMENTS

Ideas and form of presentation of the material have been borrowed from so many sources, and mathematical style has been influenced by so many authors, that to give an accurate list of acknowledgments is hardly possible. However, some obvious debts are recognized at the end of some chapters in the notes.

I am indebted to Professors Y. Murata, T. Pinfold, U. L. G. Rao, and W. R. S. Sutherland, coparticipants in the Workshop in Mathematical Economics, Econometrics and Optimization at Dalhousie University, for reading sections of the manuscript and for most useful discussions. Of course, they do not bear any responsibility for any remaining mistakes.

I am grateful to Mr. L. MacLean, who read the whole manuscript, corrected my English, and made valuable suggestions to improve the presentation of several subjects; to Dr. T. J. Huang, who prepared many of the examples included in Part II; and, to Mr. V. Voutsinas, who helped me check errors in earlier drafts of the manuscript.

My appreciation is extended to Professor Karl Shell, series editor, for his interest in the publication of this book, and to the staff of Academic Press who made my task a very pleasant experience.

Mrs. S. Haggarty typed the whole manuscript, and Mrs. P. Tomlinson and Mrs. S. Kline retyped some parts of it. For their very difficult task I express my special gratitude.

Finally, I wish to express my thanks for financial assistance to the Research Development Committee in the Social Sciences and Humanities, Dalhousie University.

NOTATION

The meaning of each symbol used in this book is explained where the symbol first appears.

Different elements of the *same set* are usually distinguished by means of superscripts, e.g., x^h, $x^k \in X$, x^0, x^1, $x^2 \in X$ or, in general $x^i \in X$, $i = 1, 2, \ldots$. For simplicity we use, on occasion, the notation x', $x'' \in X$ or, when no confusion is likely to arise, even $x, y, z \in X$.

Elements belonging to *different specified sets* are often denoted as $x \in X$, $y \in Y$, $z \in Z$, or, say, $(x, y) \in X \times Y$. When the set is indexed, we always use a subscript and write $x_i \in X_i$. The latter notation is standard, and we follow it even if X_i is itself a product set or a subset of a product set. The reader should be aware of the context in which x_i is discussed, since we may have, for instance, $X_i \subset R$, $X_i \subset R^n$, $X_i \subset R^{\Sigma n_j}$. When necessary, we emphasize the fact that X_i is a product set and write $x_i = (x_{1,i}, \ldots, x_{n,i})$.

The only exceptions to these notational rules will be found, together with the corresponding explanations, in Chapter 7, Section 7.4.

PART I

POINT SET TOPOLOGY

CHAPTER 1

SETS AND SET OPERATIONS

1.1 Introduction

Set theory lies at the very foundations of mathematics. The concepts developed and studied in the different branches of mathematics can be founded upon and derived from the basic ideas that are the object of set theory. This reason would be sufficient to justify our interest in the subject. However, there is another, perhaps stronger reason for our concern with set theory, namely, the analytical power characteristic of set-theoretical concepts. Set theory has become an invaluable tool to most theoretical disciplines, and in particular to economic theory.

The present chapter introduces the necessary preliminary notions. Chapter 2 will discuss the more restricted notion of set of points and define some metric concepts as a basis for the analysis of the remaining chapters of this part. Some topological ideas are presented in Chapters 4–7.

1.2 Preliminary Notions

The idea of a set is too elementary to be the object of a proper definition. However, we can agree on the sense in which we shall use the word. For us a set will be any well-defined list, collection, or class of objects. Examples of sets are

1. all integers,
2. all the roots of $x^2 + 5x + 6 = 0$,
3. all points of the circle $x^2 + y^2 = 9$,
4. all consumers in an economy.

It is necessary to distinguish clearly a set, in general denoted by a capital letter, for example, X, from its elements, generally denoted by a lowercase letter, for example, x. We can then state whether an element belongs to a certain set, denoted $x \in X$, or does not belong to it, denoted $x \notin X$.

A set can be described by listing its elements (tabular form):

$$X = \{1, 2, 3, 4, 5, 6, 7\},$$
$$Y = \{1, 2, 3, 4, 5, 6, \ldots\},$$

or by pointing out the properties its elements satisfy (set-builder form):

$$X = \{x \mid x \text{ is integer, } 1 \leq x \leq 7\},$$
$$Y = \{y \mid y \text{ is integer, } y \geq 1\}.$$

It is convenient to make a distinction between the set consisting of only one element, and the element considered in its "own right." While it is true that $x \in \{x\}$, we also have that $x \neq \{x\}$.

An important concept is that of the set without any elements called the empty, void, or null set, and denoted $\varnothing$. Note that $\varnothing \neq \{0\}$.

Here are some illustrations of the above notions. Consider the set,

$$X = \{x \mid x^2 - 3x + 2 = 0\}.$$

Obviously

$$1 \in X, \qquad 2 \in X \qquad \text{but} \qquad 0 \notin X, \qquad 3 \notin X.$$

We can then write

$$X = \{1, 2\}.$$

A necessary distinction is that between finite and infinite sets. A precise definition will be given later. For the present we shall say that a set is finite if it contains exactly n different elements, where n is a positive integer; otherwise we shall say that it is infinite.

Examples of finite and infinite sets are, respectively,

$$X = \{x \mid x \text{ is integer}, 1 \leq x \leq 10\}$$

and

$$Y = \{y \mid y \text{ is real}, 1 \leq y \leq 10\}.$$

Equality of Sets

Definition 1.1 Two sets A and B are said to be equal if and only if for every $x \in A$ we have $x \in B$ and for every $x \in B$ we have $x \in A$.

Example Let $A = \{x \mid x^2 - 3x + 2 = 0\}$ and $B = \{1, 2\}$. Then obviously $A = B$.

Subsets

Definition 1.2 Let X, Y be two sets. Then X is said to be a subset of Y, denoted $X \subset Y$, if and only if for every $x \in X$ we have $x \in Y$.

The above definition implies the equivalence of the statements "X is subset of Y" and "X is included in Y." We may also say "Y is a superset of X," that is, "Y includes X."

Example 1 If $A = \{y \mid 0 \leq y \leq 10\}$ and $B = \{y \mid 0 \leq y \leq 8\}$, then $B \subset A$.

Example 2 If $U = \{a, b, c, d, e\}$ and $V = \{a, b, c\}$, then $V \subset U$.

Example 3 If $W = \{1, 2, 3, 4\}$ and $Z = \{1, 2, 3, 4\}$, then $W \subset Z$ but also $Z \subset W$.

Example 3 and Definition 1.2 suggest an alternative way of defining equality of sets.

Definition 1.3 Let A, B be two sets. Then A and B are said to be equal if and only if $A \subset B$ and $B \subset A$.

The latter definition indicates that any set may be regarded as a subset of itself. We shall call such a subset an improper subset, and reserve the name proper subset for the following case.

Definition 1.4 Let X, Y be two sets. Then X is said to be a proper subset of Y if and only if $X \subset Y$ and $X \neq Y$.

In certain circumstances, it is useful to consider all the sets under study as being subsets of a certain fixed set called the universal set or the "universe of discourse." Also we shall agree at this point to regard the empty set $\varnothing$ as a subset of every set.

A set can also have other sets as elements. If $X = \{x^1, x^2\}$, $Y = \{y^1, y^2, y^3\}$ and $Z = \{z\}$, then we can also imagine a certain set $V = \{X, Y, Z\}$. If this is the case, note that the following statements are true:

$$y^1 \in Y, \qquad \{y^1, y^2\} \subset Y, \qquad X \in V, \qquad \{X, Y\} \subset V, \qquad \{X\} \subset V.$$

The following statements are not true:

$$y^1 \in V, \qquad \{y^1, y^2\} \subset V, \qquad X \subset V.$$

A collection of sets is generally referred to as a family of sets. The family of all subsets of any finite set X is usually called the power set of X, and denoted 2^x. Since for any finite X we have $X \subset X$ and $\varnothing \subset X$, the number of elements of 2^x—number of subsets of X—is given by 2^n, where n is the number of elements of X.

Example Let $X = \{1, 2, 3\}$. Then $2^x = \{\{1, 2, 3\}, \{1, 2\}, \{1, 3\}, \{2, 3\}, \{1\}, \{2\}, \{3\}, \varnothing\}$.

Definition 1.5 Two sets X, Y are said to be noncomparable sets if X is not a subset of Y and Y is not a subset of X, that is, if $X \not\subset Y$ and $Y \not\subset X$; otherwise they are said to be comparable.

Definition 1.6 Two sets X, Y are said to be disjoint if they have no common elements.

Example The sets $A = \{x \mid x > 2\}$ and $B = \{x \mid x < 2\}$ are disjoint.

Theorem 1.1 Consider the sets A, B, C such that $A \subset B$ and $B \subset C$. Then $A \subset C$.

Proof Since $A \subset B$, for any $x \in A$ we have $x \in B$. But $B \subset C$ so that for any $x \in B$ we must also have $x \in C$. Thus $A \subset C$.

With the use of Venn diagrams some of the ideas discussed above are illustrated in Fig. 1.1.

1.3 Functions

The following concept will be very useful in the context of the present and next three chapters.

Definition 1.7 A function, mapping or transformation is a rule which assigns to each element of a set X a nonempty subset of a set Y. The function is said to be single valued, denoted $f: X \to Y$, if each assigned subset of Y consists of a single element. Otherwise it is said to be multivalued and denoted $F: X \to Y$.

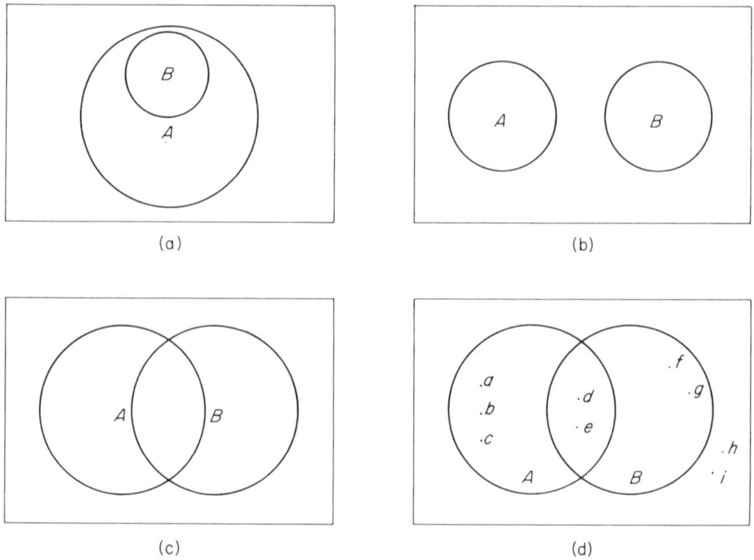

Figure 1.1. (a) Comparable sets. (b) Noncomparable disjoint sets. (c) Noncomparable nondisjoint sets. (d) Sets and elements.

In Chapters 5 and 6 we shall deal, in some detail, with single-valued and multivalued functions, respectively.

For the remainder of this and for the next three chapters, the term function will mean, without exception, a single-valued function.

It is useful to distinguish among the different components of a function $f: X \to Y$. X is called the domain, Y the codomain of the function. The unique element $y \in Y$ assigned to an $x \in X$ is the image of x under f and denoted $f(x)$. The range of the function is the subset of Y which contains all those $y \in Y$ which appear as images of at least one $x \in X$.

Example 1 A rule assigning to each commodity x in a market its unit price p is a function $f: X \to R$, where X is the set of commodities and R the set of real numbers. Since prices are usually positive numbers, the range of f is a subset of R.

Example 2 Consider $f(x) = x^3$. This function assigns to every number its cube. If x belongs to the set of real numbers then the image of $x, f(x)$ is also a real number.

A function $f: X \to Y$ is said to be one to one if (a) every element $y \in Y$ is the image of some element $x \in X$, and (b) every element $y \in Y$ is the image of no more than one $x \in X$.

1.4 Operations with Sets

We now examine the basic operations which may be performed on sets, and some of their properties. A formal definition of the word operation is not necessary at this point and will be given in another place.

Definition 1.8 Let A_i be a subset of a given set S, and $\{A_i\}$ represent a family of subsets of S. Further let I be a set of numbers, usually positive integers. Then a one-to-one function $f: I \to \{A_i\}$ that assigns to each $i \in I$ an $A_i \in \{A_i\}$ is said to define an indexed family of sets. The set I is called the index set. The indexed family of subsets of S is denoted $\{A_i\}_{i \in I}$, "A sub i, i in I."

Union of Sets

Definition 1.9 Let $\{X_i\}_{i \in I}$ be an indexed family of sets, and consider $J \subset I$. The union of X_i over J is the set of all elements which belong to at least one X_i, where $i \in J$, that is,

$$\bigcup_{i \in J} X_i = \{x \mid \text{there is one } i \in J \text{ such that } x \in X_i\}.$$

Example 1 Let us have

$$X_1 = \{a, b\}, \qquad X_2 = \{a, b, c\}, \qquad X_3 = \{d\}, \qquad X_4 = \{h, g, f\}.$$

Then

$$\bigcup_{i=1}^{4} X_i = \{a, b, c, d, f, g, h\}.$$

Example 2 Let

$$X_1 = \{x \mid 0 \le x \le 5\}, \qquad X_2 = \{x \mid 8 \le x \le 10\}.$$

Then

$$X_1 \cup X_2 = \{x \mid 0 \le x \le 5, 8 \le x \le 10\}.$$

Theorem 1.2 Consider the set $\bigcup_{i \in I} X_i$. Then for every $i \in I$, $X_i \subset \bigcup_{i \in I} X_i$.

Proof Call x_i the typical element of X_i. Since $x_i \in X_i$, by definition of union we must have $x_i \in \bigcup_{i \in I} X_i$. But since this is true for every $x_i \in X_i$, we necessarily have $X_i \subset \bigcup_{i \in I} X_i$.

Intersection of Sets

Definition 1.10 Let $\{X_i\}_{i \in I}$ be an indexed family of sets, and consider $J \subset I$. The intersection of X_i over J is the set of all elements which belong to every

X_i, where $i \in J$, that is,

$$\bigcap_{i \in J} X_i = \{x \mid x \in X_i \text{ for every, } i \in J\}.$$

Example 1 Consider

$$X_1 = \{a, b, c\}, \qquad X_2 = \{b, c, d\}, \qquad X_3 = \{c, d, e\}.$$

Then

$$\bigcap_{i=1}^{3} X_i = \{c\}.$$

Example 2 If we have

$$X_1 = \{x \mid 0 \le x \le 10\}, \qquad X_2 = \{x \mid 5 \le x \le 15\}.$$

Then

$$X_1 \cap X_2 = \{x \mid 5 \le x \le 10\}.$$

Note that if any pair of the X_i are disjoint sets, then $\bigcap_{i \in I} X_i = \varnothing$.

Theorem 1.3 Consider the set $\bigcap_{i \in I} X_i$. Then for every $i \in I$, $(\bigcap_{i \in I} X_i) \subset X_i$.

Proof Call x the typical element of $\bigcap_{i \in I} X_i$. Since $x \in \bigcap_{i \in I} X_i$, then by definition we must have $x \in X_i$ for every $i \in I$. But since this is true for every $x \in \bigcap_{i \in I} X_i$ we necessarily have $(\bigcap_{i \in I} X_i) \subset X_i$.

Difference of Sets

Definition 1.11 Let X_1, X_2 be two sets. The difference of X_1 and X_2 is the set of all elements which belong to X_1 but do not belong to X_2, that is,

$$X_1 \setminus X_2 = \{x \mid x \in X_1, x \notin X_2\}.$$

Example Consider

$$X_1 = \{x \mid 0 \le x \le 10\}, \qquad X_2 = \{x \mid 5 \le x \le 15\}.$$

Then

$$X_1 \setminus X_2 = \{x \mid 0 \le x < 5\}, \qquad X_2 \setminus X_1 = \{x \mid 10 < x \le 15\}.$$

Theorem 1.4 Let $X_1 \setminus X_2$ be a nonempty set. Then $(X_1 \setminus X_2) \subset X_1$.

Proof Call x the typical element of $X_1 \setminus X_2$. By definition $x \in X_1$, $x \notin X_2$. Since this is true for every $x \in (X_1 \setminus X_2)$ we necessarily have $(X_1 \setminus X_2) \subset X_1$.

Theorem 1.5 Let X_1, X_2 be two sets, and let $X_1 \setminus X_2$, $X_2 \setminus X_1$, and $X_1 \cap X_2$ be nonempty. Then the sets $X_1 \setminus X_2$, $X_2 \setminus X_1$, and $X_1 \cap X_2$ are disjoint (pairwise).

Proof Call x the typical element of $X_1 \setminus X_2$ and x' the typical element of $X_2 \setminus X_1$. Since by definition $x \in X_1$, $x \notin X_2$ we have $x \notin (X_2 \setminus X_1)$. Similarly, $x' \in X_2, x' \notin X_1$ so that $x' \notin (X_1 \setminus X_2)$. Since this is true for every $x \in (X_1 \setminus X_2)$ and $x' \in (X_2 \setminus X_1)$, we have that $(X_1 \setminus X_2) \cap (X_2 \setminus X_1) = \varnothing$. Now call x'' the typical element of $X_1 \cap X_2$. We must then have $x'' \in X_1$ and $x'' \in X_2$. But then, by the definition of difference, $x'' \notin (X_1 \setminus X_2)$ and $x'' \notin (X_2 \setminus X_1)$. So it results $(X_1 \setminus X_2) \cap (X_1 \cap X_2) = \varnothing$ and $(X_2 \setminus X_1) \cap (X_1 \cap X_2) = \varnothing$.

Complementation of Sets

Definition 1.12 Let X denote any set and let U represent the universal set. Then the complement of X, denoted by X^c, is the set of all elements in U which do not belong to X; that is,

$$X^c = \{x \mid x \in U, x \notin X\}$$

or, if the universe is clearly understood, simply

$$X^c = \{x \mid x \notin X\}.$$

The following four properties of set complements are easy to show and thus left as exercises.

Remark 1.1 Let X be any set, and U represent the universal set. Then $X \cup X^c = U$.

Remark 1.2 Let X be any set. Then $X \cap X^c = \varnothing$.

Remark 1.3 If U represents the universal set, then $U^c = \varnothing$ and $\varnothing^c = U$.

Remark 1.4 Let X be any set. Then $(X^c)^c = X$.

Theorem 1.6 Let X_1, X_2 be two sets. Then $X_1 \setminus X_2 = X_1 \cap X_2{}^c$.

Proof We have

$$\begin{aligned} X_1 \setminus X_2 &= \{x \mid x \in X_1, x \notin X_2\} \\ &= \{x \mid x \in X_1, x \in X_2{}^c\} = X_1 \cap X_2{}^c. \end{aligned}$$

Partition of Sets

Definition 1.13 Let $\{X_i\}_{i \in I}$ be an indexed family of nonempty subsets of a given set S. Then $\{X_i\}_{i \in I}$ is said to form a partition of S if and only if

(1) $\bigcup_{i \in I} X_i = S$;
(2) for any X_i, X_j, either $X_i \cap X_j = \varnothing$ or $X_i = X_j$.

Example Suppose we have the set of integers

$$Z = \{\ldots, -2, -1, 0, 1, 2, \ldots\}.$$

If for the index set $I = \{1, 2, 3\}$ we define $Z_1 = \{\ldots, -2, -1\}$, $Z_2 = \{0\}$, $Z_3 = \{1, 2, \ldots\}$, it is clear that $\{Z_i\}_{i \in I} = \{Z_1, Z_2, Z_3\}$ forms a partition of Z.

Operations on Comparable Sets

A special case occurs when the sets on which the discussed operations are performed are comparable. The following theorem relates to some properties of operations on comparable sets.

Theorem 1.7 Let A, B be two sets. Then $A \subset B$ implies that (a) $A \cap B = A$; (b) $A \cup B = B$; (c) $A \cup (B \setminus A) = B$; (d) $B^c \subset A^c$.

Proof (a) Consider any $x \in (A \cap B)$. Then we have $x \in A$ and $x \in B$. Assume A is a proper subset of B. Then for every $x \in A$ we have $x \in B$, but the converse is not true. Consequently $A \cap B = A$.

(b) Consider any $x \in (A \cup B)$. Then we have $x \in A$, or $x \in B$, or both. Assume now A is a proper subset of B. Note that for any element of $A \cup B$ we can only have that either $x \in A$ and $x \in B$ or $x \notin A$ and $x \in B$. Thus $A \cup B = B$.

(c) Consider $x \in (B \setminus A)$. Then we have $x \in B$ and $x \notin A$. It is obvious then that from the union, $\{x | x \in B, x \notin A\} \cup \{x | x \in A\}$ together with the condition $A \subset B$ we obtain $A \cup (B \setminus A) = B$.

(d) Left as an exercise.

1.5 Algebra of Sets

Operations with sets and subsets satisfy determinate properties known as laws of set algebra. A few of them have been studied already. Nevertheless, it will be convenient to have all of these laws grouped together. We list the properties first. Some proofs are given thereafter.

Idempotent Laws

$$X \cup X = X, \qquad X \cap X = X.$$

Commutative Laws

$$\bigcup_{i=1}^{n} X_i = \bigcup_{i=n}^{1} X_i, \qquad \bigcap_{i=1}^{n} X_i = \bigcap_{i=n}^{1} X_i.$$

Associative Laws

$$\left(\bigcup_{i=1}^{n-1} X_i\right) \cup X_n = X_1 \cup \left(\bigcup_{i=2}^{n} X_i\right), \qquad \left(\bigcap_{i=1}^{n-1} X_i\right) \cap X_n = X_1 \cap \left(\bigcap_{i=2}^{n} X_i\right).$$

Distributive Laws

$$Y \cup \left(\bigcap_{i=1}^{n} X_i \right) = (Y \cup X_1) \cap (Y \cup X_2) \cap \cdots \cap (Y \cup X_n),$$

$$Y \cap \left(\bigcup_{i=1}^{n} X_i \right) = (Y \cap X_1) \cup (Y \cap X_2) \cup \cdots \cup (Y \cap X_n).$$

Identity Laws

$$X \cup \varnothing = X, \qquad X \cup U = U, \qquad X \cap U = X, \qquad X \cap \varnothing = \varnothing.$$

Complement Laws

$$X \cup X^c = U, \qquad (X^c)^c = X, \qquad X \cap X^c = \varnothing, \qquad U^c = \varnothing, \qquad \varnothing^c = U.$$

Dualization or DeMorgan's Laws

The following property is of considerable importance and we state it in the form of a theorem.

Theorem 1.8 Let $\{X_i\}_{i \in I}$ be an indexed family of subsets of a set S. Then

(a) $(\bigcup_{i \in I} X_i)^c = \bigcap_{i \in I} X_i^c$,
(b) $(\bigcap_{i \in I} X_i)^c = \bigcup_{i \in I} X_i^c$.

Proof We prove part (a) of the theorem leaving part (b) as an exercise. Consider any $x \in (\bigcup_{i \in I} X_i)^c$; then $x \notin (\bigcup_{i \in I} X_i)$. This means that for each index $j \in I$, $x \notin X_j$. Then, for each index $j \in I$, $x \in X_j^c$, and so $x \in (\bigcap_{i \in I} X_i^c)$. Therefore

(1) $(\bigcup_{i \in I} X_i)^c \subset \bigcap_{i \in I} X_i^c$.

Conversely, take any $x \in (\bigcap_{i \in I} X_i^c)$, then, $x \in X_j^c$ for each index $j \in I$. This means that for each index $j \in I$, $x \notin X_j$ and as a result $x \notin (\bigcup_{i \in I} X_i)$. Consequently $x \in (\bigcup_{i \in I} X_i)^c$ and therefore

(2) $(\bigcap_{i \in I} X_i^c) \subset (\bigcup_{i \in I} X_i)^c$.

It is easy to see that expressions (1) and (2) together imply statement (a) of our theorem.

1.6 Duality Principle

Suppose we have a certain proposition involving a pair of sets X, Y and operators (the symbols which specify the sort of operation) $*$, $\odot$. If the given proposition is valid, then it might be possible to derive a second valid proposition from it by interchanging the two sets and the two operators as well.

When this is possible, we usually refer to the original proposition as the primal and to the second as its dual. Set-theoretical propositions allow, in general, for duality because of the symmetric nature of set-theoretical postulates.

In later chapters we shall meet the notion of duality in some different contexts. So far as set theory is concerned, we ought to point out at this stage the relevance that the duality principle has for the job of proving theorems. Assume we have proved a theorem that is the consequence of certain axioms. If the axioms have meaningful duals, then it would be possible to prove a second theorem, the dual of the original one, by only using the dual of each step in the first proof.

The following two theorems may serve to illustrate the use of the duality principle. The proposition of Theorem 1.9 is proved by means of the laws of set algebra. The proposition of Theorem 1.10 is just its dual, and is proved by using the dual of each step in the first theorem.

Theorem 1.9 Let X, Y be two sets. Then $(X \cap Y) \cup (X \cap Y^c) = X$.

Proof Consider the left-hand side of $(X \cap Y) \cup (X \cap Y^c) = X$. By the distributive law

(1) $(X \cap Y) \cup (X \cap Y^c) = X \cap (Y \cup Y^c)$.

By the complement law

(2) $Y \cup Y^c = U$ (the universal set).

We then have

(3) $(X \cap Y) \cup (X \cap Y^c) = X \cap U = X$,

and the theorem is proved.

Theorem 1.10 Let X, Y be two sets. Then $(X \cup Y) \cap (X \cup Y^c) = X$.

Proof Note that $(X \cup Y) \cap (X \cup Y^c) = X$ is simply the dual of $(X \cap Y) \cup (X \cap Y^c) = X$, proved in Theorem 1.9. We use the duality principle. The dual of (1) is

(1′) $(X \cup Y) \cap (X \cup Y^c) = X \cup (Y \cap Y^c)$.

Now the dual of (2) is

(2′) $Y \cap Y^c = \varnothing$.

Finally, the dual of (3) is

(3′) $(X \cup Y) \cap (X \cup Y^c) = X \cup \varnothing = X$,

and the theorem is proved.

EXERCISES

E.1.1 Suppose X is any finite set and determine whether each of the following propositions is true: (a) $\varnothing \subset X$; (b) $\varnothing \in X$; (c) $X \subset 2^x$; (d) $X \in 2^x$.

E.1.2 Show that if the number of elements of a set X is n, then the number of elements of its power set 2^x is 2^n.

E.1.3 Let X_1, X_2, X_3, and X_4 be sets such that $X_1 \subset X_2$, $X_2 \subset X_3$, $X_3 \subset X_4$, and $X_4 \subset X_1$. Show that $X_1 = X_2 = X_3 = X_4$.

E.1.4 Let $\{X_i\}_{i \in I}$ be an indexed family of sets and $J \subset I$. Show that

$$\bigcup_{i \in J} X_i \subset \bigcup_{i \in I} X_i \qquad \text{and} \qquad \bigcap_{i \in I} X_i \subset \bigcap_{i \in J} X_i.$$

Hint: Consider Theorems 1.2 and 1.3.

E.1.5 Show that the set $X = \{1, 2, 3, 4\}$ has 15 possible different partitions.

E.1.6 Show that (a) $(X^c)^c = X$; (b) $X \cup X^c = U$; (c) $X \cap X^c = \varnothing$.

E.1.7 Prove proposition (d) of Theorem 1.7.

E.1.8 Prove proposition (b) of Theorem 1.8.

E.1.9 Let X, Y, Z be any three sets. Prove the validity of the following identities:

(a) $X \cup (Y \cap Z) = (X \cup Y) \cap (X \cup Z)$,
(b) $X \cap (Y \cup Z) = (X \cap Y) \cup (X \cap Z)$,
(c) $(X \cap Y) \cup (X \cap Z) \cup (Y \cap Z) = (X \cup Y) \cap (X \cup Z) \cap (Y \cup Z)$.

E.1.10 Let X, Y, Z be any three sets. Prove the validity of the following identities:

(a) $(X \cap Y) \setminus Z = (X \setminus Z) \cap (Y \setminus Z)$,
(b) $(X \setminus Y) \cap Z = (X \cap Z) \setminus (Y \cap Z)$,
(c) $(X \setminus Y)^c = X^c \cup Y$.

PROBLEMS

P.1.1 Let p and q represent two propositions. If U is the set of all logical possibilities, call P and Q the subsets of U for which the propositions p and q are true, respectively. Use the symbols $\vee$, $\wedge$, and $\sim$ to express the connectives "or," "and," and "not," respectively, and show that the compound propositions below have the indicated sets of possibilities: (a) $p \vee q$: $P \cup Q$; (b) $p \wedge q$: $P \cap Q$; (c) $\sim p$: P^c.

Hint: Kemeny *et al.* [1, Chapters 1, 2].

P.1.2 As a generalization of the results of the preceding problem show that: (a) for each proposition there is a "truth set"; (b) for each connective there is a set operation; (c) connectives satisfy the laws of set algebra; (d) for each relation between propositions there is a relation between sets.

Hint: See Hint to P.1.1.

P.1.3 Russell's paradox. Suppose A is the family of all sets which do not contain themselves as elements, that is, $A = \{X \mid X \notin X\}$. Does $A \in A$?

P.1.4 The input–output table in Fig. 1.2 represents an economy of n sectors each of which produces one and only one commodity. The coefficient a_{ij} stands for the amount of the ith commodity used per unit of output in the jth sector.

Figure 1.2

Commodities \ Activity sectors				
	a_{11}	a_{12}	$\cdots$	a_{1n}
	a_{21}	a_{22}	$\cdots$	a_{2n}
	$\vdots$	$\vdots$	$\vdots$	$\vdots$
	a_{n1}	a_{n2}	$\cdots$	a_{nn}

Since $K = \{1, 2, \ldots, n\}$ is the set of indices, a proper subset $S \subset K$ represents the subeconomy of all those a_{ij} such that $i, j \in S$. A subeconomy S is said to be independent if it is possible to produce the goods in S without using any good not in S. Note that if the ith commodity is not used in the jth sector, $a_{ij} = 0$.

Use the partition $\{S_1, S_2\}$ of K and state formally the conditions for (a) S_1 to be an independent subeconomy; (b) S_1 and S_2 to be both independent subeconomies.

Hint: See Chapter 19, Definition 19.1.

P.1.5 As a generalization of the preceding problem, suppose the economy produces m commodities by means of n activities. One sector may produce more than one commodity, and the same commodity may be produced by more than one sector. We have now the index sets $K = \{1, 2, \ldots, m\}$ and $H = \{1, 2, \ldots, n\}$. A subeconomy is now a set of a_{ij} such that $i \in S \subset K$, and $j \in T \subset H$. Consider the partitions $\{S_1, S_2\}$ of K and $\{T_1, T_2\}$ of H, and state formally the conditions for: (a) (S_1, T_1) to be an independent subeconomy; (b) (S_1, T_1) and (S_2, T_2) to be both independent subeconomies.

Hint: Observe that the input–output table now has m rows and n columns; see Definition 20.4, Chapter 20.

NOTES

For an excellent introduction to set theory see Halmos [1]. At a more advanced level see Fraenkel [1] and Hausdorff [1] (see also Suppes [1]). Both Kuratowski [2] and Mendelson [1] give detailed treatments of sets as introductions to general topology. The relation between set theory and logic is discussed by Kemeny *et al.* [1]. For applications of set theory to economics Koopmans [1] gives illuminating examples.

CHAPTER 2

PRODUCTS OF SETS, METRIC SPACES, AND RELATIONS

We shall deal in this chapter with the two quite different concepts of metric space and relations. Since the idea of a product of sets underlies both concepts, it seems adequate to introduce it first.

2.1 Products of Sets

Given a set X, a sequence (x^i) of its elements is an intuitive concept. It suggests the idea of a collection of elements following a certain order.

If what we are given is a list of indexed sets $X_1, \ldots, X_n, \ldots$, we may very well extend our idea of sequence to that of a sequence of sets.

In the following pages we shall attempt to formalize these intuitive ideas and let them serve as the basis for our further investigations.

Sequences

Definition 2.1 Let N represent the set of positive integers and let X be any set. A function from N to X which assigns to any positive integer $i \in N$ an element $x^i \in X$ is called a sequence and denoted

$$(x^i) = (x^1, x^2, \ldots, x^n, \ldots).$$

A sequence may consist of a finite or infinite number of elements, and be called a finite or infinite sequence, respectively

A finite sequence which consists of exactly n elements, where n is a positive integer, is said to form an "ordered n-tuple." It ought to be clear that the order in which the elements of the sequence appear is essential. Thus, two "ordered n-tuples," $(x^1, x^2, \ldots, x^n)$ and $(y^1, y^2, \ldots, y^n)$ are said to be equal, denoted $(x^1, x^2, \ldots, x^n) = (y^1, y^2, \ldots, y^n)$, if and only if $x^1 = y^1$; $x^2 = y^2$; $\ldots$; $x^n = y^n$.

An "ordered pair" (x^1, x^2) is a finite sequence of two elements in which one x^1 has been designated as the first element, and the other x^2 has been designated as the second element.

The terms "finite sequence" and "ordered n-tuple" are thus synonymous. Consider now elements x_i such that $x_i \in X_i$, $i = 1, \ldots, n$.

Direct Products of Sets

Definition 2.2 Let $X_1, X_2, \ldots, X_n$ be a finite sequence of sets indexed by $I = \{1, 2, \ldots, n\}$. The direct product of $X_1, X_2, \ldots, X_n$ denoted $\prod_{i=1}^{i=n} X_i$ is the set of all finite sequences (ordered n-tuples) $(x_1, x_2, \ldots, x_n)$ such that $x_1 \in X_1$, $x_2 \in X_2, \ldots, x_n \in X_n$.

Note that in the preceding definition we do not simply refer to a set of sets but rather to a sequence of sets. The distinction is not trivial. The order of the sets in the product is essential to the concept. In general, for two sets X_1 and X_2, $X_1 \times X_2 \neq X_2 \times X_1$. Equality of the two direct products would require $X_1 = X_2$.

Example 1 Let us consider the sets of outcomes of the throw of three coins that we shall call X, Y, Z, in that order. Call H and T the possible outcomes with each coin. The set of possibilities is then the direct product $X \times Y \times Z$. In other words, the set,

$$\{(H, H, H), (H, H, T), (H, T, H), (T, H, H), (H, T, T), (T, H, T), (T, T, H), (T, T, T)\}.$$

Example 2 Let R be the set of real numbers, and let $X = Y = R$. The Cartesian or direct product $X \times Y$ is the set of all "ordered pairs" (x, y). It is the well-known coordinate plane of analytical geometry.

It is easy now to generalize our previous definition to the case of an infinite sequence of sets $X_1, X_2, \ldots, X_n, \ldots$.

Definition 2.3 Let $X_1, X_2, \ldots, X_n, \ldots$ be an infinite sequence of sets indexed by the positive integers. The direct product of $X_1, X_2, \ldots, X_n, \ldots$, denoted $\prod_{i=1}^{i=\infty} X_i$, is the set of all infinite sequences $(x_1, x_2, \ldots, x_n, \ldots)$ such that $x_i \in X_i$ for each positive integer.

Notationally we can describe the set of Definition 2.2 as $\prod_{i=1}^{n} X_i = \{(x_1, \ldots, x_n) | x_i \in X_i \text{ for } i = 1, \ldots, n\}$ and that of Definition 2.3 as $\prod_{i=1}^{\infty} X_i = \{(x_1, \ldots, x_n, \ldots) | x_i \in X_i \text{ for } i = 1, \ldots, n, \ldots, \infty\}$.

2.2 Sets of Points

From now on we shall focus our attention mainly on sets whose elements are numbers (particularly real numbers) or sequences of numbers (particularly of real numbers).

Operations

Definition 2.4 Let X be a set. An n-ary operation $*$ on X is a function from the product set $X \times X \times \cdots \times X$ (n times) to the set X. In particular, a binary operation $*$ on X is a function from $X \times X$ to X.

Notice that the operation assigns to each n-tuple $(x^1, \ldots, x^n) \in X^n$ a unique $x \in X$. In particular, a binary operation assigns to each pair $(x^1, x^2) \in X^2$ a unique $x \in X$. In both cases the range of the function is a subset of X.

Binary operations are of great utility to our further considerations, and it seems convenient to examine them in closer detail.

Definition 2.5 Let the set X be a subset of a set Y. X is said to be closed under a binary operation $*$ defined on Y, if $x^i * x^j = x^k \in X$ for every $x^i, x^j \in X$. Conversely, a binary operation is said to be closed in a set X if its domain is the full set $X \times X$.

Example Consider the sets

$$Y = \{x | x \geq -10\} \qquad \text{and} \qquad X = \{x | x \geq 0\},$$

so that $X \subset Y$. Define the operations $+$ and $-$ on Y. It is immediate that (a) X is closed under $+$, since for any $x^i, x^j \in X$ also $(x^i + x^j) \in X$; (b) X is not closed under $-$, since there are some $x^i, x^j \in X$ for which $(x^i - x^j) \notin X$.

Properties of Operations

We shall consider now the basic algebraic properties binary operations may satisfy. This allows us to classify operations accordingly.

Commutative Operations

Definition 2.6 A binary operation $*$ on X is said to be commutative if for every $x^i, x^j \in X$ we have $x^i * x^j = x^j * x^i$.

Example Addition of real numbers.

Associative Operations

Definition 2.7 A binary operation $*$ on X is said to be associative if for every $x^i, x^j, x^h \in X$ we have $x^i * (x^j * x^h) = (x^i * x^j) * x^h$.

Example Multiplication of real numbers.

Distributive Operations

Definition 2.8 A binary operation $*$ on X is said to be distributive over a binary operation $\odot$ on X if for every $x^i, x^j, x^h \in X$ we have $x^i * (x^j \odot x^h) = (x^i * x^j) \odot (x^i * x^h)$.

Example Multiplication of real numbers is distributive over addition of real numbers.

Dual Operations

Definition 2.9 A binary operation $*$ on X is said to be the dual of a binary operation $\odot$ on X if $*$ is distributive over $\odot$ and $\odot$ is distributive over $*$, that is, if for every $x^i, x^j, x^h \in X$

$$x^i * (x^j \odot x^h) = (x^i * x^j) \odot (x^i * x^h)$$

and also

$$x^i \odot (x^j * x^h) = (x^i \odot x^j) * (x^i \odot x^h).$$

Example Union and intersection of sets are dual operations, one being the dual of the other.

Identity and Inverse Elements

There are certain elements in a set with an operation whose properties are of special interest. Although this question will really concern us only in a latter chapter, it seems convenient to introduce the idea at this time.

Definition 2.10 Let X be a set with a binary operation $*$. An element $\bar{x} \in X$ is said to be an identity for $*$ on X if and only if for every $x \in X$, $x * \bar{x} = \bar{x} * x = x$.

Example 0 is an identity element for addition on the set of integers, while 1 is an identity element for multiplication on the set of real numbers.

Definition 2.11 Let X be a set with a binary operation $*$ with the identity element $\bar{x}$. An element $x' \in X$ is said to be an inverse of $x \in X$ if $x * x' = x' * x = \bar{x}$.

Example x is the inverse for addition of $-x$ in the set of integers, while x is the inverse for multiplication of $1/x$ in the set of real numbers.

Sets of Numbers

The concepts of operation and of a set closed under an operation permit us to put some order in the set of numbers. If we take the set of complex numbers as our universe of discourse, it is obvious that we can easily distinguish first between the set of numbers having nonzero imaginary part and the set of numbers having zero imaginary part. The latter is the set of real numbers.

But let us simplify matters and start with the set of real numbers, which we shall consider for our present purposes the universal set. Let R be the set of real numbers. The following subsets of R may be mentioned:

(a) The set of rational numbers, defined by $Q = \{x \mid x \in R;\ x = p/q$ where $p, q \in Z, q \neq 0\}$, where Z represents the set of integers. Q is closed under the operations $+$, $\times$, $-$, and $\cdot/\cdot$ (except by 0).

(b) The set of irrational numbers, which is the complement of Q in R and may be denoted Q^c.

(c) The set of integers $Z = \{\ldots, -2, -1, 0, 1, 2, \ldots\}$, which is a subset of the set of rational numbers and is closed under the operations $+$, $\times$, $-$.

(d) The set of natural numbers (positive integers N) which is a subset of Z closed under the operations $+$ and $\times$.

We intuitively know that the set of real numbers R is completely ordered by the expression $\leq$ ("less than or equal to"). Leaving the formal definition of what "totally or completely ordered" means for a later chapter, we should be able to agree at this point that for $x, y, z \in R$ the following statements are true:

(a) either $x < y$, $x = y$ or $y < x$;
(b) if $x < y$ and $y < z$, then $x < z$;
(c) if $x < y$, then $x + z < y + z$;
(d) if $x < y$ and $0 < z$ (z is positive), then $x \cdot z < y \cdot z$
(e) if $x < y$ and $z < 0$ (z is negative), then $y \cdot z < x \cdot z$.

In the following pages we shall use the notations $x \leq y$ or $y \geq x$ and $x < y$ or $y > x$.

Definition 2.12 Consider $x \in R$. The absolute value of x, denoted $|x|$, is defined by $|x| = x$ if $x \geq 0$, and $|x| = -x$ if $x < 0$.

Example Let $x = 20$; since $x > 0$, we have $|20| = 20$. Let $x = -20$; since $x < 0$, we have $|-20| = 20$.

Intervals

Definition 2.13 Let there be $x \in R$. Sets of the form $\{x \mid a \leq x \leq b\}$, $\{x \mid a < x < b\}$, $\{x \mid a \leq x < b\}$, $\{x \mid a < x \leq b\}$, $\{x \mid a \leq x\}$, and $\{x \mid x \leq b\}$ are called intervals and, respectively, denoted $[a, b]$, $]a, b[$, $[a, b[$, $]a, b]$, $[a, \rightarrow[$, and $]\leftarrow, b]$. The numbers a and b are called the lower and the upper bounds, respectively, of the intervals. A set of the form $[a, b]$ is called a closed interval.

A set $]a, b[$ is said to be an open interval. Intervals of a "mixed" sort are possible, as follows from Definition 2.13. Note that an interval is closed if it contains its bounds.

Properties of Intervals

Let R be the set of real numbers. Let V be the family of all intervals of the set R, including the void set $\varnothing$ and, for any $a \in R$, the interval $[a, a]$. Finally, consider any two intervals $A, B \in V$. It is easy to prove that the following properties are valid:

Remark 2.1 $A \cap B \in V$.

Remark 2.2 If $A \cap B \neq \varnothing$, then $A \cup B \in V$.

Remark 2.3 If $A \not\subset B$ and $B \not\subset A$ then $(A \setminus B) \in V$.

Example Consider the sets $A = \{x \mid 0 \leq x < 10\}$ and $B = \{x \mid 5 < x \leq 15\}$.

Note that

$$\begin{aligned} A \cap B &= \{x \mid 5 < x < 10\} =]5, 10[, \\ A \cup B &= \{x \mid 0 \leq x \leq 15\} = [0, 15], \\ A \setminus B &= \{x \mid 0 \leq x \leq 5\} = [0, 5]. \end{aligned}$$

Sets of Points

In Definition 2.2 we defined the product set $\prod_{i=1}^{n}{}' X_i$ as the set of n-tuples $(x_i) = (x_1, \ldots, x_n)$. Note that the sequence (x_i) is not considered as a set itself but rather as an element of a set.

If $X_i = C$, the set of complex numbers, the set of sequences (x_i) may be written C^n. Analogously, if $X_i = R$, the set of real numbers, the set of all n-tuples (x_i) may be denoted R^n.

Sets of the type C^n or R^n are generally referred to as sets of points. The elements $(x_i) = (x_1, \ldots, x_n)$, wherein the order of the components is essential, are called points. Sometimes it is necessary to require from C^n, R^n, and the elements (x_i) some additional mathematical properties. In one such special case the sets of points are called vector spaces and the points vectors. We shall study the theory of vector spaces in Part II.

Given a set of points R^n, we may regard $R^{n-1}, \ldots, R^2, R$ as subsets of it. They are, as we shall see in another place, a special kind of subset of R^n.

The important fact, anyway, is that if we regard R^n as the set of points with typical element $(x_1, \ldots, x_n)$, we may regard R as a special case, namely as a set of points with typical element x. This standpoint serves to give a quite general character to our theory.

In the remainder of this chapter and the forthcoming chapters, unless stated otherwise, sets are meant to be sets of points. We shall be particularly concerned with the set R^n, from now on our universal set, and subsets X, Y, Z with typical elements $(x_1, \ldots, x_n)$, $(y_1, \ldots, y_n)$, $(z_1, \ldots, z_n)$, respectively.

To simplify notations we shall write x instead of (x_i) to refer to the elements of X. The ith component of x will be denoted x_i; x^j, x^h, x^k will denote different points in X.

In particular, if the set under consideration is R or any subsets X, Y, Z, we shall assume that the typical elements are, respectively, the one-component points or scalars x, y, and z.

Example 1 Let $X = \{x \mid x \in R^5, |x_i| < 4 \text{ for } i = 1, 2, \ldots, 5\}$ (read "the set of all x such that x is in R^5, where the ith component of x is in absolute value less than 4, for $i = 1, 2, \ldots, 5$"). Note that the following points are in X:

$$(0, 0, 0, 0, 0), \qquad (1, 2, 3, 3, 0), \qquad (-1, 3, 2, -2, 1),$$

but the following do not belong to X:

$$(4, 0, 0, 0, 0), \qquad (-6, 2, 10, 2), \qquad (-4, 0, 0, 0, 4).$$

As defined below, Y and Z are not subsets of X

$$Y = \{x \mid x \in R^5,\ x_i > 4\}, \qquad Z = \{x \mid x \in R^7,\ |x_i| < 4\},$$

but the following set is a subset of X:

$$S = \{x \mid x \in R^5,\ |x_i| < 2\}.$$

Example 2 The set $\{x \mid x \in R,\ 1/x \leq 10\}$ is a special kind of set of points. The elements x are scalars.

Example 3 The set $\{(x, y) \mid x^2 + y^2 = 9\}$ is a set of points in the coordinate plane.

Example 4 In economic theory the name consumer is given to any agent (for example, individual) whose function is to choose and carry out a "consumption plan". A "consumption plan" is a list of all quantities of inputs (food, clothes, etc.) flowing to the consumer and outputs (labor, etc.) flowing from the consumer. Generally inputs are denoted by positive and outputs by negative numbers. A "consumption plan" is then a point $(x_1, \ldots, x_k)$ where k is the number of commodities considered. If no additional restrictions are put the consumption set $X \subset R^k$ of a consumer is the set of all "consumption plans" $x \in R^k$.

2.3 Metric Spaces

Sets of points, like those studied in Section 2.2 are frequently called spaces. In particular, the word space is used when the emphasis is put on some geometrical properties of the set under consideration. In such cases, the word space is usually applied to the universal set underlying the study.

A good example of a space, in the sense given above, is the n-dimensional space R^n. It is, of course, the same concept as that of the set of points R^n.

A more complex idea is that of a metric space. In essence, a metric space is a set of points together with a well-defined quantitative measure giving the degree of closeness of pairs of points in the set.

Metric

Definition 2.14 Let X be a nonempty set of points. A real-valued function d: $X \times X \to R$ is called a metric or distance function if and only if it satisfies, for every $x, y, z \in X$, the following axioms:

(a) $d(x, y) \geq 0$.
(b) $d(x, y) = 0$ if and only if $x = y$.
(c) $d(x, y) = d(y, x)$.
(d) $d(x, z) \leq d(x, y) + d(y, z)$.

The real number $d(x, y)$ is called the distance between the points $x, y \in X$.

It should be clear that different types of real-valued functions may satisfy the requirements of the above definition. In consequence, different kinds of distance functions may be constructed.

For instance let $a, b \in R$, the set of real numbers. A possible metric satisfying Definition 2.14 is then

$$d(a, b) = |a - b|.$$

Another well-known metric is the so called "Euclidean distance function" defined as, for every $x, y \in X = R^n$,

$$d(x, y) = \left[\sum_{i=1}^{n} (x_i - y_i)^2\right]^{1/2}.$$

A formal definition of a metric space is now straightforward.

Definition 2.15 A nonempty set of points X together with a distance function $d\colon X \times X \to R$ is called a metric space and denoted (X, d).

Example 1 The set R and the metric $d(a, b) = |a - b|$, for every $a, b \in R$ define the one-dimensional metric space (R, d).

Example 2 The set $X = R^n$ and the Euclidean distance function,

$$d(x, y) = \left[\sum_{i=1}^{n} (x_i - y_i)^2\right]^{1/2} \qquad \text{for every } x, y \in X,$$

determine the Euclidean space $(X, d) = E^n$.

In the next chapter we refer to the connections that exist between metric spaces and topological spaces. In Part II the theory of vector spaces is discussed. For the remainder of this chapter we shall consider a quite different subject.

2.4 Binary Relations

It is customary to define a relation by the capital letter R. However, to avoid any possible notational confusion with the set of real numbers R, we have decided to denote a relation by means of the symbol $\tilde{R}$.

Definition 2.16 Let X, Y be two sets. A binary relation $\tilde{R}$ from X to Y is a subset of the direct product $X \times Y$. If $(x, y) \in \tilde{R}$, where $x \in X$ and $y \in Y$, we say that x is related to y and write $x \tilde{R} y$; otherwise, we say that x is not related to y and denote $x \not\tilde{R} y$.

A Logical Concept of Relation

From a logical point of view a binary relation consists of two sets, X and Y, and a propositional function defined on $X \times Y$, that is, a statement which can be true or false for any ordered pair $(x, y) \in X \times Y$.

The set of pairs for which the propositional function is true is usually called the solution set of the relation.

Example Consider the sets $X = \{2, 3, 4\}$ and $Y = \{3, 4, 5, 6\}$, and let us have the propositional function "x exactly divides y." The product set is then

$$X \times Y = \{(2, 3), (2, 4), (2, 5), (2, 6), (3, 3), (3, 4), (3, 5), (3, 6), (4, 3), (4, 4), (4, 5), (4, 6)\}.$$

The set of pairs that satisfy the propositional function, that is, our solution set, is obviously

$$\tilde{R} = \{(2, 4), (2, 6), (3, 3), (3, 6), (4, 4)\}.$$

In purely mathematical terms, the different specifications involved in the logical description are not necessary. They may serve to illustrate what is usually meant by the term relation. However, in a mathematical sense, a relation is not more and not less than a subset of a product set. This can be seen very clearly from the foregoing example. And this is precisely what has been pointed out by Definition 2.16.

Domain and Range

Definition 2.17 Let $\tilde{R}$ be a relation from X to Y. The domain of $\tilde{R}$ is the set of all first elements of $(x, y) \in \tilde{R}$, that is the set $D = \{x \mid x \in X, (x, y) \in \tilde{R}\}$. The range of $\tilde{R}$ is the set of all second elements of $(x, y) \in \tilde{R}$, that is the set $E = \{y \mid y \in Y, (x, y) \in \tilde{R}\}$. It is obvious that $D \subset X$ and $E \subset Y$.

Example For our previous example, since $\tilde{R} = \{(2, 4), (2, 6), (3, 3), (3, 6), (4, 4)\}$ it is easy to see that the domain is $D = \{2, 3, 4\}$ and the range is $E = \{3, 4, 6\}$.

Graph of a Relation

Definition 2.18 Let $\tilde{R}$ be a relation from X to Y. The graph of $\tilde{R}$ consists of all those ordered pairs (x, y) of the coordinate plane $X \times Y$ which belong to $\tilde{R}$.

Example Figure 2.1 corresponds to the relation $\tilde{R}$ of our previous examples.

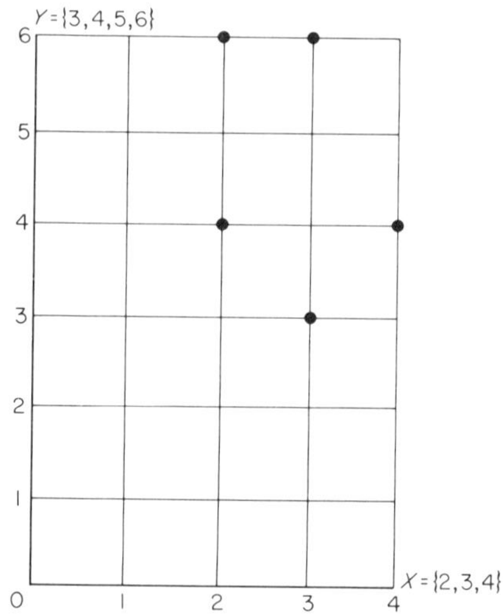

Figure 2.1

$X = \{2, 3, 4\}$
$Y = \{3, 4, 5, 6\}$
"x exactly divides y"
$\tilde{R} = \{(2, 4), (2, 6), (3, 3), (3, 6), (4, 4)\}$
$D = \{2, 3, 4\}$
$E = \{3, 4, 6\}$

2.5 Classes of Relations

Let us examine now some special types of relations.

Inverse Relation

Definition 2.19 Let $\tilde{R}$ be any relation from X to Y. The inverse of $\tilde{R}$, denoted $\tilde{R}^{-1}$, is the relation from Y to X whose elements are those ordered pairs $(y, x) \in Y \times X$ which, when reversed, yield the ordered pairs $(x, y) \in \tilde{R}$. So, for any relation $\tilde{R}$ we have $\tilde{R}^{-1} = \{(y, x) \mid (x, y) \in \tilde{R}\}$.

Example Let $X = \{1, 2, 3\}$, $Y = \{2, 3, 4\}$. Consider the proposition "$x \geq y$." Then

$$X \times Y = \{(1, 2), (1, 3), (1, 4), (2, 2), (2, 3), (2, 4), (3, 2), (3, 3), (3, 4)\}$$

and

$$\tilde{R} = \{(2, 2), (3, 2), (3, 3)\}.$$

Since

$$Y \times X = \{(2, 1), (2, 2), (2, 3), (3, 1), (3, 2), (3, 3), (4, 1), (4, 2), (4, 3)\},$$
$$\tilde{R}^{-1} = \{(2, 2), (2, 3), (3, 3)\}.$$

Reflexive Relation

Definition 2.20 Let $\tilde{R}$ be a relation on X, that is, from the set X to itself. $\tilde{R}$ is said to be reflexive if for every $x \in X$, $(x, x) \in \tilde{R}$.

Example Let X be the consumption set [set of all commodity bundles, "consumption plans" or simply points (x_i)] of a consumer of an economy. Consider now the relation on X defined by the proposition $x^k \precsim x^h$ for x^k, $x^h \in X$, where $\precsim$ means "at most as desired as." That the relation is reflexive is obvious from the fact that, for $k = h$, we have $(x^k, x^k) \in \tilde{R}$ or $(x^h, x^h) \in \tilde{R}$.

Symmetric Relation

Definition 2.21 Let $\tilde{R}$ be a relation on X. $\tilde{R}$ is said to be symmetric if $(x, y) \in \tilde{R}$ implies $(y, x) \in \tilde{R}$, for $x, y \in X$.

Example 1 Consider the coordinate plane and the proposition "triangle x is similar to triangle y." For any triangles x, y, "x similar to y" also implies "y similar to x." So if $(x, y) \in \tilde{R}$ then $(y, x) \in \tilde{R}$.

Example 2 A nonsymmetric relation is involved in the proposition $x < y$ ("x less than y") defined on the set of positive integers N.

Antisymmetric Relation

Definition 2.22 Consider the relation $\tilde{R}$ on X. $\tilde{R}$ is said to define an antisymmetric relation if for $x, y \in X$, $(x, y) \in \tilde{R}$ and $(y, x) \in \tilde{R}$ imply $x = y$.

Example Let N be the set of natural numbers. Consider the relation $\tilde{R}$ as "m exactly divides n." It is obvious that $(m, n) \in \tilde{R}$ and $(n, m) \in \tilde{R}$ imply $m = n$, for any numbers $m, n \in N$.

Nonsymmetric, Nonantisymmetric Relations

A brief comment on Definitions 2.21 and 2.22 is necessary at this point, in particular to avoid some very common misunderstandings.

A first glance at Definitions 2.21 and 2.22 would seem to suggest that the concepts symmetric relation and antisymmetric relation are mutually exclusive, that is, in set-theoretical terminology, that the set of antisymmetric relations and that of symmetric relations are disjoint sets. However, that this is not the case can be illustrated in the following way.

Consider the set $X = \{1, 2, 3\}$. The direct product is

$$X^2 = \{(1, 1), (1, 2)\ (1, 3), (2, 1), (2, 2), (2, 3), (3, 1), (3, 2), (3, 3)\}.$$

A good example of a symmetric relation is, according to Definition 2.21,

$$\tilde{R}_S = \{(1, 2), (1, 3), (2, 1), (3, 1)\}.$$

On the other hand, a relation which satisfies Definition 2.22 is antisymmetric,

$$\tilde{R}_A = \{(1, 2), (2, 2), (3, 1)\}.$$

The following relation satisfies the definition of symmetric relation and the definition of antisymmetric relation as well:

$$\tilde{R}_{S,A} = \{(1, 1), (2, 2), (3, 3)\}.$$

From the latter example it becomes clear that relations whose only elements are the ordered pairs of the "main diagonal" of the direct product are both symmetric and antisymmetric. That would obviously be the case for the relation defined on N, the set of positive integers, by the proposition $m = n$, for $m, n \in N$.

Finally, a type of relation that is neither symmetric nor antisymmetric can be defined. This is the case with

$$\tilde{R}_{NS,NA} = \{(1, 2), (2, 1), (1, 3)\}.$$

Here, the fact that (1, 3) belongs to the relation but not so (3, 1) does exclude it from Definition 2.21. On the other hand it is excluded from Definition 2.22 by the fact that it contains both (1, 2) and (2, 1).

The preceding discussion is illustrated by the use of Venn diagrams in Fig. 2.2. Let $\{\tilde{R}_S\}$, $\{\tilde{R}_A\}$, $\{\tilde{R}_{S,A}\}$, and $\{\tilde{R}_{NS,NA}\}$ represent, respectively, the sets of all symmetric, antisymmetric, symmetric–antisymmetric, and nonsymmetric nonantisymmetric relations.

In a finer classification the following sets would have to be distinguished:

(a) the set of symmetric relations $\{\tilde{R}_S\}$,
(b) the set of nonsymmetric relations $\{\tilde{R}_S\}^c$,
(c) the set of antisymmetric relations $\{\tilde{R}_A\}$,
(d) the set of nonantisymmetric relations $\{\tilde{R}_A\}^c$,
(e) the set of symmetric–antisymmetric relations $\{\tilde{R}_S\} \cap \{\tilde{R}_A\}$,
(f) the set of nonsymmetric, nonantisymmetric relations $[\{\tilde{R}_S\} \cup \{\tilde{R}_A\}]^c$.

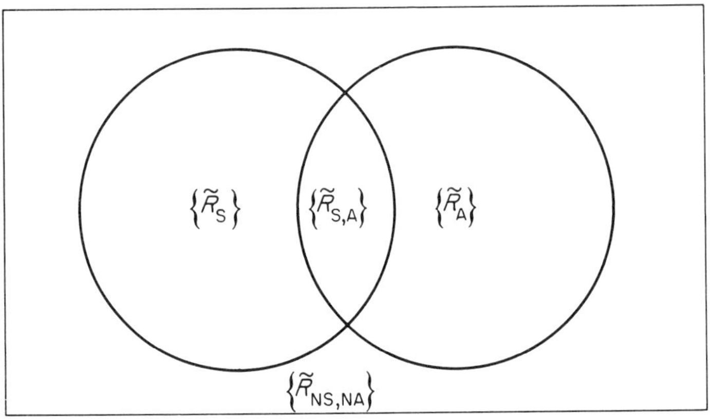

Figure 2.2

Transitive Relation

Definition 2.23 A relation $\tilde{R}$ from X to X is said to be transitive if $(x, y) \in \tilde{R}$ and $(y, z) \in \tilde{R}$ imply $(x, z) \in \tilde{R}$ for $x, y, z \in X$.

Example Let N be the set of positive integers. Note that the relation defined by the proposition "$x < y$" is transitive, since for any three numbers $m, n, p \in N$, $m < n$ together with $n < p$ imply $m < p$.

Equivalence Relation

A particularly important class of relations is formed by those relations which simultaneously satisfy Definitions 2.20, 2.21, and 2.23. A formal definition is given next.

Definition 2.24 Let $\tilde{R}$ be a relation on X. Then $\tilde{R}$ is said to be an equivalence relation if and only if it satisfies the following properties:

(a) $\tilde{R}$ is reflexive; that is, for any $x \in X$, $(x, x) \in \tilde{R}$,
(b) $\tilde{R}$ is symmetric; that is, for any $x, y \in X$, $(x, y) \in \tilde{R}$ implies $(y, x) \in \tilde{R}$,
(c) $\tilde{R}$ is transitive; that is, for any $x, y, z \in X$, $(x, y) \in \tilde{R}$ and $(y, z) \in \tilde{R}$ imply $(x, z) \in \tilde{R}$.

Example 1 Let X be the consumption set of the example of Definition 2.20. Consider now the relation defined by the proposition $x^k \sim x^h$, where $\sim$ means "is indifferent to" for $x^k, x^h \in X$. That the relation so defined is an equivalence relation results from the fact that

(a) for $k = h$ every $(x^k, x^k) \in \tilde{R}$;
(b) if $x^k \sim x^h$ then $x^h \sim x^k$ for any k, h;
(c) $x^k \sim x^h$ and $x^h \sim x^m$ imply $x^k \sim x^m$ for any k, h, m.

Example 2 A special case of an equivalence relation is "equality." For any three objects x, y, z we have (a) $x = x$, (b) $x = y$ implies $y = x$, (c) $x = y$ and $y = z$ imply $x = z$.

EXERCISES

Products of Sets

E.2.1 Let X_1, X_2, X_3 be three sets. Show that $X_1 \times (X_2 \times X_3) = (X_1 \times X_2) \times X_3$ and state under which condition is the product $\prod_{i=1}^{3} X_i$ equal to the product $\prod_{i=3}^{1} X_i$.

E.2.2 Let $A \subset X$, $B \subset Y$, $C \subset X$, $D \subset Y$, and show the validity of the following identities:

(a) $(A \times B) \subset (X \times Y)$,
(b) $(X \times Y) \setminus (A \times B) = ((X \setminus A) \times B) \cup (X \times (Y \setminus B)) \cup ((X \setminus A) \times (Y \setminus B))$,
(c) $(A \times B) \cap (C \times D) = (A \cap C) \times (B \cap D)$.

E.2.3 Let $A \subset X$. Show that $A \times A = (A \times X) \cap (X \times A)$.

E.2.4 Let X, Y be any two sets. Prove that $(X \times Y) \cap (Y \times X) \neq \varnothing$ if and only if $X \cap Y \neq \varnothing$.

E.2.5 Let $\{X_i\}_{i \in I}$ be an indexed family of sets with $X = \bigcup_{i \in I} X_i$ and define a function $f: I \to X$ such that $f(i) \in X_i$ for every $i \in I$. Show that the Cartesian product $\prod_{i \in I} X_i$ can now be defined as the set of functions $\{f | f: I \to X$ such that $f(i) \in X_i\}$.

Hint: Consider Definition 1.8, Chapter 1.

Intervals

E.2.6 Let V represent the family of all intervals of R, and let us take $A = [-1, 1[$, $B = [1, 2]$, $C =]1, 2]$. Tell why $A \cap B \in V$, $A \cap C \in V$, $B \cap C \in V$, $A \cup C \notin V$, $B \cup C \in V$.

Hint: Perform operations and also compare with Remarks 2.1 and 2.2.

E.2.7 Let V represent the family of all intervals of R, and take $D = [1, 4]$, $E =]2, 3[$, $F =]5, 6[$ and $G = [3, 5]$. Answer why $D \setminus E \notin V$, $D \setminus F \in V$, $D \setminus G \in V$, $E \setminus F \in V$.

Hint: Use Remark 2.3.

Metric Spaces

E.2.8 Let X be any set and, for $x, y \in X$, define a function d such that $d(x, y) = 0$ if $x = y$, and $d(x, y) = 1$ if $x \neq y$. Show that (X, d) is a metric space.

Hint: Show that d is a metric.

E.2.9 Using Definition 2.14 show that the following sets and functions form metric spaces: (a) (R, d) with $d(x, y) = |x - y|$; (b) (R^2, d) with $d(x, y) = |x_1 - y_1| + |x_2 - y_2|$; (c) (R^n, d) with $[\sum_{i=1}^n (x_i - y_i)^2]^{1/2}$.

E.2.10 Show that (R, d) is not a metric space when d is defined by $d(x, y) = |x^2 - y^2|$.

Relations

E.2.11 Draw the graphs of the following relations on R:

$$\{(x, y) \mid x^2 + y^2 = 9\}, \qquad \{(x, y) \mid x^2 + y^2 < 9\}.$$

E.2.12 Determine domain, range and inverse of the following relations on the set of natural numbers:

$$\{(x, y) \mid 2x + y = 10\}, \qquad \{(x, y) \mid x + y = 10\}.$$

E.2.13 Why does the relation on R defined by $x^2 + y^2 = r^2$ not satisfy the definition of single-valued function?

Hint: See Definition 1.7, Chapter 1, and draw a graph for $r = 2$.

E.2.14 Classify as reflexive, symmetric, transitive, and equivalence each of the following relations:

(a) Let $A = \{1, 2, 3, 4\}$ and consider

$$\tilde{R}_1 = \{(1, 3)\}, \quad \tilde{R}_2 = \{(1, 1), (1, 2)\}, \quad \tilde{R}_3 = \{(1, 1), (2, 2), (3, 3)\},$$
$$\tilde{R}_4 = \{(1, 1), (2, 2), (3, 3), (4, 4)\}, \quad \tilde{R}_5 = \{(2, 2), (2, 3), (3, 2)\}.$$

(b) Let X represent the set of all straight lines in R^2 so that

$$\tilde{R} = \{(a, b) \mid a, b \in X, a \text{ is perpendicular to } b\}.$$

(c) Let N be the set of natural numbers and

$$\tilde{R} = \{(x, y) \mid x, y \in N, x \leq y\}.$$

(d) Let $\{A_i\}_{i \in I}$ represent an indexed family of sets so that

$$\tilde{R} = \{(A_k, A_h) \mid A_k, A_h \in \{A_i\}_{i \in I}, A_k \cap A_h = \emptyset\}.$$

E.2.15 Let $\tilde{R}_1$ and $\tilde{R}_2$ be two relations on a set X. Prove: If $\tilde{R}_1$ and $\tilde{R}_2$ are equivalence relations, then $\tilde{R}_1 \cap \tilde{R}_2$ is an equivalence relation.

Hint: Observe that $\tilde{R}_1$, $\tilde{R}_2$, and therefore $\tilde{R}_1 \cap \tilde{R}_2$, are subsets of X^2 and, furthermore, that $\tilde{R}_1 \cap \tilde{R}_2 \subset \tilde{R}_1$ and $\tilde{R}_1 \cap \tilde{R}_2 \subset \tilde{R}_2$.

PROBLEMS

P.2.1 Prove Remarks 2.1 and 2.2.

P.2.2 Prove Remark 2.3.

P.2.3 Let $X \neq \varnothing$ be a set under an equivalence relation $\tilde{R}$. For every $a \in X$, the set $X_a = \{x \mid x \in X, x \tilde{R} a\}$ is called the equivalence class of a. Show that $\tilde{R}$ can be chosen such that the collection of all equivalence classes forms a partition of X.

Hint: See also P.3.2.

P.2.4 An economy of k commodities, m consumers and n producers is a system consisting of (a) for each $i = 1, \ldots, m$ a consumption set $X_i \subset R^k$ under a (preference–indifference) relation $\tilde{R}_i$; (b) for each $j = 1, \ldots, n$ a production set $Y_j \subset R^k$; (c) a bundle of initial resources $r \in R^k$.

If $(x_{1,i}, \ldots, x_{k,i}) = x_i \in X_i$ represents the demand of the ith consumer, and $(y_{1,j}, \ldots, y_{k,j}) = y_j \in Y_j$ the supply of the jth producer, then the points

$$x = \sum_{i=1}^{m} x_i = \left(\sum_{i=1}^{m} x_{1,i}, \ldots, \sum_{i=1}^{m} x_{k,i}\right)$$

and

$$y = \sum_{j=1}^{n} y_j = \left(\sum_{j=1}^{n} y_{1,j}, \ldots, \sum_{j=1}^{n} y_{k,j}\right)$$

represent total demand and total supply, respectively.

The difference $x - y = \sum_{i=1}^{m} x_i - \sum_{j=1}^{n} y_j$ is called the net demand; $x - y - r$ is the excess demand. A state of the economy is an $(m+n)$-tuple $((x_i), (y_j))$ of points in R^k; it is called a market equilibrium when $x - y = r$; it is said to be attainable when $x_i \in X_i$, for every i, $y_j \in Y_j$, for every j, and $x - y = r$.

Call M the set of all market equilibria, and A the set of all attainable states. Prove that $A = ((\prod_i X_i) \times (\prod_j Y_j)) \cap M$.

Hint: Debreu [4].

P.2.5 Let $x, y, z \in X$. A relation on X is said to be circular if it satisfies the following condition: If $z \tilde{R} y$ and $y \tilde{R} x$, then $x \tilde{R} z$. (a) Give examples of circular relations; (b) prove that a necessary and sufficient condition for $\tilde{R}$ to be an equivalence relation is to be both reflexive and circular.

NOTES

Most subjects of this chapter are discussed by Kuratowski [2] and Mendelson [1]. In particular, for a complete discussion of metric spaces see Berge [2] and Franz [1]. An excellent advanced book is that by Dunford and Schwartz [1].

CHAPTER 3

PREORDERED AND ORDERED SETS

The idea that the elements of a set may be subject to a certain order has already been introduced in the previous chapter. What we shall do now is to translate our intuitive concept of order into a formal one.

3.1 Preorderings and Orderings

A formal concept of order can be introduced in a simple way. Let X be a set with elements $x, y, z, \ldots$. Suppose a relation $\tilde{R}$ satisfying ordering axioms has been defined on X which any two elements $x, y \in X$ satisfy or do not satisfy. Notationally we shall express this by any of the following statements:

$x \precsim y$ (read "x precedes y"),
$x \prec y$ (read "x strictly precedes y"),
$x \succsim y$ (read "x dominates y"),
$x \succ y$ (read "x strictly dominates y"),

or in the negative case:

$x \not\precsim y$ (read "x does not precede y"),
$x \nprec y$ (read "x does not strictly precede y"),
$x \not\succsim y$ (read "x does not dominate y"),
$x \nsucc y$ (read "x does not strictly dominate y").

If it happens that for $x, y \in X$ we have $x \not\precsim y$, $x \not\succsim y$, $x \nprec y$, and $x \nsucc y$, then we shall say that x and y are not comparable elements for the relation $\tilde{R}$ and write $x \| y$.

For the remainder of this chapter we shall concentrate our attention on "weak" orderings ($\precsim$, $\succsim$) rather than on "strong" ones ($\prec$, $\succ$).

The set X is said to be ordered if its elements satisfy some of the following axioms:

A1 $x \precsim y$ and $y \precsim x$ imply $x = y$.
A2 $x \precsim y$ and $y \precsim z$ imply $x \precsim z$.
A3 For any $x, y \in X$ either $x \precsim y$ or $y \precsim x$.
A4 Let S represent any nonempty subset of X. There is $s^1 \in S$ such that $s^1 \precsim s$ for every $s \in S$.

In particular, Axioms A1 and A2 are said to define a partial order, whereas Axioms A1, A2, and A3 are said to determine a total order. The four axioms taken together constitute the so-called well-ordering principle.

An interesting case occurs when Axiom A1 is not satisfied, that is, when it is possible to have at once $x \precsim y$ and $y \precsim x$ without the implication $x = y$. Such a situation is called a quasi-ordering or preordering. If it is true that $x \precsim y$ and $y \precsim x$, we shall say that "x is equivalent to y" and write $x \sim y$. It is clear that a preordering may be total or partial, and satisfy Axiom A4 as well. Preorderings are of particular interest in economic theory as one of the examples below will show.

Example 1 Let $N = \{1, 2, 3, \ldots\}$ be the set of natural numbers. N is ordered by $\leq$, satisfying Axioms A1, A2, and A3. Also any subset of N has a first element s^1, as required by Axiom A4. For instance, the subsets $\{1, 2, 3\}$, $\{6, 7, 8\}$, $\{100, 101, 102\}$ have, respectively, the first elements 1, 6, 100 according to Axiom A4. So N is a well-ordered set.

Example 2 Let R be the set of real numbers with the "natural order" $\leq$. R is a totally ordered set, but not a well-ordered set.

Example 3 Consider the set $X = \{1, 2, 3, 4, 5\}$ ordered by "x is an exact divisor of y." The order is only partial since several elements are not comparable. For instance

$$2 \| 3, \qquad 2 \| 5, \qquad \text{and} \qquad 4 \| 5.$$

Example 4 Let X represent the consumption set of a consumer, with the preference relation $x \precsim y$ (read "x is at most as desired as y"). For two commodity bundles x, y it is possible to have $x \precsim y$ and $y \precsim x$, that is, $x \sim y$. If x is equivalent to y we usually say that, from the point of view of the consumer, x and y are "indifferent." Such a relation on X defines a preordering.

As orderings are the results of relations defined on given sets, it is usual to take account of this fact when formulating the corresponding definitions.

Preordered Sets

Definition 3.1 A set X is said to be preordered or quasi-ordered by a relation $\tilde{R}$, if $\tilde{R}$ is reflexive and transitive; that is, for any $x, y, z \in X$; (1) $x \tilde{R} x$; (2) $x \tilde{R} y$ and $y \tilde{R} z$ imply $x \tilde{R} z$.

The consumption set X of Example 4 above is a totally preordered set, since it includes all possible commodity bundles. Noncomparable bundles are thus not allowed. Two commodity bundles may, however, be indifferent.

Partially Ordered Sets

Definition 3.2 A set X is said to be partially ordered by a relation $\tilde{R}$, if $\tilde{R}$ is reflexive, transitive, and antisymmetric; that is, for any $x, y, z \in X$, (1) $x \tilde{R} x$; (2) $x \tilde{R} y$ and $y \tilde{R} z$ imply $x \tilde{R} z$; (3) $x \tilde{R} y$ and $y \tilde{R} x$ imply $x = y$.

Example Consider the set $X = \{1, 2, 3, 4, \ldots\}$ ordered by "x is multiple of y." Then we have $2 \precsim 1$, $2 \precsim 2$, $4 \precsim 1$, $4 \precsim 4$, $8 \precsim 4$, etc., but also $5 \| 4$, $8 \| 6$, $9 \| 5$, etc.

Totally Ordered Sets

Definition 3.3 A set X is said to be totally ordered under a relation $\tilde{R}$ if it is partially ordered and if it satisfies, additionally, the following axiom: For any $x, y \in X$ either $x \tilde{R} y$ or $y \tilde{R} x$.

Example Consider the set $X = \{10, 20, 40, 80\}$. X is totally ordered by "x exactly divides y."

Note that a totally ordered set cannot have noncomparable elements.

Well-Ordered Sets

Definition 3.4 A set X is said to be well ordered under a relation $\tilde{R}$ if it is totally ordered and if it satisfies, additionally, the following property: for any set $S \subset X$ there is an $s^1 \in S$ such that $s^1 \precsim s$ for every $s \in S$.

In other words, a well-ordered set is a totally ordered set which has the property that any of its subsets possesses a "first element," that is, an element which precedes any other element in the subset.

In Example 1 above we saw that the set of positive integers N is well ordered. It is easy to see that any subset of N satisfies Definition 3.4 in the sense of having a "first element." Furthermore, it is intuitive that any subset of a well-ordered set is by itself a well-ordered set.

3.2 Preordered and Ordered Subsets

Definitions 3.1 and 3.2 suggest the idea that an ordering is a special case of a preordering, namely a preordering with the additional quality of being antisymmetric. For the remainder of this chapter we shall be studying properties of preordered sets with the understanding, if not stated otherwise, that they apply to ordered sets as well.

Definition 3.5 Let a set X be preordered under a relation $\tilde{R}$, and let S be a subset of X. Assume that $\tilde{R}^S = \tilde{R} \cap (S \times S)$, called the restriction of $\tilde{R}$ to S, is not empty. Then $\tilde{R}^S$ defines a preordering on S.

The above definition implies that subsets of preordered sets may very well be preordered ($\tilde{R}^S$ nonempty) or not preordered at all ($\tilde{R}^S$ empty). In particular, a subset of a partially preordered set may fall in any of the following three categories: totally preordered, partially preordered, not preordered. On the other hand, if the preordering defined on the original set is total, then any of its subsets will of course be totally preordered too.

The following examples illustrate the definition.

Example 1 Consider the set $X = \{2, 4, 6, 8, 10, 12\}$ strongly ordered by "x is an exact divisor of y and at most half as large as y." We then have

$$\tilde{R} = \{(2, 4), (2, 6), (2, 8), (2, 10), (2, 12), (4, 8), (4, 12), (6, 12)\}.$$

Consider now the following subsets of X;

(1) $W = \{8, 10, 12\}$.

We have

$$W \times W = \{(8, 8), (8, 10), (8, 12), (10, 8), (10, 10), (10, 12), (12, 8), (12, 10), (12, 12)\}.$$

Therefore $\tilde{R}^W = \tilde{R} \cap (W \times W) = \varnothing$ and W is not an ordered subset of X.

(2) $Z = \{2, 4, 6\}$.

The direct product is

$$Z \times Z = \{(2, 2), (2, 4), (2, 6), (4, 2), (4, 4), (4, 6), (6, 2), (6, 4), (6, 6)\}.$$

The restriction is now $\tilde{R}^Z = \tilde{R} \cap (Z \times Z) = \{(2, 4), (2, 6)\}$ and Z is partially ordered but not totally ordered for $4 \| 6$.

(3) $Y = \{2, 4, 8\}$.

We have now

$$Y \times Y = \{(2, 2), (2, 4), (2, 8), (4, 2), (4, 4), (4, 8), (8, 2), (8, 4), (8, 8)\}$$

so that $\tilde{R}^Y = \tilde{R} \cap (Y \times Y) = \{(2, 4), (2, 8), (4, 8)\}$. Since all the elements of Y are comparable under $\tilde{R}^Y$, it is totally ordered by definition.

Example 2 Consider again the case of a given consumption set with a preference preordering defined on it. Since a preference preordering is always total, every subset of the consumption set is also totally preordered.

3.3 Elements of Preordered and Ordered Sets

There are some interesting elements in preordered sets which deserve our special attention. We have already seen the "first element" of a set. We take now this together with other elements as the subject of our next definition. But we should not forget that what we are saying here regarding the elements of a preordered set is also valid, if not stated otherwise, for the elements of an ordered set.

Definition 3.6 Let X be a preordered set and consider $a^0, b^0 \in X$. Then,

(1) a^0 is called a first or smallest element of X if and only if a^0 precedes any element in X; that is, $a^0 \precsim x$ for every $x \in X$;

(2) b^0 is called a last or largest element of X if and only if b^0 dominates any element in X; that is, $x \precsim b^0$ for every $x \in X$;

(3) a^0 is called a minimal element of X if and only if no other element strictly precedes it; that is, for every $x \in X$, $x \precsim a^0$ implies $x \sim a^0$ if X is preordered and $x = a^0$ if X is ordered;

(4) b^0 is called a maximal element of X if and only if no other element strictly dominates it; that is for every $x \in X$, $b^0 \precsim x$ implies $x \sim b^0$ if X is preordered and $x = b^0$ if X is ordered.

The concepts of first, last, minimal, and maximal element of a preordered set are important enough to justify the following lengthy but simple examples.

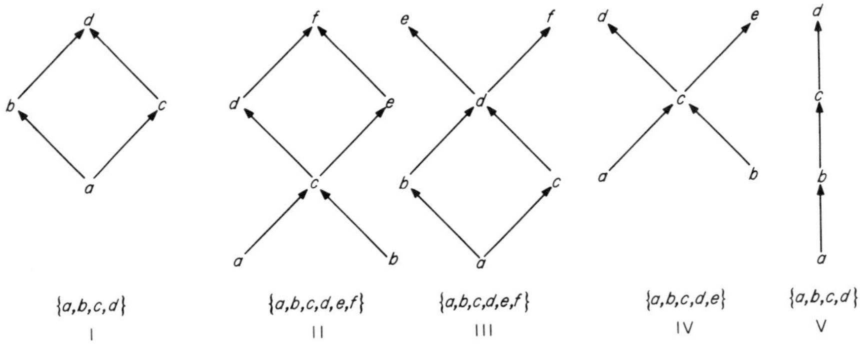

Figure 3.1

Example 1 Consider the set X ordered as in the diagrams in Fig. 3.1 where an arrow $a \to b$ indicates that "a precedes b" or "b dominates a". For each of the five cases shown, the first, last, minimal, and maximal elements are given in Table 3.1.

Table 3.1

Ordering of Elements in Sets Shown in Fig. 3.1

	First element	Last element	Minimal element	Maximal element
I	a	d	a	d
II	none	f	a and b	f
III	a	none	a	e and f
IV	none	none	a and b	d and e
V	a	d	a	d

It ought to be understood at this point that in order to decide whether or not an element is first (last) we have to determine whether or not it precedes (dominates) every other element in the set, whereas for classifying an element as minimal (maximal) we only have to consider whether or not it is strictly preceded (strictly dominated) by another element.

Example 2 Let X be as defined below. In each case find first, last, minimal, and maximal elements.

I. $X = \{x \mid 10 < x < 100\}$ ordered by $\leq$. X does not have any of the four studied elements.

II. $X = \{x \mid 10 \leq x \leq 100\}$ ordered by $\leq$. 10 is the first and minimal element. 100 is the last and maximal element.

III. $X = \{x \mid 10 \leq x < 100\}$ ordered by $\leq$. 10 is the first and minimal element. There are neither last nor maximal elements.

IV. $X = \{x \mid 10 < x \leq 100\}$ ordered by $\leq$. There is neither a first nor minimal element. 100 is the last and maximal element.

From Definition 3.6 and the examples discussed above we can easily derive the following properties.

Remark 3.1 Let X be a preordered set. If the preordering is partial, then a first element is a minimal element and a last element is a maximal element. If the preordering is total then the converse is also true.

Remark 3.2 Let X be partially ordered. Then X has at most one first element and at most one last element. A first element, if it exists, is the only minimal element, and a last element, if it exists, is the only maximal element.

Remark 3.3 Let X be totally ordered. Then X has at most one minimal element, its first element, and at most one maximal element, its last element.

Remark 3.4 A partially ordered finite set X has at least one minimal element and at least one maximal element, A infinite set X, partially or totally ordered, does not need to have any minimal or maximal element.

The last three remarks do not present special difficulties. The first remark should be clearly understood. Think that the first (last) element of a preordered set may be equivalent to or "indifferent" of other elements of the set, without being equal or not comparable to these. Hence a preordered set may have more than one first or last element.

A special situation arises when X is a subset of R, the set of real number. R is obviously totally ordered by $\leq$ and so is X. By Remark 3.3 we know that first and last elements, if they exist, are unique. We have then the following case.

Definition 3.7 Consider the set of real numbers R, totally ordered by $\leq$ and let X be a subset of R. A first element of X, when it exists, is said to be the minimum of X, and denoted min X. A last element of X, when it exists, is said to be the maximum of X, and denoted max X.

3.4 Preorderings on Product Sets

Until now we have considered preorderings defined on sets in a quite general manner, that is without paying much attention to the characteristics of the underlying sets. The case of a set of points was thus included in our considerations. Obviously, nothing will impede us in defining a preordering on a set whose elements are sequences, and the general principles would still hold.

Nevertheless, a closer look at the case of preordered set of points will serve to stress some specific problems.

Suppose we are given the sets X_i, for $i = 1, \ldots, n$ each of them preordered by the preordering $\precsim_i$ defined on X_i. Call x_i the typical element of X_i.

Consider next the product set $X = \prod_{i=1}^{n} X_i$ with the sequence $x = (x_1, \ldots, x_n)$ as typical element.

A preordering $\precsim$ on the product set X has been indirectly defined, assuming that for $x, x' \in X$ we have $x \precsim x'$ if $x_i \precsim_i x_i'$ for every i. But we must be careful about the meaning of this new relation, and expansion of our notation seems convenient:

(a) $x \precsim x'$ means $x_i \precsim_i x_i'$ for every i;
(b) $x \prec x'$ means $x_i \precsim_i x_i'$ for all i and $x_i \prec_i x_i'$ for at least one i;
(c) $x \lll x'$ will mean $x_i \prec_i x_i'$ for every i.

A special case arises when instead of preordering we have just an ordering. Consider, for instance, the Euclidean space E^n. The underlying set of the direct product E^n is obviously the set of real numbers R, ordered by the "natural order" $\leq$. We have now that also E^n is ordered by the relation "less than or equal," but to avoid ambiguity we must expand our notation. For any x, $x' \in E^n$ it is useful to distinguish among the following cases:

(a) $x \leq x'$ means $x_i \leq x_i'$ for every i;
(b) $x < x'$ means $x_i \leq x_i'$ for all i with at least one $x_i < x_i'$;
(c) $x \ll x'$ means $x_i < x_i'$ for every i.

3.5 Lower and Upper Bounds

Definition 3.8 Let S be a preordered set and X a subset of S. An element $y \in S$ is called a lower bound of X if for any $x \in X$ we have $y \precsim x$. Similarly an element $z \in S$ is called an upper bound of X if for any $x \in X$ we have $x \precsim z$.

Note that neither y nor z needs to be unique, so that we may perfectly well imagine y as an element of a set of lower bounds Y, and z as an element of a set of upper bounds Z. The example below illustrates what we are saying.

Example Let S and X be preordered as indicated in Fig. 3.2, where, again, $a \to b$ should be understood to mean "$a \precsim b$." A double arrow, $a \leftrightarrow b$, means "$a \sim b$"; that is, "a is indifferent of b."

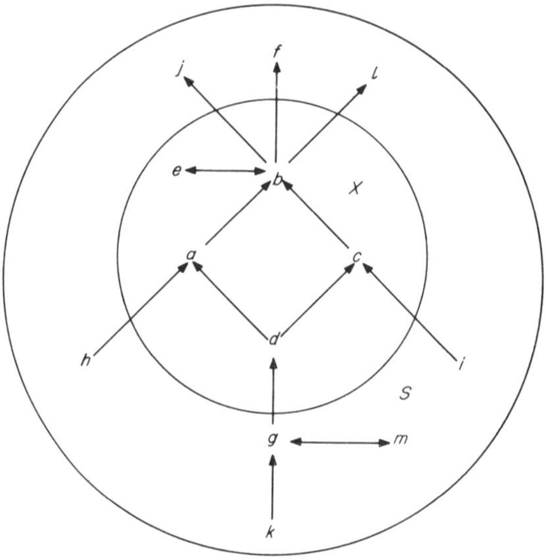

Figure 3.2

We have, according to Definition 3.8, that the lower bounds of X are d, g, m, k and the upper bounds of X are b, e, j, f, l. The set of lower bounds is then $Y = \{d, g, m, k\}$, and that of upper bounds $Z = \{b, e, j, f, l\}$.

Definition 3.9 Let X be a subset of a preordered set S. Let Y and Z represent, respectively, the set of all lower bounds and the set of all upper bounds of X. Then a last element of Y, if it exists, is called a greatest lower bound (*glb*) of X, and a first element of Z, if it exists, is called a least upper bound (*lub*) of X.

Example In the case of the previous example, given the preordering defined on S, we have that d dominates any other element in Y, and b and e ($b \sim e$) precede any element in Z. Consequently, d is the glb of X and b, e are the lubs.

Note that because of the possibility of two or more elements being equivalent, a preordering does not guarantee the uniqueness of either glb or lub. This lack of uniqueness is a direct consequence of the lack of uniqueness of last and first elements of preordered sets. The following remarks contemplate a different case.

Remark 3.5 If X is a subset of an ordered set S, and a glb exists, then it is unique and called the infimum of X, denoted $\inf(X)$. Similarly, if a lub exists, then it is unique, and called the supremum of X, denoted $\sup(X)$.

It is immediate that the uniqueness of glb and lub is a consequence of the uniqueness of last and first elements of ordered sets.

Example Figure 3.3 is a modified version of the example used after Definition 3.8, and is aimed at illustrating the case of an ordered set. It is obvious that double arrows $\leftrightarrow$ cannot appear in our diagram now. The set of lower bounds of X is now $Y = \{d, g, k\}$, that of upper bounds $Z = \{b, j, f, l\}$.

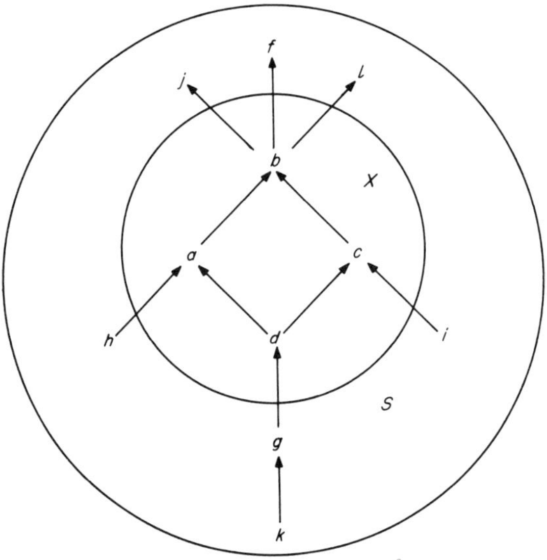

Figure 3.3

The glb and lub are necessarily unique; that is, $\inf(X) = d$, $\sup(X) = b$. Note that although in the two last examples the glb and the lub of X do belong to the set itself, this is not a requirement.

EXERCISES

E.3.1 Let $A = \{a, b, c\}$ and C be the collection of all nonempty subsets of A. Order the elements of C by set inclusion ($\subset$) and show whether this ordering is total or partial. Is it a preordering?

Hint: $\{a\} \not\subset \{b, c\}$, and $\subset$ together with $\supset$ imply $=$.

E.3.2 Let R be the set of real numbers, and C the collection of all nonempty subsets of R. Order the elements of C by set inclusion ($\subset$) and determine whether the ordering is partial or total.

Hint: Observe, for instance, that $[0, 1] \not\subset Q$.

E.3.3 Order the space R^2 by a lexicographic ordering.

Hint: Assume A and B are totally ordered sets. Then $A \times B$ is said to satisfy a lexicographic ordering if $(a, b) \succ (a', b')$ whenever $a \succ a'$, or $a = a'$ and $b \succ b'$.

Is the ordering on R^2 total?

E.3.4 Let C represent the collection of all subsets of $]0, 1[$, and consider any subcollection $\{X_i\}_{i \in I}$ of C. If C is ordered by inclusion ($\subset$), does the collection $\{X_i\}_{i \in I}$ have a lub and/or a glb?

Hint: Study the sets $\bigcup_{i \in I} X_i$ and $\bigcap_{i \in I} X_i$.

E.3.5 In what way will our previous results be modified if we ask for the lub and glb of the family C instead of the subcollection $\{X_i\}_{i \in I}$?

E.3.6 Find first and last element of C and $\{X_i\}_{i \in I}$ in E.3.4.

E.3.7 Prove the following proposition: If a totally ordered set X has two maximal elements, x^1 and x^2, then $x^1 = x^2$.

Hint: All elements of a totally ordered set are comparable.

E.3.8 What is the result in E.3.7 if X is totally preordered but not totally ordered?

E.3.9 Indicate the appropriate conclusion from the following statement: "X is totally preordered and has a and b as its first elements." (a) $a = b$; (b) $a \sim b$; (c) $a \| b$; (d) the statement is incorrect.

E.3.10 Find an example for each of the propositions of Remarks 3.1 to 3.4. Try to distinguish between orderings and preorderings.

PROBLEMS

P.3.1 Consider the economy of Problem P.2.4. It is intuitive that the preference–indifference relation $\precsim_i$ defined on each consumption set X_i, $i = 1, \ldots, m$, induces a preordering $\precsim$ on the set of attainable states $A = \{((x_i), (y_j)) \mid x_i \in X_i$, for every i, $y_j \in Y_j$, for every j, and $x - y = r\}$. Show that this induced preference–indifference preordering is only partial.

P.3.2 Let $(X_i, \precsim_i)$ represent the preordered consumption set of the ith consumer. For every $x_i' \in X_i$, the set $\{x_i \mid x_i \in X_i, x_i \sim_i x_i'\}$ represents an indifference class. Show that the collection of all indifference classes of X_i is a partition of X_i.

Hint: See P.2.3.

P.3.3 In consumption theory it is customary to assume that the preordering on a consumption set satisfies some convexity condition. For every x^1, x^2 in the consumption set of a certain consumer (we drop the index i to simplify notation), and $\lambda \in]0, 1[$, we may alternatively assume:

(1) if $x^2 \succsim x^1$, then $\lambda x^2 + (1 - \lambda)x^1 \succsim x^1$;
(2) if $x^2 \succ x^1$, then $\lambda x^2 + (1 - \lambda)x^1 \succ x^1$;
(3) if $x^2 \sim x^1$, then $\lambda x^2 + (1 - \lambda)x^1 \succ x^1$.

Suppose the consumption set is the nonnegative orthant of R^2, and illustrate three alternative shapes of an indifference class $\{x \mid x \in R^2,\ x \sim x'\}$ which would satisfy the preceding conditions. Draw graphs.

P.3.4 Figure 3.4 is drawn under the assumption that: (a) the preordering on X satisfies Assumption (3) of P.3.3; (b) x^S is a satiation consumption.

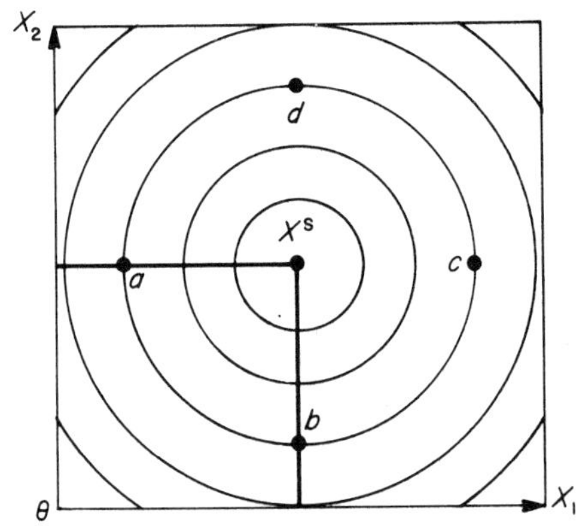

Figure 3.4

Give a meaningful economic interpretation of each of the sections *ab*, *bc*, *cd*, and *ad* of a typical "closed" indifference curve.

P.3.5 Give economic examples which would conflict with the assumption of a convex preordering of a consumption set.

NOTES

Orderings are dealt with by Kuratowski [2]. For a slightly more general definition of preordering than the one given here see Berge [2]. Preorderings are called quasi-orderings by Birkhoff [1]. The notation followed in this chapter is very much the same as that of Debreu [4] and Herstein and Milnor [1]. Differences are minor.

CHAPTER 4

PROPERTIES OF SETS OF POINTS

The present chapter is somewhat more difficult than the previous ones, and we ought to make sure of our ability to understand the different concepts discussed in it.

We shall be concerned with the study of some very important properties of sets of points. They are basic for a great many of the topics we will discuss later on. Besides, they are essential analytical pieces which support a good part of the building of modern theoretical economics.

4.1 Denumerable and Countable Sets

Let us introduce first the notion of equivalence between sets.

Definition 4.1 Two sets X, Y are said to be equivalent, denoted $X \sim Y$, if there exists a one-to-one function from X to Y.

Note that a one-to-one function, according to what has been said in Section 1.3, associates each $x \in X$ with one and only one $y \in Y$. In other words, each $y \in Y$ is the image of one and only one element of X.

We are now able to redefine a finite set.

Definition 4.2 A set X is said to be finite if and only if it is empty or equivalent to a set $S = \{1, 2, \ldots, n\}$ for some $n \in N$, the set of natural numbers; otherwise it is said to be infinite.

Take for some $n \in N$ the subset $L = \{1, 2, \ldots, n\}$. A set $X = \{x^1, x^2, \ldots, x^n\}$ is finite, according to Definition 4.2, if a one-to-one function from L to X can be defined or, in other words if $X \sim L$.

That two equivalent, finite sets have the same number of elements follows immediately from the above definition.

The concept of set equivalence is used once more in the following definition.

Definition 4.3 A set X is said to be denumerable if and only if it is equivalent to N. It is said to be countable if and only if it is denumerable or finite.

In other words, a countable set has the property that its elements can be put in a one-to-one correspondence with the elements of N (denumerable) or with the elements of a proper subset of N (finite). By Definition 4.2 the empty set is countable too.

Example 1 The set of letters of the English alphabet is obviously countable, since it is finite.

Example 2 The set of terms of an infinite sequence $x^1, x^2, x^3, \ldots$ is a typical example of a denumerable, and thus countable set. If we call it X, and define then a one-to-one function $f\colon N \to X$ such that the nth term of the sequence becomes the image of $n \in N$, that is, $f(n) = x^n$, it is easy to see that X is denumerable.

Four properties related to countable sets will be examined next.

Theorem 4.1 Let X, Y be nonempty sets. If X is countable and there is a one-to-one function $g : X \to Y$, then Y is countable.

Proof Since X is countable we have a one-to-one function $f : L \to X$, where $L \subset N$ (note we could have $L = N$). On the other hand, we know by hypothesis that the function $g\colon X \to Y$ is one to one. It is possible then to define a function $h\colon L \to Y$ such that, for each $n \in L$, and $x^n \in X$, $y^n \in Y$, $h(n) = g[f(n)] = g(x^n) = y^n$. The function $h\colon L \to Y$ is obviously one to one, so that Y is countable. Functions defined like h are called composite functions. We shall study them in the next chapter.

Theorem 4.2 Let N be the set of natural numbers. Then $N \times N$ is countable.

Proof That N is countable is obvious, since it is possible to define a one-to-one function $i : N \to N$. Now take the elements of $N \times N$ and arrange them as follows:

$$\begin{matrix} (1, 1)(1, 2) & \cdots (1, n) & \cdots \\ (2, 1)(2, 2) & \cdots (2, n) & \cdots \\ \vdots \quad \vdots & \vdots & \\ (m, 1)(m, 2) & \cdots (m, n) & \cdots \\ \vdots \quad \vdots & \vdots & \end{matrix}$$

It is easy to see that we may display them as an infinite sequence, (1, 1), (2, 1), (1, 2), (1, 3), (2, 2), ... so that a one-to-one function $f : N \to N \times N$ exists, such that

$$\begin{aligned} (1, 1) &= f(1) \\ (2, 1) &= f(2) \\ (1, 2) &= f(3) \\ &\vdots \end{aligned}$$

and the theorem is proved.

Generalizing the result of Theorem 4.2, we get Theorem 4.3. The proof is left as an exercise.

Theorem 4.3 The direct product of n countable sets is countable.

Theorem 4.4 Let $X_1, X_2, \ldots, X_n, \ldots$ be a sequence of countable sets. Then $\bigcup_{i=1}^{\infty} X_i$ is also countable.

Proof Since every set in the sequence is countable, we write

$$\begin{aligned} X_1 &= \{x_1^1, x_1^2, \ldots, x_1^n, \ldots\} \\ X_2 &= \{x_2^1, x_2^2, \ldots, x_2^n, \ldots\} \\ &\vdots \\ X_n &= \{x_n^1, x_n^2, \ldots, x_n^n, \ldots\}. \end{aligned}$$

The union of all X_i is clearly the set of elements whose indices, as ordered pairs, belong to the product set $N \times N$. That is $\bigcup_{i=1}^{\infty} X_i = \{x_i^j \mid (i, j) \in N \times N\}$. The function

$$f : \bigcup_{i=1}^{\infty} X_i \to N \times N, \qquad \text{such that} \quad (i, j) = f(x_i^j),$$

is one to one. But $N \times N$ is countable by Theorem 4.2. Then, according to Theorem 4.1 $\bigcup_{i=1}^{\infty} X_i$ is countable.

Example The set of rational numbers Q is denumerable. Partitioning Q into the subsets of positive rationals, zero, and negative rationals, we have $Q = Q^+ \cup \{0\} \cup Q^-$.

We can define a one-to-one function from Q^+ to $N \times N$ by $f(p/q) = (p, q)$, where p/q is any rational in Q^+. Similarly we can define a one-to-one function from Q^- to $N \times N$. By Theorem 4.2, $N \times N$ is denumerable, so that Q^+ and Q^- are denumerable too. Furthermore, $Q = Q^+ \cup \{0\} \cup Q^-$ is denumerable by Theorem 4.4.

Note that by Theorem 4.3 this result can be generalized for Q^n.

4.2 Closedness

Our attention will now be almost exclusively focused on the space R^n and on subsets of it. As a special case, and mostly for simplicity, we shall frequently use R in our examples. But note that R can be regarded as a subset of R^n.

Convergent Sequences and Limits

Definition 4.4 Let $(a^m \mid m \in N) = a^1, a^2, \ldots$ represent a sequence of real numbers. The sequence is said to be convergent, or more precisely to converge to $a^0 \in R$, denoted $a^m \to a^0$, if for every $\varepsilon > 0$ there exists a positive integer m' such that if $m > m'$ then $|a^m - a^0| < \varepsilon$. The number $a^0 \in R$ is called the limit of the sequence, and denoted $\lim_{m \to \infty} a^m = a^0$.

Consider our definition. If we define the open interval $]a^0 - \varepsilon, a^0 + \varepsilon[$, it is easy to see that the expression $|a^m - a^0| < \varepsilon$ simply means $a^m \in]a^0 - \varepsilon, a^0 + \varepsilon[$.

We can now generalize our concept of convergence for the case of a sequence $(x^m \mid m \in N)$ of points of R^n. Note that our sequence can also be written $x^1, x^2, \ldots$. Every term of the sequence is now a point or n-tuple of R^n.

Definition 4.5 Let $(x^m \mid m \in N) = x^1, x^2, \ldots$ represent a sequence of points of R^n. The sequence is said to be convergent, or more precisely to converge to a point $x^0 \in R^n$, denoted $x^m \to x^0$, if for every coordinate i $(i = 1, \ldots, n)$, we have $x_i^m \to x_i^0$. The point $x^0 \in R^n$ is called the limit of the sequence, and denoted $\lim_{m \to \infty} x^m = x^0$.

Accumulation Points, Derived Set and Closure

Consider now two sets of points, say S and X, such that $X \subset S$. Without further consequence to our reasoning we may regard S either as identical to R^n or as a subset of it.

Definition 4.6 Let X be a subset of a set of points S. Then $x' \in S$ is said to be an accumulation or limit point of X if x' is the limit of a sequence of points belonging to X but distinct from x'.

According to the above definition an accumulation point of X is the limit of a sequence, all the terms of which, with perhaps the exception of the accumulation point itself, belong to X. The limit may or may not belong to X.

The concept of accumulation point is important enough to justify an additional effort to approach the question from a slightly different angle.

Let us consider a δ-neighborhood of a point $x' \in S$, that is, the set of all points $x \in S$ such that their distance is $d(x', x) < \delta$, where δ is a positive real number. We can then formulate the following alternative definition.

Definition 4.7 Let X be a subset of a set of points S. Then $x' \in S$ is called an accumulation or limit point of X if and only if every δ-neighborhood of x' contains some point of X other than x'.

Example 1 Consider $S = R$ and $X = \{1, 2, 3, 4, 5\}$. It is easy to see that X has no accumulation points. There is no sequence satisfying the conditions mentioned in Definition 4.6. Alternatively, for δ sufficiently small, there is no δ-neighborhood of an element $x \in S$ containing another element $y \in X$.

Example 2 Consider $S = R$ and $X = \{x \mid 5 < x < 10\}$. Note that every δ-neighborhood of 10, with δ as small as desired, will contain elements of X other than 10. The same is true, of course, for 5.

The points 5 and 10, which in our case do not belong to X, are accumulation points of X. But note also that any other point of X does satisfy Definitions 4.6 and 4.7 and is an accumulation point of X.

Example 3 Consider $S = R$ and $X = \{x \mid 5 \leq x \leq 10\}$. With the same reasoning as in the previous example we conclude that every $x \in X$ is an accumulation point of X. Obviously X contains all its accumulation points, including 5 and 10.

Example 4 Consider $S = R$ and $X = \{1, 1/2, 1/3, \ldots, 1/n, \ldots\}$. We can see that in this case only 0 satisfies the two definitions above. In consequence 0 is the only accumulation or limit point of X.

We have seen four classes of sets. A set without limit points at all, which happened to be a finite set. A set which did not contain all its limit points but whose points were all accumulation points. A set which did contain all its accumulation points and whose elements were all accumulation points. Finally, a set having only one accumulation point and whose elements were not accumulation points. The latter case suggests to us the following definition.

Definition 4.8 Let X be a subset of a set of points S. Any point of X which is not an accumulation point of X is said to be an isolated point of X.

Example In the case $X = \{1, 1/2, 1/3, \ldots, 1/n, \ldots\}$ it is clear that 0 is the only accumulation point. Any element $x \in X$ is an isolated point.

Definition 4.9 Let X be a set of points. The set of all accumulation points of X is called the derived set of X, and denoted X^{d}.

Example Consider our previous four examples. We have

if $X = \{1, 2, 3, 4, 5\}$, $X^{\mathrm{d}} = \varnothing$;
if $X = \{x \mid 5 < x < 10\}$, $X^{\mathrm{d}} = \{x \mid 5 \leq x \leq 10\}$;
if $X = \{x \mid 5 \leq x \leq 10\}$, $X^{\mathrm{d}} = \{x \mid 5 \leq x \leq 10\}$;
if $X = \{1, 1/2, \ldots, 1/n, \ldots\}$, $X^{\mathrm{d}} = \{0\}$.

Let us introduce a new concept.

Definition 4.10 Let X be a subset of a set of points S. Then $x \in S$ is called an adherent point of X if there is a sequence of points of X converging to x.

It is necessary to distinguish clearly between the notion of adherent point and that of accumulation point. Comparing Definitions 4.10 and 4.6 it can be seen that the former is more general than the latter. Both concepts are defined in terms of a point and a sequence converging to it. However, for an accumulation point, the sequence must contain an element of the set under consideration other than the accumulation point itself. The latter is not required for an adherent point. An accumulation point is always an adherent point but the converse is not necessarily true.

Example 1 Take $X = \{1, 2, 3, 4, 5\} \subset S$ where $S = R$. Any $x \in X$ is adherent to X. To show that this is true, it is enough to take the sequence whose points are all equal to x. That already satisfies Definition 4.10. So 1, 2, 3, 4, 5 are all adherent to X but none is an accumulation point of X.

Example 2 Take $\{x \mid 5 < x < 10\}$ or $\{x \mid 5 \leq x \leq 10\}$. Here, obviously all adherent points are also accumulation points.

Example 3 Consider $\{1, 1/2, 1/3, \ldots, 1/n, \ldots\}$. We already know the only accumulation point is 0. The other elements are called isolated points. Now following the same reasoning as in previous examples, the set of adherent points could be expressed as $\{1, 1/2, 1/3, \ldots, 1/n, \ldots, 0\}$.

We are now able to introduce the following idea.

Definition 4.11 Let X be a set of points. The set of all adherent points of X is called the adherence or closure of X, and denoted $\overline{X}$.

Example Writing X^{d} for the derived set of X and $\overline{X}$ for its closure, and using the same four sets of previous examples, we have the following.

(1) $X = \{1, 2, 3, 4, 5\}$, $X^{\mathrm{d}} = \varnothing$; $\overline{X} = \{1, 2, 3, 4, 5\}$;
(2) $X = \{x \mid 5 < x < 10\}$, $X^{\mathrm{d}} = \{x \mid 5 \leq x \leq 10\}$; $\overline{X} = \{x \mid 5 \leq x \leq 10\}$;
(3) $X = \{x \mid 5 \leq x \leq 10\}$, $X^{\mathrm{d}} = \{x \mid 5 \leq x \leq 10\}$; $\overline{X} = \{x \mid 5 \leq x \leq 10\}$;
(4) $X = \{1, 1/2, \ldots, 1/n, \ldots\}$, $X^{\mathrm{d}} = \{0\}$; $\overline{X} = \{1, 1/2, \ldots, 1/n, \ldots 0\}$.

Note that from the four sets above only the first and the third are equal to their respective closures. They contain all their adherent points. Later we shall come back and look closer at sets with this property.

We shall study next some properties of the closure of a set.

Theorem 4.5 Let X be a set of points and let $\overline{X}$ be its closure. Then $X \subset \overline{X}$.

Proof Take a sequence of terms (x^n) with the property that $x^n = x$ for $n = 1, 2, \ldots$. That means $\lim_{n \to \infty} x^n = x$. Note then, if $x \in X$ then $x = \lim_{n \to \infty} x^n$, where $x^n = x$. So $x \in \overline{X}$ and Definitions 4.10 and 4.11 imply that $X \subset \overline{X}$.

Theorem 4.6 Let X_1, X_2 be two sets. Then $X_1 \subset X_2$ implies $\overline{X}_1 \subset \overline{X}_2$.

Proof Consider an element $x \in \overline{X}_1$. Then there is a sequence (x^n) in X_1 which converges to x, that is, $x^n \to x$. But $X_1 \subset X_2$, so (x^n) is a sequence in X_2 and $x \in \overline{X}_2$. Therefore $\overline{X}_1 \subset \overline{X}_2$.

Theorem 4.7 Let X_1, X_2 be two sets. Then $\overline{X_1 \cup X_2} = \overline{X}_1 \cup \overline{X}_2$.

Proof We prove first

$$\overline{X_1 \cup X_2} \subset \overline{X}_1 \cup \overline{X}_2. \tag{1}$$

If

$$x \in \overline{X_1 \cup X_2} \quad \text{and} \quad x = \lim_{n \to \infty} x^n,$$

we must have for any n that either $x^n \in X_1$ or $x^n \in X_2$ so that either $x \in \overline{X}_1$ or $x \in \overline{X}_2$ and so $x \in \overline{X}_1 \cup \overline{X}_2$. It must then be

$$\overline{X_1 \cup X_2} \subset \overline{X}_1 \cup \overline{X}_2.$$

Next we prove

$$\overline{X}_1 \cup \overline{X}_2 \subset \overline{X_1 \cup X_2}. \tag{2}$$

From the previous theorem we know that if $X_1 \subset X_2$ then $\overline{X}_1 \subset \overline{X}_2$. On the other hand, from Chapter 1

$$X_1 \subset X_1 \cup X_2 \quad \text{and} \quad X_2 \subset X_1 \cup X_2$$

so that

$$\overline{X}_1 \subset \overline{X_1 \cup X_2} \quad \text{and} \quad X_2 \subset \overline{X_1 \cup X_2}.$$

Joining memberwise the latter expression we have

$$\overline{X}_1 \cup \overline{X}_2 \subset \overline{X_1 \cup X_2}.$$

Now from expressions (1) and (2) we obtain

$$\overline{X_1 \cup X_2} = \overline{X}_1 \cup \overline{X}_2.$$

Theorem 4.8 Let X_1, X_2 be two sets. Then

$$\overline{X_1 \cap X_2} \subset \overline{X}_1 \cap \overline{X}_2.$$

Proof From Chapter 1 we know that $X_1 \cap X_2 \subset X_1$ and $X_1 \cap X_2 \subset X_2$, but by Theorem 4.6 this implies

$$\overline{X_1 \cap X_2} \subset \overline{X}_1 \qquad \text{and} \qquad \overline{X_1 \cap X_2} \subset \overline{X}_2.$$

Therefore

$$\overline{X_1 \cap X_2} \subset \overline{X}_1 \cap \overline{X}_2.$$

Theorem 4.9 Let $\{X_i\}_{i \in I}$ be an indexed family of sets and I an arbitrary set of indices. Then

$$\bigcup_{i \in I} \overline{X}_i \subset \overline{\bigcup_{i \in I} X_i}.$$

Proof For every $j \in I$ we have

$$X_j \subset \bigcup_{i \in I} X_i.$$

From Theorem 4.6

$$\overline{X}_j \subset \overline{\bigcup_{i \in I} X_i} \qquad \text{for every} \quad j \in I.$$

Therefore

$$\bigcup_{i \in I} \overline{X}_j \subset \overline{\bigcup_{i \in I} X_i}.$$

Replacing j by i we now obtain

$$\bigcup_{i \in I} \overline{X}_i \subset \overline{\bigcup_{i \in I} X_i}.$$

Theorem 4.10 Let $\{X_i\}_{i \in I}$ be an indexed family of sets. Then

$$\overline{\bigcap_{i \in I} X_i} \subset \bigcap_{i \in I} \overline{X}_i.$$

Proof For every $j \in I$, it is

$$\bigcap_{i \in I} X_i \subset X_j.$$

From Theorem 4.6

$$\overline{\bigcap_{i\in I} X_i} \subset \overline{X}_j \qquad \text{for every} \quad j \in I.$$

Therefore

$$\overline{\bigcap_{i\in I} X_i} \subset \bigcap_{j\in I} \overline{X}_j .$$

Now replacing j by i yields

$$\overline{\bigcap_{i\in I} X_i} \subset \bigcap_{i\in I} \overline{X}_i .$$

Closed and Open Sets

The concept of closure or adherence of a set makes it possible to classify sets into two fundamental categories.

Definition 4.12 Let X be a subset of S. Then X is said to be closed if and only if $X = \overline{X}$.

It is clear from this definition that a closed set $\{x \mid x \in R^n, |x_i| \leq a, i = 1, \ldots, n\}$ contains all its accumulation points. In other words, if X^{d} represents the derived set of X, then $X^{\mathrm{d}} \subset X$. A definition of an open set immediately follows from that of a closed set. However, it is illustrative to describe an open set in a slightly different manner.

Definition 4.13 Let X be a subset of S. Then X is said to be open if and only if its complement in S is closed, that is $\overline{X}^{\mathrm{c}} = X^{\mathrm{c}}$.

The complement of X in S is of course the difference $S \setminus X$. That $\overline{X}^{\mathrm{c}} = X^{\mathrm{c}}$ simply means $S \setminus X = \overline{S \setminus X}$. The set "complement of X" contains all its own adherent points if X is open.

Example Note that the following sets are closed: $R^n, R, X = \{x \mid 0 \leq x \leq 10\}$, $X = \{x \mid x \geq 10\}$, $X = \{x \mid x \in R^n,\ x_i \geq 0 \text{ for } i = 1, \ldots, n\}$. The following sets are open: $X = \{x \mid 0 < x < 10\}$, $X = \{x \mid x > 10\}$, $X = \{x \mid x \in R^n,\ x_i < 10 \text{ for } i = 1, \ldots, n\}$.

The above examples do not distinguish between sets of n-component points and "one-dimensional" points, that is, elements of R. Such a distinction is not necessary and, as we have already pointed out, R is a special subset of R^n. The next remark generalizes this principle. The proof is left as an exercise.

Remark 4.1 Let $X_1, X_2, \ldots, X_m$ be closed subsets of the spaces R^{n_1}, $R^{n_2}, \ldots, R^{n_m}$. Then $X_1 \times X_2 \times \cdots \times X_m$ is a closed subset of $R^{\Sigma n_j}$.

Operations with Closed and Open Sets

Let us consider here the union and intersection of closed sets and of open sets.

Theorem 4.11 Let $X_1, X_2, \ldots, X_n$ be a finite number of closed sets. Then $\bigcup_{i=1}^n X_i$ is a closed set.

Proof Take the first two sets, X_1 and X_2. Since they are closed, we have $X_1 = \bar{X}_1$ and $X_2 = \bar{X}_2$. We then have

$$X_1 \cup X_2 = \bar{X}_1 \cup \bar{X}_2 = \overline{X_1 \cup X_2}.$$

Now taking X_3, since $X_3 = \bar{X}_3$, we can write

$$X_1 \cup X_2 \cup X_3 = \bar{X}_1 \cup \bar{X}_2 \cup \bar{X}_3 = \overline{X_1 \cup X_2 \cup X_3}.$$

Using the same reasoning on n sets, we get

$$\bigcup_{i=1}^{n} X_i = \bigcup_{i=1}^{n} \bar{X}_i = \overline{\bigcup_{i=1}^{n} X_i}$$

and the theorem is proved.

Theorem 4.12 Let $X_1, X_2, \ldots$ be an arbitrary number of closed sets. Then $\bigcap_i X_i$ is a closed set.

Proof Since the sets are closed we have $X_i = \bar{X}_i$ and

$$\overline{\bigcap_i X_i} \subset \bigcap_i \bar{X}_i = \bigcap_i X_i$$

by Theorem 4.8. Therefore $\bigcap_i X_i$ is closed.

Theorem 4.13 Let $X_1, X_2, \ldots$ be an arbitrary number of open sets. Then $\bigcup_i X_i$ is an open set.

Proof By hypothesis all X_i are open. Then, by Definition 4.13, all X_i^c are closed. Applying Theorem 4.12, we find also that $\bigcap_i X_i^c$ is closed. Now by De Morgan's law (see Theorem 1.8 in Chapter 1) we have $(\bigcup_i X_i)^c = \bigcap_i X_i^c$. It is clear that since $(\bigcup_i X_i)^c$ is closed $\bigcup_i X_i$ must be open, and the theorem is proved.

Theorem 4.14 Let $X_1, X_2, \ldots, X_n$ be a finite number of open sets. Then $\bigcap_{i=1}^n X_i$ is an open set.

Proof By hypothesis all X_i are open. Then all X_i^c are closed. Theorem 4.11 indicates that $\bigcup_{i=1}^n X_i^c$ must be closed. By De Morgan's law (Chapter 1, Theorem 1.8) we have $(\bigcap_{i=1}^n X_i)^c = \bigcup_{i=1}^n X_i^c$. Since $(\bigcap_{i=1}^n X_i)^c$ is closed we must have $\bigcap_{i=1}^n X_i$ is open.

Dense and Boundary Sets

Related to the notion of closedness is the concept of a dense set. For our further investigations we shall continue making use of the analytical tools developed in previous pages.

Definition 4.14 Let S be a set of points and let X be a subset of S. Then X is said to be dense in S (or simply dense) if $\overline{X} = S$.

The idea contained in the preceding definition can be described as follows: For any point $x \in S$ there are points of X which are as close to x as desired.

On the other hand we have the following:

Definition 4.15 A set X is called a boundary set if its complement X^c is a dense set.

The latter concept merely means that a set is a boundary set when $\overline{X}^c = S$.

Example Let R be the set of real numbers. Then Q, the set of rational numbers, is dense in R. Every real number $x \in R$ is an accumulation point of Q, for every δ-neighborhood of x will contain rational numbers other than x. We have then $\overline{Q} = R$. Note that Q is a boundary set too.

An interesting relationship between dense and countable subsets of R^n is the following:

Theorem 4.15 Let S be a subset of R^n. Then S contains a countable set X which is dense in S.

Proof To prove the theorem it is enough to show that any $S \subset R^n$ contains "at least" one subset that is countable and dense in S. In the example following Theorem 4.4 we proved that Q is a countable set and so is Q^n.

From the fact that $\overline{Q} = R$ we immediately discover that $\overline{Q}^n = R^n$. If, for simplicity, we take $S = R$ $(S = R^n)$ it is obvious that Q (respectively, Q^n) is dense in S and countable. The same is of course valid for $S \subset R$ $(S \subset R^n)$.

A different concept is that of a set dense in itself. We define it as follows.

Definition 4.16 Let X be a set of points. Then X is said to be dense in itself if each of its points is an accumulation point of X, that is, $X \subset X^d$. Furthermore, X is said to be a perfect set if it is dense in itself and closed, that is, $X = X^d$ $(X = \overline{X})$.

From the definition it is clear that a set dense in itself does not contain any isolated point. A perfect set has the additional property that it contains all its accumulation or limit points.

Example 1 $X = \{1, 1/2, 1/3, \ldots, 1/n, \ldots\}$ is obviously neither dense in R nor dense in itself. In particular note that $X^d = \{0\}$ so that $X \not\subset X^d$.

Example 2 Note that $X = \{x \mid 10 < x < 100\}$ is dense in itself, while $X = \{x \mid 10 \leq x \leq 100\}$ is a perfect set.

Interior, Boundary, and Exterior of a Set

In the following definitions we focus our attention on X subset of R^n. We refer then to the "exterior of X," "complement of X," etc. Had we considered, instead, a set $X \subset S$, where $S \subset R^n$, we would have to refer to the "exterior of X in S," the "complement of X in S" [that is, $(S \setminus X)$], and so on. The conclusions, however, are fundamentally similar.

Definition 4.17 Let X be a subset of R^n and consider $x \in R^n$. Then

(a) x is called an interior point of X if it is not adherent to X^c, the complement of X;
(b) x is called a boundary point of X if it is adherent to both X and its complement X^c;
(c) x is called an exterior point of X if it is neither an interior nor a boundary point of X.

Example Let there be $X = \{x \mid x \in R^3, |x_i| \leq 5 \text{ for } i = 1, 2, 3\}$. Then $(2, 3, 1)$, $(-3, 4, 0)$ are interior points of X. The points $(2, 3, 5)$, $(5, 4, 0)$, $(5, 5, 5)$ are boundary points of X. Finally $(0, 0, 12)$, $(2, 9, -1)$, $(-8, 9, 6)$ are all exterior points of X. Note also that $X^c = \{x \mid x \in R^3, |x_i| > 5 \text{ for } i = 1, 2, 3\}$.

Definition 4.18 Let X be a subset of R^n. Then

(a) the interior of X, denoted $\text{int}(X)$, is the set of all points interior to X;
(b) the boundary of X, denoted $\text{b}(X)$, is the set of all boundary points of X;
(c) the exterior of X, denoted $\text{ext}(X)$, is the set of all points exterior to X.

Example Again take $X = \{x \mid x \in R^3, |x_i| \leq 5 \text{ for } i = 1, 2, 3\}$. The interior of X is the set $\text{int}(X) = \{x \mid x \in R^3, |x_i| < 5 \text{ for } i = 1, 2, 3\}$. Its boundary is the set of all points on the surface of the cube X. Finally, the exterior of X is the set $X = \{x \mid x \in R^3, |x_i| > 5 \text{ for } i = 1, 2, 3\}$.

The following two properties are direct results of the above definitions.

Remark 4.2 A set $X \subset R^n$ is closed if and only if it contains its boundary, that is, $\text{b}(X) \subset X$.

Remark 4.3 Let X be a subset of R^n. Then the sets $\text{int}(X)$, $\text{b}(X)$, and $\text{ext}(X)$ form a partition of R^n.

Topological Spaces

We shall complete this part of our study with the introduction of the concept of a topological space. The introduction could have been made before now, for it is possible to study several of the notions previously discussed within the framework of a topological space. Topological concepts have been and will continue to be present in our study.

Our aim at this point is, however, a very limited one. The concept of a metric space, that of a Euclidean space in particular, did provide us with a framework which is sufficiently general to deal with most of the problems we had proposed; and we shall be able to find answers to our questions without major departures from it.

However, the distance separating the concept of a topological space and that of a metric space is a short one. It seems worthwhile to bridge the gap between them, and define a more general frame in which to study some of the questions we shall soon be facing.

We approach the problem from two angles.

Definition 4.19 A nonempty set of points X together with a family τ of its subsets is called a topological space, and denoted (X, τ), if and only if the following axioms are satisfied:

A1 $X \in \tau$.
A2 $\varnothing \in \tau$.
A3 If $O_1, O_2, \ldots, O_n \in \tau$, then $O_1 \cap O_2 \cap \cdots \cap O_n \in \tau$.
A4 If for each $i \in I$, $O_i \in \tau$, then $\bigcup_{i \in I} O_i \in \tau$.

The family τ is called the topology on X while the subsets $O_1, O_2, \ldots, O_n \subset X$ are the τ-open sets, or simply, the open sets.

As can be read in the above definition, the notion of an open set constitutes a basic ingredient of the concept of a topology.

A different, though equivalent, concept of a topological space can be formulated in terms of the following principles, known as the four Kuratowski closure axioms for $O_i, O_j \subset X$;

K1 $\overline{O_i \cup O_j} = \bar{O}_i \cup \bar{O}_j$.
K2 $O_i \subset \bar{O}_i$.
K3 $\varnothing = \bar{\varnothing}$.
K4 $\overline{(\bar{O}_i)} = \bar{O}_i$.

The validity of the first two expressions has already been considered. That of the last two is obvious for $\varnothing$ and $\bar{O}_i$ are closed, and thus equal to their closures.

Thinking of closure as an operator that assigns to each subset $O_i \subset X$ its closure $\bar{O}_i \subset X$, such that the Kuratowski axioms are satisfied, we find an alternative way of defining a topology on X. A topological space may then be thought of as a set X in which the operation closure has been defined such that K1, K2, K3, and K4 are satisfied.

The two definitions are equivalent.

A topological space is an extremely general concept and we can distinguish several special classes.

A metric space satisfies, of course, the conditions contained in the definition of a topological space. We are pointing out here a special class of a topological space. However, it would be convenient not to identify the two spaces, but refer to the topological space "associated" with a given metric space. In this manner, we would be able to emphasize the link between the two concepts without losing sight of the definitional difference involved.

Let us be precise in this point.

Definition 4.20 Let (X, d) be a metric space, and τ the family of open sets of X satisfying the four axioms of Definition 4.19. Then (X, d) is said to give rise to the topological space (X, τ), while (X, τ) is called the topological space associated with (X, d).

Example 1 A vector space, the theory of which will be studied in Part II, is a type of metric space. There is then a topological space associated with it.

In general, in agreement with the terminology we have been using, there is a Euclidean topological space associated with the Euclidean metric space (R^n, d). It is generally called the "Euclidean n-dimensional space with the usual topology."

Example 2 As a special case let us mention the real line. As a topological space it is associated with the metric space (R, d) where $d(x, y) = |x - y|$. The open sets are the open intervals.

Example 3 Let N be the set of natural numbers and, for each $n \in N$, define the set $O_n = \{n, n+1, n+2, \ldots\}$. Furthermore, let $\tau = \{\varnothing, O_1, O_2, \ldots, O_n, \ldots\}$. Then (N, τ) is a topological space.

4.3 Boundedness

The property of sets of points we are going to consider now does not present any particular difficulty. We shall be sure to understand it correctly, however, and not to confuse it with the notion of closedness. They are two different concepts.

Consider a point $x \in R^n$, where of course $x = (x_1, \ldots, x_n)$. For the present the coordinate of x with the largest absolute value will be called the norm of x, and denoted $|x|$. In a more formal way, the norm of the point x is $|x| = \max(|x_1|, \ldots, |x_n|)$. (Note there are other possible norms.)

The following definition is straightforward.

Definition 4.21 A subset K of R^n such that $K = \{x \mid x \in R^n, |x| \leq k$ where k is real and positive$\}$ is called a closed cube with center θ and edge $2k$.

Note that the n-dimensional set K is a generalization of the usual concept of a cube in R^3. That K is a closed set is obvious from the definition.

The closed square $K = \{x \mid x \in R^2, |x| \leq k, k$ real and positive$\}$ and closed segment $K = \{x \mid x \in R, |x| \leq k, k$ real and positive$\}$ are, of course, special cases of Definition 4.21. We shall use, however, the more general term "cube" in our next definition.

Definition 4.22 Let X be a subset of R^n. X is said to be bounded if and only if it is contained in a closed cube K.

Example 1 The sets

$$X = \{x \mid 10 < x < 100\}, \qquad X = \{x \mid 10 \leq x \leq 100\},$$
$$X = \{x \mid x \in R^5, |x| < 10\}, \qquad X = \{x \mid x \in R^5, |x| \leq 10\},$$

are obviously bounded. The first two are contained in a closed segment $K = [10, 100]$. Analogously, the last two are contained in a closed cube of R^5 with edge $2k = 20$.

Example 2 Note that as a consequence of Definition 4.22 the following sets must be regarded as unbounded:

$$X = \{\ldots, -2, -1, 0, 1, 2, \ldots\},$$
$$X = \{x \mid x \geq 1\},$$
$$X = \{x \mid x \in R^5, |x| \geq 5\},$$
$$X = \{x \mid x \in R^5, x_i \leq 5, i = 1, 2, \ldots, 5\},$$
$$X = R^n.$$

All sets of the latter example do not satisfy our definition, and therefore are unbounded. None of them can be contained in a closed cube.

However, within the class of unbounded sets, some qualifications may be very useful.

Definition 4.23 Let X be a subset of R^n. X is said to be bounded from below (above) if it has a lower bound (upper bound).

As we see, an unbounded set may still be bounded from below or bounded from above, but not both. The latter conclusion is obvious and we shall put it in the form of a remark.

Remark 4.4 Let X be a subset of R^n. If X is bounded from below and bounded from above, then X is bounded.

Example 1 The set $X = \{1, 1/2, 1/3, \ldots, 1/n, \ldots\}$ is bounded since it is contained in the closed interval $K = [0, 1]$. The set $X = \{x \mid x \in Q^n\}$ or simply $X = Q^n$ is unbounded. The sets $X = \{0, 1, 2, 3, \ldots\}$, $X = \{x \mid x \in R^n, x_i \geq k\}$ and $X = \{x \mid x \in R^n, x_i > k\}$ are all unbounded, but bounded from below. The sets $X = \{\ldots, -2, -1, 0\}$, $x = \{x \mid x \in R^n, x_i \leq k\}$ and $X = \{x \mid x \in R^n, x_i < k\}$ are all unbounded, but bounded from above.

Example 2 We have already described the consumption set for a consumer in an economy as the set of all commodity bundles (points) the consumer would be willing to buy.

If no other constraint is put on the consumer's behavior (for instance no wealth constraint is imposed), then it is only logical to assume that his consumption set is unbounded. Nevertheless, as it is realistic to suppose that the consumer has a minimum amount of commodities he must consume if he wants to survive, the additional assumption is usually made that the consumption set is bounded from below.

Totally Bounded Sets

We shall introduce now the concept of a more restricted kind of boundedness, the importance of which will become clear in Section 4.4. We begin by defining an ε-net of a set of points X.

Definition 4.24 Let X be a subset of a metric space S, $E = \{e^1, \ldots, e^n\}$ a finite set of points, and ε a positive real number. E is said to be an ε-net of X if for every $x \in X$ there is an $e^{i_0} \in E$ such that $d(x, e^{i_0}) < \varepsilon$.

Now the notion of total boundedness can be easily defined.

Definition 4.25 Let X be a subset of a metric space S. Then X is said to be totally bounded if it has an ε-net for every $\varepsilon > 0$.

A first result from the preceding definition is immediate. Let $x, x' \in X$ represent any two points. Define then the distance $d(x, x') = |x - x'|$. We have thus determined a set of distances. We take next the least upper bound of the distances, call it the diameter of the set X, and denote it $d(X)$. Definition 4.22 tells us $d(X) < \infty$.

What Definition 4.25 tells can now be interpreted as follows. If a set X is totally bounded, then it is possible to decompose it into a finite number of sets, each with diameter less than ε. And this holds for every $\varepsilon > 0$.

Clearly, a totally bounded set is necessarily bounded in the sense of Definition 4.22. The converse, however, is not always true.

4.4 Compactness

Compactness is an important property a set of points may possess. The study of it usually occupies a very distinguished place in any topology book.

Insofar as we are concerned, compactness will play an essential role later, when we shall focus our attention on the theory of mappings and discuss the question of the existence of fixed points.

Let us also point out that in theoretical economics several important theorems strongly depend, more or less directly, on assumptions about the compactness of certain sets—consumption sets, production sets, and so forth —which may be crucial pieces of specific economic models.

To the term compactness as we shall see, it is possible to attach more than one meaning. In our study, for instance, we shall distinguish between "sequential compactness" and "bicompactness," and define them quite differently.

For extremely general topological spaces, neither does bicompactness imply sequential compactness nor does sequential compactness imply bicompactness. However, at the lower level of generality we shall allow for in this book (see the Notes at the end of the chapter), we shall work under the postulate that a bicompact set is always sequentially compact. The converse, however, need not be true. In the special case of a metric space, on the other hand, the situation is different. A bicompact subset of a metric space is necessarily sequentially compact, and conversely, every sequentially compact subset of a metric space is bicompact.

We are interested here in the two notions of compactness we have just mentioned. They will be given by means of definitions, each of which will be useful later in different contexts. But, since our study is almost exclusively related to metric spaces, we will be able to drop the distinction and, in following chapters, refer simply to the compactness of a set. For the special case of a Euclidean space, and this class of space is most important in economic theory, "closedness and boundedness" and "compactness" will appear to be equivalent properties.

Sequential Compactness

The notion of a convergent subsequence is essential for our next definition, so let us consider it first.

Take a sequence of points $x^1, x^2, \ldots$ and assume (i_n) represents a sequence of positive integers with $i_1 < i_2 < \cdots$. Now the sequence $x^{i_1}, x^{i_2}, \ldots$ is said to form a subsequence of $x^1, x^2, \ldots$.

Of course, like a sequence, a subsequence may be convergent or not. A property we state here without proof is that every bounded sequence has a convergent subsequence.

Definition 4.26 Let X be a subset of a topological space S. X is said to be sequentially compact if for each sequence $x^1, x^2, \ldots$ in X, there is a convergent subsequence the limit of which is some point $x \in X$.

In other words, every sequence of points of the set X contains a subsequence which is convergent to a point of X.

Example According to our definition the following sets are sequentially compact: $X = \{x \mid x \in R, a \leq x \leq b\}$, $X = \{(x, y) \mid (x, y) \in R^2, x^2 + y^2 \leq 16\}$, and $X = \{x \mid x \in R^n, |x| \leq k\}$.

Suppose now S is a sequentially compact space and consider a closed subset X of it.

Theorem 4.16 Let S be a sequentially compact space and X a closed subset of S. Then X is sequentially compact.

Proof Take a sequence $x^1, x^2, \ldots$ in X. Since S is sequentially compact, there exists a subsequence of $x^1, x^2, \ldots$ which converges to $x \in S$. But, by hypothesis, X is closed. So, X contains its limit points and therefore $x \in X$. Therefore, X is sequentially compact.

Example The closed interval [0, 10] is obviously a sequentially compact set. It is easy to see that any closed subset of [0, 10] is a sequentially compact set, whereas any open subset is not.

The next two theorems deal with the relation which exist between the properties of sequential compactness, on the one hand, and boundedness and closedness, on the other.

Theorem 4.17 Let X be a sequentially compact set. Then X is bounded.

Proof By definition, a sequentially compact set possesses the following property: With each sequence of points in X we can select a subsequence which converges to a point in X. In other words, for any sequence $x^1, x^2, \ldots$ we can find a subsequence $x^{i_1}, x^{i_2}, \ldots$ such that $\lim_{n \to \infty} x^{i_n} = x$, and $x \in X$. That means that there is always a positive number δ such that the distance $d(x, x^{i_j}) < \delta$. But assume the contrary, namely, that there is always some $\delta > 0$ such that $d(x, x^{i_j}) \geq \delta$. This would mean that there is a sequence of points in X for which no convergent subsequence could be found with the result that X would not be sequentially compact, contrary to hypothesis.

The latter result can also be formulated as follows: Let X be sequentially compact and x, x' any two points of X. Then the distance $d(x, x') = |x - x'|$ is finite.

Alternatively, define the diameter $d(X)$ of X as the lub of the set of all distances $|x - x'|$. Obviously, if X is sequentially compact, then its diameter $d(X)$ is finite.

Theorem 4.18 Let X be a sequentially compact set. Then X is closed.

Proof Suppose X is not closed. We can then find a sequence of points of X which converges to $x' \notin X$. Furthermore, every subsequence of the sequence converges to x'. So X is not sequentially compact, contrary to hypothesis.

The results of the last two theorems are particularly interesting when referred to subsets of the Euclidean space $E^n = (R^n, d)$. In this case, the converses of Theorem 4.17 and 4.18 also hold. Let us formulate this result as a remark. The proof is left as an exercise.

Remark 4.5 Let X be a subset of the Euclidean space E^n. Then the statements "X is sequentially compact" and "X is closed and bounded" are equivalent.

At the end of Section 4.3 (see in particular Definition 4.25), the idea of a totally bounded set was discussed. Let us consider now an important relation that exists between the properties of sequential compactness and total boundedness. Observe that, in the next theorem, our frame of reference is a metric space.

Theorem 4.19 Let X be a sequentially compact subset of a metric space S. Then X is totally bounded.

Proof Suppose X is not totally bounded. Thus, there is an $\varepsilon > 0$ for which X does not have an ε-net. In other words, there is a sequence of points $x^1, x^2, \ldots$ in X with $d(x^i, x^j) \geq \varepsilon$ for $i \neq j$. Since such a sequence cannot contain a convergent subsequence, it results that X is not sequentially compact, thus violating our assumption.

All that has been said is valid for sets of points. Let us study now what happens with the direct product of a fixed number of subsets X_i of the spaces R^{n_j}, where $j = 1, \ldots, m$. The following theorem shows a very important result.

Theorem 4.20 Let $X_1, X_2, \ldots, X_m$ be sequentially compact subsets of $R^{n_1}, R^{n_2}, \ldots, R^{n_m}$. Then the direct product $\prod_{j=1}^m X_j$ is a sequentially compact subset of $R^{\Sigma n_j}$.

Proof Let $x_1, x_2, \ldots, x_m$ denote elements of $X_1, X_2, \ldots, X_m$, respectively, and let $(x_1, x_2, \ldots, x_m)$ be an element of $\prod_{j=1}^m X_j$. Let $((x_1^1, x_1^2, \ldots), (x_2^1, x_2^2, \ldots), \ldots, (x_m^1, x_m^2, \ldots))$ denote a sequence of $\prod_{j=1}^m X_j$, where $(x_j^1, x_j^2, \ldots) \in X_j$ for $j = 1, 2, \ldots, m$. Since each X_j is sequentially compact, we can find a convergent subsequence for $(x_j^1, x_j^2, \ldots), j = 1, 2, \ldots, m$, such that for $i_j^1 < i_j^2 < \cdots$,

$$\lim_{n\to\infty} x_j^{i_j^n} = x_j;$$

that is,

$$\lim_{n\to\infty} x_1^{i_1^n} = x_1$$

$$\lim_{n\to\infty} x_2^{i_2^n} = x_2$$

$$\vdots$$

$$\lim_{n\to\infty} x_m^{i_m^n} = x_m.$$

This amounts to saying that

$$\lim_{n\to\infty} (x_1^{i_1^n}, x_2^{i_2^n}, \ldots, x_m^{i_m^n}) = (x_1, x_2, \ldots, x_m);$$

that is, $\prod_{j=1}^m X_j$ is sequentially compact.

Bicompactness

The property we are going to consider now is referred to, by many authors, simply as compactness. The term bicompactness, however, has a long tradition in the discipline and, in addition, will help to avoid confusion. This is the main reason we prefer to use it here.

Before defining bicompactness let us begin with a most important concept, namely that of a covering of a set.

Definition 4.27 Let $\{G_i\}$ be a collection of nonempty subsets of S. Let X be a nonempty subset of S. The collection $\{G_i\}$ is said to be a covering of X if $X \subset \bigcup_i G_i$. In particular, we call $\{G_i\}$ a finite covering if it is a finite collection; and we say that $\{G_i\}$ is an open covering if each $G_i \in \{G_i\}$ is an open set. Furthermore, if $\{G_i\}$ is a covering of X and $\{G_i\} \subset \{J_i\}$, we shall say that $\{G_i\}$ is a subcovering of $\{J_i\}$.

The concept of a finite covering is most significant for it will allow us to define bicompactness. But there is another place in which the concept is also relevant. Let us look at this first.

In Section 4.3, two ideas were introduced which we shall recall now. First, we defined an ε-net of a set and then, by means of it, we were able to talk about total boundedness as a special property of a metric space.

Remember that, if an ε-net exists for X, then we can decompose X into a finite number of sets having diameter less than ε for every $\varepsilon > 0$.

The ε-net of a totally bounded set is, by definition, a finite set of points. For every $x \in X$ it contains a point $e^{i_0} \in \{e^1, \ldots, e^n\}$ such that $d(x, e^{i_0}) < \varepsilon$.

If instead of considering the finite set of points $E = \{e^1, \ldots, e^n\}$ and the distance $d(x, e^{i_0}) < \varepsilon$, we think of the finite collection of neighborhoods $\{G_i\} = \{\alpha(x^1), \ldots, \alpha(x^n)\}$ with diameter $d(\alpha(x^i)) < \varepsilon$, we would have the following situation: If $x^i \in A$, where A is an open set in X, there will be a neighborhood $\alpha(x^i)$ such that $x^i \in \alpha(x^i) \subset A$, and, thus, $A = \bigcup \alpha(x^i)$, absorbing all $x^i \in A$.

In E^n, we immediately think in a collection of spherical neighborhoods with rational diameter less than ε, for, as we know, Q^n is a countable set dense in R^n. For any fixed ε we would then have a countable collection of neighborhoods which would constitute a "countable covering."

Now, in our particular case, we have a set X which by hypothesis is totally bounded. In other words, we would be able to construct a finite covering $\{G_i\}$ for each $\varepsilon > 0$.

It is apparent now that, applying the very useful notion of a finite covering, Definition 4.25 can be conveniently reformulated as follows.

Definition 4.28 Let X be a subset of a metric space S. X is said to be totally bounded if it has a finite open covering $\{G_i\}$ with $d(G_i) < \varepsilon$ for every $\varepsilon > 0$.

Let us come back to our point of departure and define bicompactness with help of Definition 4.28.

Definition 4.29 Let X be a subset of a topological space. Then X is said to be a bicompact set if and only if each open covering of X contains a finite subcovering.

What the new definition means is the following: If $X \subset \bigcup_i G_i$ and we can select a finite number of open sets $G_{i_1}, \ldots, G_{i_r}$ of $\{G_i\}$ such that $X \subset G_{i_1} \cup \cdots \cup G_{i_r}$, then X is bicompact. Each element of X belongs to at least one open set and the number of these sets is finite.

The concept of a covering will reappear later in Chapter 7, in particular in relation to Definition 7.10.

Let us study now some interesting properties of bicompact spaces.

Theorem 4.21 Let S be a bicompact space. Then every infinite subset of S contains at least one accumulation point in S.

Proof Take an infinite subset $A \subset S$ and suppose it has no accumulation point in S. Then, each $x \in S$ belongs to an open set, say G_x, with at most one point of A. Now, the collection $\{G_x | x \in S\}$ forms an open covering of S. Since S is bicompact, $\{G_x\}$ has a finite subcovering, say $\{G_{x_1}, \ldots, G_{x_m}\}$. Therefore, $A \subset S \subset G_{x_1} \cup \cdots \cup G_{x_m}$. Each G_{x_i} contains only one point of A.

Thus, $A \subset G_{x_1} \cup \cdots \cup G_{x_m}$ can contain at most m points. This means A is a finite set, which violates our assumption. Therefore, every infinite subset of S contains at least one accumulation point in S.

The next property is most important, for it establishes a first link between a bicompact and a sequentially compact set at our restricted level of generality (see Notes).

Theorem 4.22 Every bicompact space is sequentially compact.

Proof Assume X is bicompact and choose a sequence $x^1, x^2, \ldots$. In regard to $x^1, x^2, \ldots$ one of two situations is possible. First, it is possible that the sequence contains only a finite number of different terms. In this case, one of them must occur an infinite number of times. It would constitute thus a convergent subsequence of (x^i). Second, it is possible that an infinite number of the terms of the sequence are different. In this case, they obviously form an infinite subset of X. Thus, since X is bicompact, the subset has at least one accumulation point. This means that there is a subsequence converging to it. Therefore, X is sequentially compact.

The next property represents, for a bicompact space, the analog of what Theorem 4.16 meant for a sequentially compact one.

Theorem 4.23 Let S be a bicompact space and X a closed subset of S. Then X is bicompact.

Proof Since S is bicompact, it has by definition an open covering $\{G_i\}$. And this must contain a finite subcovering, which in turn is an open finite covering of the subset X. Therefore, the bicompactness of X is immediate.

Union and Intersection of Compact Sets

From our previous study of sequential compactness and bicompactness we are now able to derive certain useful results concerning union and intersection of sets.

Theorem 4.24 Let $X_1, \ldots, X_n$ be a finite number of sequentially compact sets. Then $\bigcup_{i=1}^n X_i$ is sequentially compact.

Proof From Theorem 4.11 the union is a closed set. Besides, since we have only a finite number of bounded sets, the union is a bounded set, that is, has a finite diameter. Therefore every sequence in $\bigcup_{i=1}^n X_i$ has a convergent subsequence with limit $x \in \bigcup_{i=1}^n X_i$.

Theorem 4.25 Let $X_1, X_2, \ldots$ be an arbitrary number of sequentially compact sets. Then $\bigcap_i X_i$ is sequentially compact.

Proof From Theorem 4.12 the intersection is a closed set and, by definition of intersection, a subset of each X_i. But each X_i is sequentially compact. Therefore, by Theorem 4.16, $\bigcap_i X_i$ is sequentially compact.

Let us consider now the case of bicompact sets. The results as we are going to see, are similar.

Theorem 4.26 Let $X_1, \ldots, X_n$ be a finite number of bicompact sets. Then $\bigcup_{i=1}^n X_i$ is bicompact.

Proof Let $\{G_i\}$ represent an open covering of the union $\bigcup_{i=1}^n X_i$. Then the sets $\{G_i\}$ cover each of the bicompact sets X_i and thus contain a finite subcovering for each X_i. There are a finite number of X_i, so $\{G_i\}$ contains a finite covering for $\bigcup_{i=1}^n X_i$ and the latter is bicompact.

Theorem 4.27 Let $X_1, X_2, \ldots$ be an arbitrary number of bicompact sets. Then $\bigcap_i X_i$ is bicompact.

Proof From Theorem 4.12 the intersection is a closed set and, by definition of intersection, a subset of each X_i. But each X_i is bicompact. Therefore, by Theorem 4.23, $\bigcap_i X_i$ is bicompact.

Compactness in Metric Spaces

Most of what has been said so far regarding compactness has a very general character, for it has been referred to the concept of a topological space.

In general while it is true that any bicompact space is also sequentially compact it is not possible to ensure that the converse is also true, namely that any sequentially compact space is also bicompact. However, there is to the latter a most important exception which will be discussed at this point. It concerns, as was already mentioned in another place, metric spaces.

Theorem 4.28 Let X be a subset of a metric space S. Then, if X is bicompact, it is sequentially compact and conversely.

Proof We have only to prove the second part of the theorem. That every bicompact space is sequentially compact has already been proved in Theorem 4.22. This result, proved for a topological space, is obviously also valid for the special case of a metric space. Let us show that in a metric space any sequentially compact set is also bicompact. Assume X, according to the theorem, is sequentially compact. As a consequence, it is totally bounded, that is, it allows for a finite open covering for every number $\varepsilon > 0$. Therefore, it is bicompact.

A more elegant but lengthy proof than that given here might be constructed by means of the famous Heine–Borel theorem of the calculus. In our case, we simply direct the reader to the definitions of total boundedness presented before. We have shown how, from the definition of an ε-net, it is possible to derive and construct a "countable covering" for X. But since a sequentially compact set is always totally bounded, the covering will only be a finite one, thus satisfying the definition of bicompactness.

At this stage, we might also expand the result of Remark 4.5.

Remark 4.6 Let X be a subset of the Euclidean space E^n. Then the statements "X is sequentially compact," "X is bicompact," and "X is closed and bounded" are equivalent.

We know now that, in metric spaces, the concepts of bicompactness and sequential compactness are equivalent. Since in our investigations we are almost exclusively concerned with metric spaces, it seems reasonable to simplify our terminology and speak in the future of compactness, with the understading that sets of points which possess this property are both bicompact and sequentially compact.

Furthermore, we should remember that for sets in Euclidean space closedness together with boundedness means compactness. This relation is most important because of the particular interest that E^n will have in our further discussions.

4.5 Connectedness

For the following definition we make use of the notion of partition as introduced in Section 1.4, Chapter 1.

Definition 4.30 Let X be a subset of R^n. Then X is said to be connected if and only if it cannot be partitioned into two nonvoid subsets closed in X. Otherwise it is said to be disconnected.

The above definition can be interpreted as follows: Let X be a set of points in R^n. We find two nonempty proper subsets of X, say Y and Z, such that the first condition for a partition of X is satisfied, that is,

(1) $X = Y \cup Z.$

Now if it also happens that

(2) $Y \cap Z = \varnothing,$

then the second condition for a partition of X is also satisfied and X is a disconnected set. On the other hand, if for any closed subsets Y, Z, for which condition (1) is satisfied, we have

(3) $Y \cap Z \neq \varnothing,$

then we say that X is connected.

In other words, a set $X \subset R^n$ is connected when it is not the union of two disjoint, nonvoid closed subsets.

The following remark results directly from Definition 4.24. The proof is left as an exercise.

Remark 4.7 Let R be the set of real numbers. A subset $X \subset R$, containing at least two points, is connected if and only if X is an interval.

Example 1 Figure 4.1 shows the connected set X. Note that $Y \cup Z = X$ while $Y \cap Z \neq \emptyset$.

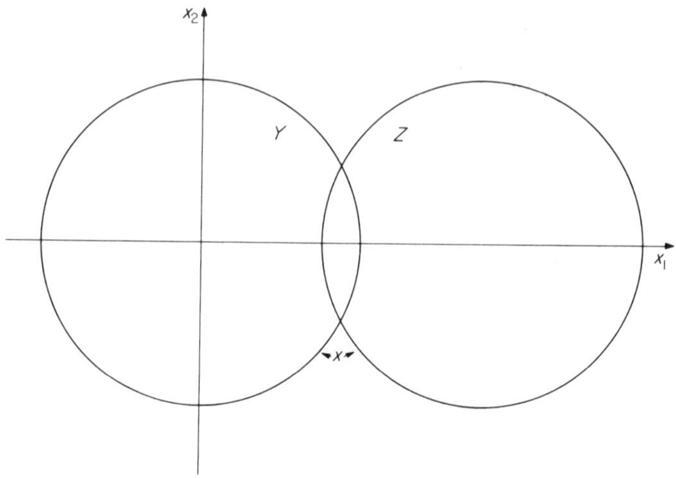

Figure 4.1

Example 2 In Fig. 4.2 X is a disconnected set. $Y \cup Z = X$ but $Y \cap Z = \emptyset$.

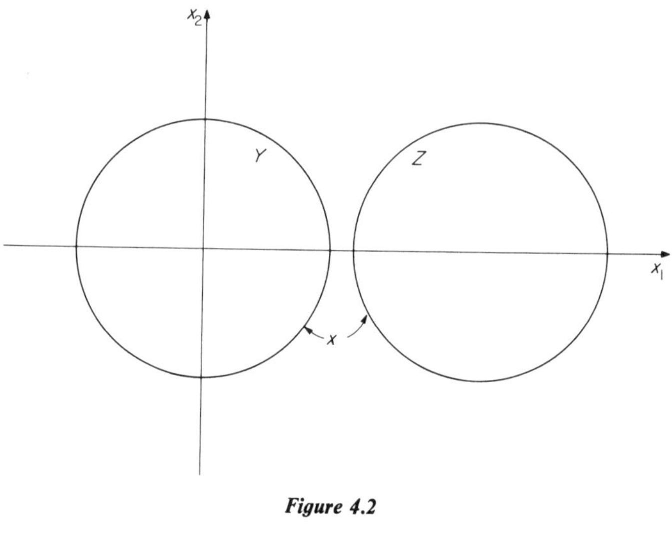

Figure 4.2

4.6 Convexity

Convexity is the last property of sets we shall consider in this chapter. For the time being we will content ourselves with introducing the concept in a very general manner. Later, in Part II, we shall meet the notion of a convex set in a more specific situation.

Convexity plays an important role in economic analysis in very different contexts: In production theory it is generally assumed that the production set is convex; in consumption theory it is not only assumed that the consumption set of each consumer satisfies this property, but also that the preference ordering of each consumer adjusts to some convexity requirements. The roles played by such assumptions are usually those of facilitating the proofs of some basic economic theorems.

Before entering the discussion about convexity of sets, it is necessary to describe some operations which can be performed with points of R^n.

For two points $x^r, x^s \in R^n$ we understand the sum to be the point $y = x^r + x^s$ such that $y_i = x_i^r + x_i^s$ for each component $i = 1, \ldots, n$. If x^r and x^s belong to a subset X of R^n, it is clear that the sum y may or may not belong to the subset X.

The product of a scalar λ and a point $x \in R^n$ is defined as the new point $\lambda x = (\lambda x_1, \ldots, \lambda x_n) = (x_1 \lambda, \ldots, x_n \lambda) = x\lambda$.

If λ is an element of the closed interval [0, 1], we define a convex combination or weighted average of x^r and x^s as $\lambda x^r + (1 - \lambda)x^s$. Notice that $\lambda x^r + (1 - \lambda)x^s$ belongs to the closed segment $[x^r, x^s]$, that is, lies on the segment joining the two points.

In general, for m points $x^1, x^2, \ldots, x^m$ and weights $\lambda_1, \lambda_2, \ldots, \lambda_m$ such that $\lambda_j \in [0, 1]$ and $\sum_{j=1}^m \lambda_j = 1$, we define a convex combination or weighted average as $\lambda_1 x^1 + \lambda_2 x^2 + \cdots + \lambda_m x^m$.

Finally, though the utility of this operation will only appear later, let us for the sake of completeness define the product of two points $x^r, x^s \in R^n$ as follows: $x^r x^s = \sum_{i=1}^n x_i^r x_i^s$. It is clear from the definition that while $x^r, x^s \in R^n$, the product $x^r x^s \in R$.

The description of a convex set is now straightforward.

Definition 4.31 Let X be a subset of R^n and x^r, x^s any two points of X. Then X is said to be a convex set if and only if, for $\lambda \in [0, 1]$, $\lambda x^r + (1 - \lambda)x^s \in X$.

The definition of convexity can also be interpreted as follows. For any two elements of the set, the points on the line segment joining the elements must also belong to the set.

Example Observe that X in Fig. 4.3 is a convex set whereas Y is not.

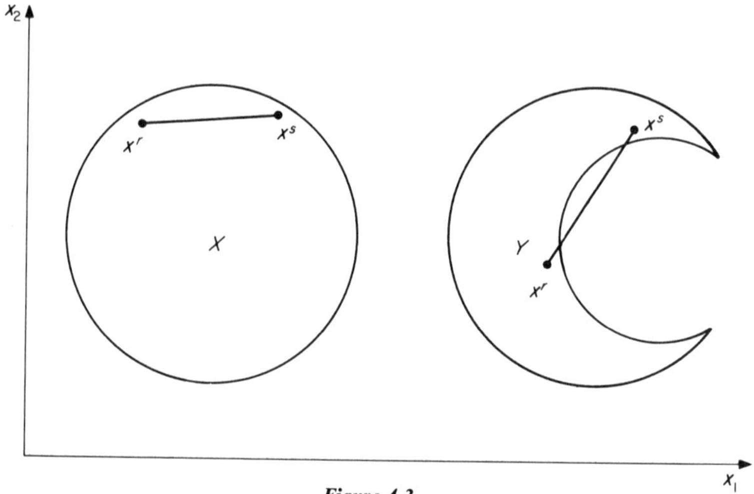

Figure 4.3

Theorem 4.29 Let $X_1, X_2, \ldots, X_m$ be convex subsets of $R^{n_1}, R^{n_2}, \ldots, R^{n_m}$. Then the direct product $\prod_{j=1}^m X_j$ is a convex subset of $R^{\Sigma n_j}$.

Proof Let $(x_1^r, x_2^r, \ldots, x_m^r)$, and $(x_1^s, x_2^s, \ldots, x_m^s)$, be two points of $\prod_{j=1}^m X_j$, where $x_1^r, x_1^s \in X_1, \ldots, x_m^r, x_m^s \in X_m$. For $\lambda \in [0, 1]$ we have

$$\lambda(x_1^r, x_2^r, \ldots, x_m^r) + (1-\lambda)(x_1^s, x_2^s, \ldots, x_m^s) = \\ (\lambda x_1^r + (1-\lambda)x_1^s, \ldots, \lambda x_m^r + (1-\lambda)x_m^s).$$

Now the new point, since the sets X_j are convex for $j = 1, 2, \ldots, m$, has the following components,

$$\begin{gathered} \lambda x_1^r + (1-\lambda)x_1^s \in X_1 \\ \lambda x_2^r + (1-\lambda)x_2^s \in X_2 \\ \vdots \\ \lambda x_m^r + (1-\lambda)x_m^s \in X_m \end{gathered}$$

and as a result lies in $\prod_{j=1}^m X_j$. In other words $\prod_{j=1}^m X_j$ is convex.

Consider now a set of convex sets.

Theorem 4.30 Let $\{X_i\}_{i \in I}$ be a family of convex sets. Then $\bigcap_{i \in I} X_i$ is a convex set.

Proof Take any two points of the intersection, say $x^r, x^s \in \bigcap_{i \in I} X_i$. We then have that $x^r, x^s \in X_i$ for every $i \in I$. But since every set is convex, we also have for $\lambda \in [0, 1]$, $\lambda x^r + (1-\lambda)x^s \in X_i$ for every $i \in I$. Thus, it also belongs to the intersection. Hence $\bigcap_{i \in I} X_i$ is convex.

The following theorem, concerning the closure of a convex set is easy to prove, and left as an exercise.

Theorem 4.31 Let X be a convex set. Then its closure $\overline{X}$ is also convex.

An interesting result can be achieved by comparing the properties of connectedness and convexity.

From the definition of convexity, it is obvious that any convex set is connected. In effect, the conditions for set convexity are much stronger than those for connectedness. On the other hand, a connected set may or may not be convex. Let us write this in the form of a remark.

Remark 4.8 If a set X is convex, then it is necessarily connected.

An important concept is that of the convex hull or convex closure of a set. We must be careful not to confuse a convex closure with the convexity property of a closure as formulated in Theorem 4.31. The latter merely refers to a property of the closure or adherence of a convex set. The former denotes a new concept.

Definition 4.32 Let X be a set of points. The convex hull or convex closure of X, denoted $\dot{X}$, is the intersection of all convex sets containing X.

From Theorem 4.30 we know that the intersection of convex sets is a convex set. We have then the following remark.

Remark 4.9 The convex hull or convex closure $\dot{X}$ of a set X is convex.

Since $\dot{X}$ has been defined as an intersection of convex sets, it only contains those points common to all sets entering the intersection operation. We have then the following two important properties.

Remark 4.10 The convex hull or convex closure $\dot{X}$ of a set X is the smallest convex set containing X.

Remark 4.11 Let X be a closed, convex set. Then $X = \dot{X}$.

The notion of convexity has a very important place in the study of linear spaces. In Part II we shall have to come back to it and pay attention to some of the questions it gives rise to.

4.7 Sum of Sets in E^n

A concept which we shall find specially useful in Part II is that of a direct sum of subsets of the Euclidean space E^n. In principle it is a generalization of the idea of a sum of points, introduced in Section 4.6.

Definition 4.33 Let $X_1, \ldots, X_r$ be r subsets of R^n. The direct sum $\sum_{j=1}^{r} X_j$ is the set of all points $y \in R^n$ such that $y = \sum_{j=1}^{r} x_j$, where $x_j \in X_j, j = 1, \ldots, r$.

In substance $\sum_{j=1}^{r} X_j$ is the set of all points which result from adding, in all possible ways, r points each of which belongs to a different X_j.

Example Let $X_1, X_2 \subset R^2$ be described, respectively, by $X_1 = \{x_1 \mid x_1 \in R^2, 0 \le x_{1,1} \le 1, x_{2,1} = 0\}$ and $X_2 = \{x_2 \mid x_2 \in R^2, x_{1,2} = 0, 0 \le x_{2,2} \le 1\}$. Then $X_1 + X_2 = \{y \mid y \in R^2, 0 \le y_i \le 1, i = 1, 2\}$. Observe that each y in the latter set has component $y_i = x_{i,1} + x_{i,2}$.

Some notational clarification is convenient at this point. For a set $X_j \subset R^n$ we can define $-X_j$ as the set of all points $-x_j$ where $x_j \in X_j$.

In particular, for the sets X_i and $-X_j$ we can define the sum $X_i + (-X_j) = X_i - X_j$ as the set of points $y \in R^n$ such that $y = x_i + (-x_j) = x_i - x_j$, where clearly $x_i \in X_i$ and $-x_j \in -X_j$.

It is very important not to mix up this idea with the concept of a set-theoretical difference of two sets as presented in Definition 1.11. We use a \, for convenience, to represent the latter. Notational differences, as shown below, correspond to conceptual differences. Notation and meaning of the difference of X_1 and X_2 is

$$X_1 \setminus X_2 = \{x \mid x \in X_1, x \notin X_2\},$$

whereas for the direct sum of X_1 and $-X_2$ we have

$$X_1 - X_2 = \{y \mid y \in R^n, y = x_1 - x_2, x_1 \in X_1, x_2 \in X_2\}.$$

Let us finally refer to three properties of the direct sum of sets. The first is immediate so we can leave the proof.

Theorem 4.32 Let $X_1, \ldots, X_r$ be r subsets of R^n. Then $\sum \bar{X}_j \subset \overline{\sum X_j}$.

Theorem 4.33 Let $X_1, \ldots, X_r$ be r compact subsets of R^n. Then $\sum_{j=1}^{r} X_j$ is compact.

Proof Suppose $Y = \sum X_j$ represents the sum of the r sets X_j. Consider a sequence $\{y^i\}$ in $Y = \sum_{j=1}^{r} X_j$. Since there are only a finite number of X_j, there must be an infinite number of terms of the sequence $\{y^i\}$ in some X_j, say X_k. Then the y^i terms in that X_k form a sequence $\{x^i\}$. Since X_k is compact, $\{x^i\}$ has a convergent subsequence. But this means $\{y^i\}$ also has a convergent subsequence and so Y is compact.

Theorem 4.34 Let $X_1, \ldots, X_r$ be r convex subsets of R^n. Then $\sum X_j$ is convex.

Proof Call the sum $Y = \sum X_j$, and take two points $y \in Y$ and $y' \in Y$. Clearly we must have $x_j \in X_j$ and $x'_j \in X_j$ such that

$$y = \sum x_j, \qquad y' = \sum x_j'.$$

Now consider the convex combination, $\lambda y + (1 - \lambda)y'$. It is obvious that we have

$$\lambda y + (1 - \lambda)y' = \lambda \sum x_j + (1 - \lambda)\sum x_j' = \sum [\lambda x_j + (1 - \lambda)x_j'].$$

Now, for $j = 1, \ldots, r$ we have $[\lambda x_j + (1 - \lambda)x_j'] \in X_j$, for each X_j is convex. Therefore $\lambda y + (1 - \lambda)y' \in Y$ and Y is convex.

EXERCISES

Countable and Denumerable Sets

E.4.1 Show that any subset of a countable set is countable.

E.4.2 Let S represent the set of all circles with rational radii and centers in E^2. Show that S is countable.

Closed and Open Sets

E.4.3 Determine all the accumulation points of the following sets: (a) Z, the set of all integers; (b) Q, the set of all rationals; (c) Q^c, the set of all irrationals; (d) $[a, b[$; (e) $\{x \mid x = (-1)^m + (1/n);\ m, n = 1, 2, \ldots\}$; (f) $\{x \mid x = (1/m) + (1/n);\ m, n = 1, 2, \ldots\}$.

E.4.4 In E.4.3 determine the corresponding derived sets and closures, and state whether each set is closed or open.

E.4.5 Prove that a set X is closed if and only if X^c is open.

Hint: Note that $(X^c)^c = X$.

E.4.6 Show with an example that the union of an arbitrary number of closed sets need not be closed.

Hint: Study the union over N of the sets of $\{A_n\}_{n \in N}$, where $A_n = \{x \mid 1 \geq x \geq 1/n\}$.

E.4.7 Give an example to show that the intersection of an arbitrary number of open sets need not be open.

Topological Spaces

E.4.8 Show that a set X with a topology $\tau = \{\varnothing, X\}$ define a topological space (X, τ).

Hint: The above topology is called "indiscrete" topology. Show that Definition 4.19 is satisfied.

E.4.9 Show that $X = \{1, 2, 3, 4\}$ together with the "indiscrete" topology $\tau = \{\varnothing, \{1, 2, 3, 4\}\}$ form a topological space.

E.4.10 Let $X = \{1, 2, 3, 4, 5\}$ with the topology $\tau = \{\varnothing, \{1\}, \{1, 2\}, \{1, 3, 4\}, \{1, 2, 5\}, \{1, 2, 3, 4\}, \{1, 2, 3, 4, 5\}\}$.

(a) Show that (X, τ) is a topological space,
(b) Determine all closed subsets of X,
(c) Specify the closures of the sets $\{1\}, \{2, 3\}, \{2, 3, 4\}, \{4, 5\}$, and find which of these sets, if any, is dense in X.

Hint: (a) Apply definition 4.19. (b) The closed subsets in X are the complements of the open subsets in τ, for example, X, $\{2, 3, 4, 5\}, \ldots$. (c) Define the closure $\bar{S}$ of $S \subset X$ as the intersection of all closed supersets of S. Observe that $\overline{\{1\}} = \{1, 2, 3, 4, 5\}$. Thus, $\{1\}$ is dense in X. What about the other three sets?

Totally Bounded Sets

E.4.11 Let $X \subset R^n$ be a totally bounded set. Prove following propositions: (a) $\bar{X}$ is totally bounded. (b) Any $S \subset X$ is totally bounded. (c) X is bounded.

Compactness

E.4.12 Show that if X is closed and Y is compact, then $X \cap Y$ is compact.

E.4.13 Show that a subset of R is compact if and only if it is closed and bounded.

E.4.14 Let $d(x, y)$ represent the distance between two points in S, and let $X \subset S$ and $Y \subset S$. Then the distance between the sets X and Y is defined as $d(X, Y) = \inf\{d(x, y) \mid x \in X, y \in Y\}$. Prove: if X and Y are compact sets, then there exist points $x' \in X$ and $y' \in Y$ such that $d(x', y') = d(X, Y)$.

E.4.15 Prove: if X is a compact subset of a metric space, then X^d and $\bar{X}$ are compact.

E.4.16 Prove Remark 4.5.

Hint: Generalized version of E.4.13.

Connectedness

E.4.17 Prove that the set $R \setminus \{0\}$ is not connected.

E.4.18 Determine which of the following subsets of R^2 is disconnected and which is connected: $A = \{(x, y) \mid x^2 - y^2 \geq 9\}$; $B = \{(x, y) \mid x^2 + y^2 \leq 9\}$.

Hint: Observe that the lower boundary of A and the upper boundary of B are given by the equations of a hyperbola and a circumference, respectively.

E.4.19 Prove Remark 4.7.

Convexity

E.4.20 Which of the following subsets of R^2 is convex?

$$A = \{(x, y) \mid x^2 - y^2 \geq 9\}; \qquad B = \{(x, y) \mid x^2 - y^2 \leq 9\};$$
$$C = \{(x, y) \mid x^2 + y^2 \geq 9\}; \qquad D = \{(x, y) \mid x^2 + y^2 \leq 9\};$$
$$E = \{(x, y) \mid x^2 + y^2 = 9\}.$$

E.4.21 Let $X \subset R^n$ be a convex set. Show that X is connected.

E.4.22 Prove Theorem 4.31.

PROBLEMS

P.4.1 Prove that the set P of all mth degree polynomials $a_0 + a_1 x + \cdots + a_m x^m$ with integral coefficients is denumerable.

Hint: For each pair $(n, m) \in N \times N$, the set P_{nm} of polynomials of mth degree and satisfying $|a_0| + |a_1| + \cdots + |a_m| = n$ is countable. Study this and analyze $P = \bigcup P_{nm}$.

P.4.2 Let $X \subset R^n$. Show that X^{d} is closed.

Hint: Show that every accumulation point of X^{d} is also an accumulation point of X, that is, $(X^{\mathrm{d}})^{\mathrm{d}} = X^{\mathrm{d}}$.

P.4.3 Show that a set X having as a topology the collection of all subsets of X forms a topological space ("discrete" topology).

P.4.4 Prove that any compact metric space (X, d) is complete.

Hint: A metric space is said to be complete if every Cauchy sequence in X converges to a point in X; a sequence $x^1, x^2, \ldots$ in X is a Cauchy sequence if for every $\eta > 0$ there is an integer $n > 0$ such that $d(x^i, x^j) > \eta$ whenever $i, j > n$.

P.4.5 In the framework of Problem P.4.2 define (a) the aggregate consumption set $X = \sum_{i=1}^m X_i$; (b) the aggregate production set $Y = \sum_{j=1}^n Y_j$; (c) the net demand set $X - Y$; (d) the excess demand set $X - Y - \{r\} = Z$. Let $A = \{((x_i), (y_j)) \mid x_i \in X_i, \text{ every } i, y_j \in Y_j, \text{ every } j, x - y = r\}$ be, once again, the set of attainable states.

Prove: (a) For $A \neq \varnothing$ it is necessary and sufficient that $r \in (X - Y)$; (b) if each X_i and each Y_j is closed and convex, A is closed and convex; (c) if each X_i and each Y_j is convex, both $X - Y$ and Z are convex.

Hint: Debreu [4].

NOTES

Franz [1] and Kuratowski [2] give excellent discussions of spaces and set properties. The terminology followed here is essentially that of Kuratowski [2]. For a complete and advanced treatment see Bourbaki [1], Dunford and Schwartz [1], and Kelley [1]. Given the level and the character of this book we have only distinguished between two concepts of compactness. Further conceptual refinements are possible. For instance: A space is called countably compact when every countable open covering has a finite subcovering. Both bicompact and sequentially compact space are countably compact. However, the converses of the two propositions involved do not hold in general. On the other hand, for the class of "first countable spaces" (see definition in Lipschutz [1]) sequential and countable compactness coincide, that is, are equivalent concepts. These spaces are general enough for the purposes of this book and they constitute, indeed, the "upper limit" of generality we allow ourselves in the current discussions. We should be aware that beyond this limit the statement "in principle, a bicompact set is always sequentially compact" and Theorem 4.22 would not hold in general. See also Vaidyanathaswamy [1]. For the economic theoretical relevance of the topics discussed in this chapter the reader should consult Debreu [4] and Nikaido [4]. See also Arrow and Hahn [1] and Nikaido [3].

CHAPTER 5

POINT-TO-POINT MAPPINGS

The concept of a function from a set X to a set Y was introduced in Chapter 1 in an instrumental sense, namely as an aid to studying certain sets, set operations and set properties in which we were especially interested. In this and in Chapter 6, the idea of a function will be the direct subject of our analysis.

The terms mapping, function, and transformation are used here with the same meaning. However, to have a consistent terminology, a classification will be useful.

Mappings may be grouped into two wide categories. First, there is the class of mappings in which an element of a set is associated with an element of another set. Mappings of this class are usually called point-to-point mappings or single-valued functions and are the subject of study in the present chapter. We shall adhere to this rather convenient terminology, but without completely excluding the use of the terms transformation or, simply, function as synonyms.

Second, there is the class of mappings in which an element of a set is associated with a nonempty subset of another set. Mappings of this class are generally referred to as point-to-set mappings or multivalued functions. We shall study mappings of this class in Chapter 6, and use the mentioned terms or the alternative designations, multivalued transformations or correspondences, as synonyms.

5.1 Single-Valued Functions

A function has been described in Chapter 1, Definition 1.7. An alternative, precise definition is the following:

Definition 5.1 Let X and Y be two sets. A function from X into or onto Y is a binary relation from X to Y satisfying the following additional properties: (a) $X \neq \varnothing$; (b) if $y^1 = f(x)$ and $y^2 = f(x)$, then $y^1 = y^2$.

Domain and Range

Our definition points out that a function is a special case of a binary relation. Not only is the domain X assumed to be nonempty but also the uniqueness of the image of every element $x \in X$ is required.

Remember that in Chapter 1 we distinguished between the codomain Y and the range of the function, the latter being the subset of Y which contains all those elements which are images of at least one element in X. The range will be denoted $f(X)$.

To symbolically describe a function, it is usual to write f: $X \to Y$ (the domain set at the left of the arrow) together with some explicit formulas summarizing the association rule in terms of operations.

In calculus, functions are most frequently described by means of a rule.

Example 1 Consider the function f: $R \to R$ given by $f(x) = x^2$. Here both domain and codomain are identical, namely, the set of real numbers. Images of -4, -2, 0, 2, 4 are, respectively, $f(-4) = 16$, $f(-2) = 4$, $f(0) = 0$, $f(2) = 4$ and $f(4) = 16$. The range of the function is $f(X) = \{x \mid x \in R,\ x \geq 0\}$.

In the usual calculus notation, as we know, the above function is simply described $y = x^2$.

Example 2 An example of a nonnumerical function is given by the rule that assigns to each country in the world its capital city.

Graph of a Function

Associated with a function is its graph. We can define it as follows:

Definition 5.2 Let f: $X \to Y$ be a function. The subset of the direct product $X \times Y$ which contains all ordered pairs $(x, f(x))$ is called the graph of f:$X \to Y$.

The preceding definition stresses once more the special character of a function when compared to a relation. For each element $x \in X$ there is one and only one ordered pair $(x, f(x))$ appearing in the graph of f: $X \to Y$.

Example In Fig. 5.1 we have drawn the well known graph of $y = x^2$ which, of course, satisfies Definition 5.2.

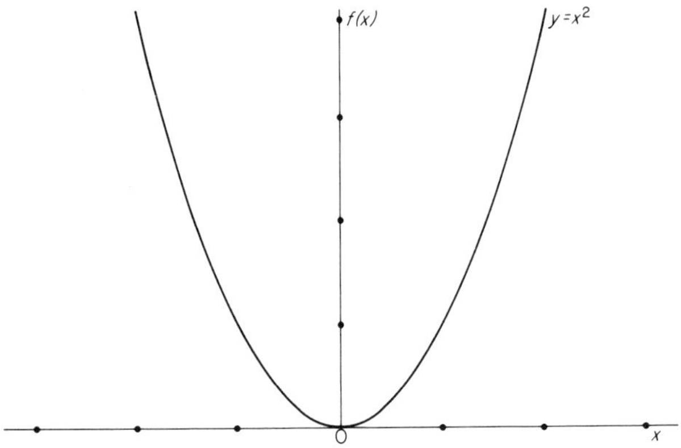

Figure 5.1

For the remainder of this chapter our interest will be concentrated on functions whose domain and codomain are sets of points. In particular, X and Y will be regarded as subsets of the set R or the product set R^n.

Scalar and Vector-Valued Functions

Consider the expression f: $X \to Y$. According to how the sets X and Y have been defined we will obtain quite different kinds of mappings. From the different possibilities there are four general types of mappings which deserve special interest.

(a) Let both X and Y be equal to R. The mapping is then f: $R \to R$. The image of each number $x \in R$ is a number $f(x) = y \in R$. A function of this sort is said to be a scalar-valued function of a scalar. In calculus terminology one refers to "functions of one variable."

(b) Let X be equal to R^n and Y to R. We have then $f\colon R^n \to R$. The image of each n-component point (or vector) $(x_1, \ldots, x_n) = x \in R^n$ is a number $f(x) = f(x_1, \ldots, x_n) = y \in R$. Such a mapping is called a scalar-valued function of a vector. The so-called "function of several variables" studied in the calculus falls within this category.

(c) Let X be equal to R and Y to R^m. The mapping becomes $f\colon R \to R^m$. Note that the image of each number (or scalar) $x \in R$ is an m-component point (or vector) $f(x) = (y_1, \ldots, y_m) = y \in R^m$. We shall call this mapping a vector-valued function of a scalar. Functional equations like differential or difference equations belong to this class.

(d) Finally let X and Y be equal to R^n and R^m, respectively. We have $f\colon R^n \to R^m$. Here the image of each n-component point (or vector) $(x_1, \ldots, x_n) = x \in R_n$ is $f(x_1, \ldots, x_n) = (y_1, \ldots, y_m) = y \in R^m$. We have here a vector-valued function of a vector. Linear transformations such as those studied in Part II belong to this class of mappings.

It is convenient to take notice of the fact that each of the described function types defines a special class of point-to-point mappings or single-valued functions. They all satisfy Definition 5.1, and we might, if we like, order them according to there decreasing generality. We would then get $f\colon R^n \to R^m$ at one end and $f\colon R \to R$ at the other end of the scale.

In particular, if R^n is thought as being an n-dimensional coordinate system, we can make the following interpretation:

Definition 5.3 Let X_j be a subset of R for $j = 1, 2, \ldots, n$ such that $\prod_{j=1}^n X_j$ is a subset of R^n. The function that associates with each $x \in \prod_{j=1}^n X_j$ its ith coordinate $x_i \in X_i$ is called the projection function on X_i, for $i = 1, 2, \ldots, n$.

A projection function then is a special type of point-to-point mapping.

There are at least two ways of describing more general function classes than those examined in (a)–(d).

First, we can replace the underlying set R by the set of complex numbers C. Second, we may define our functions on an infinite product set $X_1 \times X_2 \times \ldots$ rather than on $X_1 \times \ldots \times X_n$. In the present chapter, however, our interest will be focused almost exclusively on mappings belonging to the classes described in the preceding.

Equal Functions, Restrictions and Extensions of Functions

Consider now two functions, say f_1 and f_2, defined on the same sets, say $f_1\colon X \to Y$, $f_2\colon X \to Y$.

Definition 5.4 Two functions f_1 and f_2 defined on the same domain and having the same range are said to be equal if and only if for every $x \in X$ we have $f_1(x) = f_2(x)$.

In other words, equality of functions requires equality of images for every point in the common domain.

Take a subset A of X and redefine function f_1 as follows:

Definition 5.5 Consider functions f_1: $A \to Y$ and f_2: $X \to Y$, where $A \subset X$. Then f_1 is said to be a restriction of f_2 and f_2 is said to be an extension of f_1 if and only if for each $x \in A$, $f_1(x) = f_2(x)$.

Example 1 Let $X = \{\alpha \mid 0 \leq \alpha \leq \pi/2\}$ and $Y = [0, 1]$. Let f_1 and f_2 be two functions from X to Y defined by $f_1 = \sin \alpha$ and $f_2 = (1 - \cos^2 \alpha)^{1/2}$, respectively. Then $f_1 = f_2$.

Example 2 Let X, $Y = R$ and $A = \{x \mid x \in R,\ x \geq 0\}$. Then f_1: $A \to Y$ defined by $f(x) = x^3$ is a restriction of f_2: $X \to Y$ defined by $f(x) = x^3$, and the latter is an extension of the former.

5.2 Classes of Mappings

In Section 5.1 we classified functions into two broad categories, according to whether their values take the form of a scalar or an m-component point. A quite different classification will result if, instead, we take into consideration the character of the association rule.

Surjective Mappings

Definition 5.6 Let f: $X \to Y$ be a function. Then f is said to be surjective or onto if every $y \in Y$ is the image of at least one $x \in X$.

Example Consider $X = \{\alpha \mid 0° \leq \alpha \leq 360°\}$ and $Y = [-1, 1]$. Then $y = \sin \alpha$ is certainly an onto function.

Injective Mappings

Definition 5.7 Let f: $X \to Y$ be a function. Then f is said to be injective or into if no $y \in Y$ is the image of more than one $x \in X$.

Example A typical into function is the exponential function $y = e^x$.

Bijective Mappings

A function may satisfy the condition of both a surjective and an injective mapping. The concept of a bijective function, already introduced in Section 1.3 of Chapter 1, can be formulated as follows.

Definition 5.8 A function f: $X \to Y$ is said to be bijective, onto–into or one to one, if it is surjective and injective.

Example Let f: $X \to Y$ be defined by the formula $y = \sin x$, where $|y| < 1$, $|x| < \pi/2$. The mapping is one to one.

It would be convenient to pay some attention to the terminology introduced here. Some authors use the term one-to-one function to refer to what we have called an injective function. The same authors generally call our bijective or one-to-one function a "one-to-one and onto" mapping.

The following special functions also deserve our special consideration.

Identity Mappings

Definition 5.9 A function i: $X \to X$ is called an identity mapping on X if $i(x) = x$ for each $x \in X$.

Constant Mappings

Definition 5.10 Let f: $X \to Y$ denote a function. Then f is said to be a constant function if there is some $y' \in Y$ such that for each $x \in X$ we have $f(x) = y'$.

Image, Inverse Image, and Single-Valued Inverse Function

Consider a function f: $X \to Y$. The image of a point $x \in X$ under the mapping f has been defined as the point $y \in Y$ such that $f(x) = y$.

In a similar way, we can define the image of a subset $X_i \subset X$ as the subset $Y_i \subset Y$ such that $f(X_i) = Y_i$. Naturally Y_i is nothing but the set of those points of Y which are images of the points of X_i.

Some interesting relations between sets can be derived here.

Theorem 5.1 Let f: $X \to Y$ be a mapping and $\{X_i\}_{i \in I}$ an indexed family of subsets of X. Then $f(\bigcup_{i \in I} X_i) = \bigcup_{i \in I} f(X_i)$.

Proof For every $x \in X_i$, $i \in I$, we have $f(x) \in f(X_i)$ so that $x \in \bigcup_{i \in I} X_i$ means $f(x) \in \bigcup_{i \in I} f(X_i)$. Thus $f(\bigcup_{i \in I} X_i) \subset \bigcup_{i \in I} f(X_i)$. Conversely, suppose $f(x) \in \bigcup_{i \in I} f(X_i)$. Then $f(x) \in f(X_k)$ for some k. So $x \in X_k$, which means $x \in \bigcup_{i \in I} X_i$. Therefore, $f(x) \in f(\bigcup_{i \in I} X_i)$ and $\bigcup_{i \in I} (f(X_i) \subset f(\bigcup_{i \in I} X_i)$; so $f(\bigcup_{i \in I} X_i) = \bigcup_{i \in I} f(X_i)$.

Theorem 5.2 Let f: $X \to Y$ be a mapping and $\{X_i\}_{i \in I}$ an indexed family of subset of X. Then $f(\bigcap_{i \in I} X_i) \subset \bigcap_{i \in I} f(X_i)$.

Proof For every $i \in I$ we clearly have $\bigcap_{i \in I} X_i \subset X_i$, so that $f(\bigcap_{i \in I} X_i) \subset f(X_i)$ also. Then, if it is true that for every $i \in I$, $f(\bigcap_{i \in I} X_i) \subset f(X_i)$, we must also have $f(\bigcap_{i \in I} X_i) \subset \bigcap_{i \in I} f(X_i)$ and the theorem is proved.

As a consequence of Theorem 5.2 we have the following property.

Remark 5.1 If f: $X \to Y$ is a one-to-one mapping and $\{X_i\}_{i \in I}$ an indexed family of subsets of X, then $f(\bigcap_{i \in I} X_i) = \bigcap_{i \in I} f(X_i)$.

Note that here each $f(x) \in f(X_i)$, $i \in I$, is the image of one and only one $x \in X_i$.

The function f: $X \to Y$ associates with each element of X its image in Y. Using an arrow to denote the "direction" of the transformation, we may indicate this fact by

$$x \to f(x) = y.$$

Now take the point $f(x) = y \in Y$ and reverse the original "direction" of the mapping; that is, associate with each $y \in Y$ the set of those elements in X of which y is the image. We may indicate this by

$$f^{-1}(y) = f^{-1}(f(x)) = \{x \mid x \in X, y = f(x)\}.$$

Note, however, that depending on the properties of our mapping, $f^{-1}(y)$ might consist of only one element, of several elements, or even be empty. If the function f: $X \to Y$ is bijective, that is, one to one, the first case will necessarily occur. If the mapping is only onto, then the inverse image of y, as the set $f^{-1}(y)$ is called, might very well consist of several elements since y could simultaneously be the image of different $x \in X$. Finally, if the transformation is only injective, then the possibility that the inverse image of a point $y \in Y$ is empty cannot be excluded.

Now let $Y_i \subset Y$ be a subset of images mapped under f: $X \to Y$.

The inverse image of Y_i can be described in an analogous way, denoting it by

$$f^{-1}(Y_i) = f^{-1}(f(X_i)) = \{x \mid x \in X_i, f(X_i) \subset Y\}.$$

Here again we can see that the inverse image of a subset Y_i will depend on the properties of f: $X \to Y$. It might even be the empty set.

Note that in general $f^{-1}(Y)$ will not determine a single-valued function from Y to X. There might be points in Y which are not images of points in X. Other points in Y might even be images of several points in X. In other words, two basic conditions for the existence of a single-valued mapping f^{-1}: $Y \to X$ will not be satisfied in general. That is, not every $y \in Y$ will have an image $f^{-1}(y) \in X$ or this image will not be unique.

There is, of course, an obvious exception to what has been said above.

Definition 5.11 Let f: $X \to Y$ be a bijective function. Then the inverse image of Y under the transformation f is a bijective mapping from Y onto X called the single-valued inverse function of f and denoted f^{-1}: $Y \to X$.

In other words, if f: $X \to Y$ is one to one then its inverse is also a one-to-one function. Note that each $y \in Y$ will have a unique image $f^{-1}(y) \in X$. Each $x \in X$ will be the image of one and only one $y \in Y$. Finally, $f^{-1}(Y) = X$.

The following results, on the contrary, are quite general and apply to any function.

Remark 5.2 Let f: $X \to Y$ be any function; then

$$f^{-1}(\bigcup_{i \in I} Y_i) = \bigcup_{i \in I} f^{-1}(Y_i).$$

Remark 5.3 Let f: $X \to Y$ be any function; then

$$f^{-1}(\bigcap_{i \in I} Y_i) = \bigcap_{i \in I} f^{-1}(Y_i).$$

Composite Functions

For convenience, let us make a minor notational change, namely, let us write f: $X_i \to X_j$ instead of f: $X \to Y$ for denoting a function.

Definition 5.12 Let f_1: $X_1 \to X_2$, f_2: $X_2 \to X_3, \ldots, f_n$: $X_n \to X_{n+1}$ be given functions. The composition or composite function of $f_1, f_2, \ldots, f_n$, denoted $f_n \cdots f_2 f_1$: $X_1 \to X_{n+1}$ is the rule which assigns to each $x \in X_1$ a unique element $f_n(\ldots f_2(f_1(x)) \ldots) \in X_{n+1}$.

Example 1 Assume f_1: $X_1 \to X_2$, f_2: $X_2 \to X_3$ are the functions defined as in Fig. 5.2.

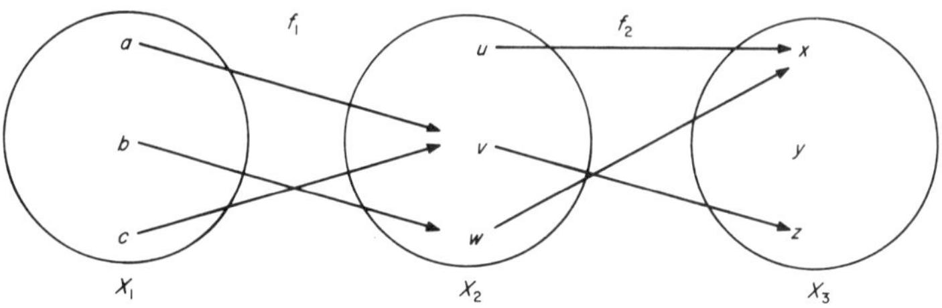

Figure 5.2

By definition, the composition of f_1 and f_2 is the function $f_2 f_1$: $X_1 \to X_3$ such that

$$f_2 f_1(a) = f_2(f_1(a)) = f_2(v) = z,$$
$$f_2 f_1(b) = f_2(f_1(b)) = f_2(w) = x,$$
$$f_2 f_1(c) = f_2(f_1(c)) = f_2(v) = z.$$

Example 2 Let $f_1: R \to R$ and $f_2: R \to R$ be defined by the formulas $f_1(x) = x^2$ and $f_2(x) = x + 5$, respectively. Find the compositions $f_2 f_1$ and $f_1 f_2$.

Let us find $f_2 f_1$ first. By definition

$$f_2 f_1(x) = f_2(f_1(x)) = f_2(x^2) = x^2 + 5.$$

Next let us form $f_1 f_2$. Using Definition 5.12 yields

$$f_1 f_2(x) = f_1(f_2(x)) = f_1(x + 5) = (x + 5)^2.$$

We see then that our composite functions are defined by the formulas

$$f_2 f_1(x) = x^2 + 5, \qquad f_1 f_2(x) = x^2 + 10x + 25.$$

As can be easily seen from Example 2, composition of functions is not, in general, a commutative operation; that is, $f_n \cdots f_2 f_1 \neq f_1 f_2 \cdots f_n$. On the other hand, as is shown in the following theorem, composition of functions is associative.

Theorem 5.3 Let $f_1: X_1 \to X_2, f_2: X_2 \to X_3, \ldots, f_n: X_n \to X_{n+1}$ be functions, such that $f_n \cdots f_2 f_1: X_1 \to X_{n+1}$. Then $(f_n \cdots f_2) f_1 = f_n (f_{n-1} \cdots f_1)$.

Proof We have that

(1) $$(f_n \cdots f_2) f_1(x) = f_n(\cdots f_2(f_1(x)) \cdots).$$

Also

(2) $$f_n(f_{n-1} \cdots f_1)(x) = f_n(\cdots f_2(f_1(x)) \cdots).$$

From (1) and (2), then

$$(f_n \cdots f_2) f_1 = f_n(f_{n-1} \cdots f_1).$$

A special case occurs when, given the composite of two functions, one function entering the composition happens to be the inverse of the other.

Theorem 5.4 Let $f: X \to Y$ be a bijective function and let $f^{-1}: Y \to X$ be its inverse. Then the composite functions $f^{-1}f$ and ff^{-1} represent identity functions on X and Y, respectively.

Proof For the composition $f^{-1}f$ we have

$$f^{-1}f(x) = f^{-1}[f(x)] = f^{-1}(y) = x;$$

$f^{-1}f$ then gives the identity function $i: X \to X$. It is sometimes denoted i_x. Analogously, for the composition ff^{-1}, we get

$$ff^{-1}(y) = f[f^{-1}(y)] = f(x) = y.$$

Thus ff^{-1} is the identity function $i: Y \to Y$ also denoted i_y.

5.3 Continuous Mappings

A particularly important class of functions is those mappings which possess the continuity property. The solutions of quite a few problems occurring in theoretical economics are based upon crucial assumptions about the continuity of some of the functions involved in the formulation of those problems. The matter, thus, seems relevant enough to justify our looking at it in some detail.

Definition 5.13 Let f: $X \to Y$ be a function, and let X and Y be subsets of R^n and R^m, respectively. Then f is said to be continuous at a point $x^0 \in X$ if $\lim_{k\to\infty} x^k = x^0$ implies $\lim_{k\to\infty} f(x^k) = f(x^0)$ for every sequence $(x^k \mid k \in N)$, where $x^k \in X$.

In consequence, a function is continuous at a point if the image of a sequence of points tending to it is a sequence tending to the image of the point.

We may also say that a function is continuous at

$$x^0 = (x_1{}^0, x_2{}^0, \ldots, x_n{}^0) \in X$$

if the system of equalities

$$\lim_{k\to\infty} x_1{}^k = x_1{}^0$$

$$\lim_{k\to\infty} x_2{}^k = x_2{}^0$$

$$\vdots$$

$$\lim_{k\to\infty} x_n{}^k = x_n{}^0$$

implies

$$\lim_{k\to\infty} f(x^k) = f(x^0).$$

Finally, we may consider that $f: X \to Y$ is continuous at x^0 if, for a given positive number η, there is some positive number δ such that $|f(x) - f(x^0)| < \eta$ whenever $|x_i - x_i{}^0| < \delta$, for $i = 1, 2, \ldots, n$.

The last two concepts are of course equivalent to Definition 5.13.

Once continuity at a point has been established, the notion of continuity on a set of points follows immediately.

Definition 5.14 Let f: $X \to Y$ be the function of the previous definition; f is said to be continuous on X (or, simply, to be continuous) if it is continuous at every point $x \in X$.

Now the following properties can be derived from the given definitions.

Theorem 5.5 Let $f_1 : X \to Y$ be a function continuous at $x^0 \in X$ and $f_2 : Y \to Z$ a function continuous at $f_1(x^0) \in Y$. Then the composite function $f_2 f_1$ is also continuous at $x^0 \in X$.

Proof Take a suitable number $\eta > 0$. For continuity of $f_2 f_1$ it is necessary to find a positive number δ such that if $x \in X$ and $d(x, x^0) < \delta$, then

$$d''\big(f_2(f_1(x)), f_2(f_1(x^0))\big) < \eta.$$

Consider first f_2. By hypothesis it is continuous at point $f_1(x^0)$. Thus, a positive number θ exists such that for $y \in Y$, $d'\big(y, f_1(x^0)\big) < \theta$, we get

$$d''\big(f_2(y), f_2(f_1(x^0))\big) < \eta.$$

But f_1 is continuous at point x^0 by hypothesis. This means that for given $\theta > 0$ a positive δ must exist such that whenever $x \in X$,

$$d(x, x^0) < \delta;$$

then

$$d'\big(f_1(x), f_1(x^0)\big) < \theta$$

and thus

$$d''\big(f_2(f_1(x)), f_2(f_1(x^0))\big) < \eta.$$

Theorem 5.5 together with Definition 5.14 lead us to recognize the following property:

Remark 5.4 If f_1: $X \to Y$ and f_2 : $Y \to Z$ are continuous functions on X and Y, respectively, then the composition $f_2 f_1$: $X \to Z$ is a continuous function on X.

A continuous function f: $X \to Y$ has the fundamental property that for every closed subset in Y, its inverse image is also a closed subset in X. In other words, if $f(X) \subset Y$ is closed, then $f^{-1}f(X) \subset X$ is closed too.

Strictly speaking, we have here a necessary and sufficient condition for function continuity, formulated in terms of the closedness of images and their inverses.

Theorem 5.6 Let f: $X \to Y$ be a function. A necessary and sufficient condition for f to be continuous is that the inverse image of every set closed in Y be a set closed in X.

Proof Let us show first that if f is discontinuous, then there is a set $B \subset Y$ such that $B = \bar{B}$ and $f^{-1}(B) \neq \overline{f^{-1}(B)}$. Choose $A \subset X$ and suppose f

is discontinuous at a point $x \in \bar{A}$. Then $f(x) \notin \overline{f(A)}$. The set $B = \overline{f(A)}$ is obviously closed. Since $f(A) \subset \overline{f(A)} = B$ we must have $A \subset f^{-1}(B)$ and, since $x \in \bar{A}$ and $f(x) \notin \overline{f(A)}$, we have $x \in \overline{f^{-1}(B)}$ but $x \notin f^{-1}(B)$. Thus, the inverse image of the closed set B is open.

Let us show next that if there exists a set $B \subset Y$ such that $B = \bar{B}$ and $f^{-1}(B) \neq \overline{f^{-1}(B)}$, then the function f is discontinuous. Suppose, thus, $B = \bar{B}$ and $f^{-1}(B) \neq \overline{f^{-1}(B)}$, so that $x \in \overline{f^{-1}(B)}$ but $x \notin f^{-1}(B)$. Hence, $f(x) \notin B = \bar{B}$ although B is the image set of $f^{-1}(B)$. Thus f is discontinuous.

A corollary of the previous theorem can now be given in the form of a remark.

Remark 5.5 Let f: $X \to Y$ be a function. A necessary and sufficient condition for f to be continuous is that the inverse image of every open set in Y be an open set in X.

Example Let there be the real-valued function defined by $f(x) = 5x + 3$. Its inverse is then the function given by the formula $f^{-1}(x) = (x - 3)/5$.

Note that the image of the closed set $X = [1, 5]$ is the closed set $f(x) = [8, 28]$, while the inverse image of the latter is also a closed set, namely, $f^{-1}(f(X)) = [1, 5]$. An open set would be obtained as a result, had we taken $X =]1, 5[$.

An important particular case of Theorem 5.6 can be formulated now for real-valued functions.

Remark 5.6 Let f: $X \to R$ be a real-valued function. A necessary and sufficient condition for f to be continuous is that the inverse image of any interval in R of the form $]-\infty, y]$ or $[y, \infty[$ be closed in X.

In Definition 5.3 the concept of projection was introduced. The following two remarks take account of its definition. The proofs are straightforward and hence left as an exercise.

Remark 5.7 Let $X_j, j = 1, 2, \ldots, r$ be a subset of R^{n_j} and define the direct product $X = \prod_{j=1}^{r} X_j$ (subset of $R^{\Sigma n_j}$). Then, the projection function of X on X_j is continuous on X.

Remark 5.8 Let X be a subset of R^m and $Y = \prod_{j=1}^{r} Y_j$ be a subset of $R^{\Sigma n_j}(Y_j \subset R^{n_j})$. Furthermore, let $f_j : X \to Y_j, j = 1, 2, \ldots, r$, be a function. If every f_j is continuous at the point $x \in X$, then the function f: $X \to Y$, defined by $f(x) = (f_1(x), \ldots, f_r(x))$, is also continuous at $x \in X$.

5.4 Mappings and Set Properties

Much of modern mathematics is concerned with the study of properties which are common to spaces under particular classes of mappings.

A fundamental case in topology is that of a one-to-one continuous function which maps a space X onto a space Y. Such a mapping possesses, of course, an inverse, and this inverse may also be continuous. It can be shown that the relation established by a mapping with these characteristics has the properties of reflexivity, symmetry, and transitivity; in other words it is an equivalence relation.

A mapping with the property that f and f^{-1} are continuous is called a homeomorphism. The spaces under the homeomorphism relation are said to be homeomorphic. Topology may be defined as the study of the properties of homeomorphic spaces or, more properly, of the invariants of homeomorphisms.

In this part of our study, without attempting to present any formal theory of homeomorphisms, we shall analyze mappings which preserve some of the properties of the sets, or spaces, on which they are defined.

Increasing Functions and Similar Sets

The first case which is of interest for us is that of an increasing function, that is, a function which maps a preordered set onto a preordered set.

Definition 5.15 Let X and Y be two sets preordered by $\precsim_X$ and $\precsim_Y$, respectively. A mapping f: $X \to Y$ is said to be increasing if for any two points $x, x' \in X$ and their images $f(x), f(x') \in Y$ we have

(a) $f(x) \underset{Y}{\sim} f(x')$ if $x \underset{X}{\sim} x'$,

(b) $f(x) \underset{Y}{\prec} f(x')$ if $x \underset{X}{\prec} x'$.

An increasing mapping need not be continuous. Increasing continuous mappings, however, play an important role in economic theory as the following example will illustrate.

Example Let X be the consumption set of an individual where $x = (x_1, \ldots, x_k)$ represents a commodity bundle.

Customary assumptions about X are that it is closed ("continuous"), has a lower bound (minimum-consumption requirement), is connected (made of "one piece"), or even convex [if x^1 and x^2 can be consumed, so can $\lambda x^1 + (1 - \lambda)x^2$].

Futhermore, it is assumed that x is totally preordered by a relation (preference preordering) so that, for a given $x' \in X$, the sets $\{x \mid x \in X, x \sim x'\}$,

$\{x | x \in X, x' \precsim x\}$, and $\{x | x \in X, x \precsim x'\}$ represent, respectively, the classes of those consumptions that are indifferent to x', at least as desired as x', and at most as desired as x'.

The preference preordering of a consumer is supposed to be continuous, and to satisfy some convexity assumption (for instance, if $x^2 \succ x^1$, then $\lambda x^2 + (1 - \lambda)x^1 \succ x^1$; $x^1, x^2 \in X$; $\lambda \in]0, 1[$).

Finally X is supposed to have no last or largest element for the preordering $\precsim$, so that the possibility of a satiation consumption (a consumption point that is preferred, that is, dominates any other consumption point in X) is explicitly excluded.

Under the above assumptions, an increasing function that maps the set X, totally preordered by $\precsim$, into R, the set of real numbers, can be defined. Such a continuous function $f: X \to R$, frequently denoted $u = u(x)$, is called the utility function of a given consumer and, as we know, it plays a considerable role in the theory of consumer behavior. The problem of the existence of such a function will be considered in another place.

The idea of an increasing function is thus easy to understand. A more general definition is the following:

Definition 5.16 Two preordered sets X and Y are said to be similar if there exists a bijective function $f: X \to Y$ such that for $x, x' \in X$, and $f(x), f(x') \in Y$ we have

(a) if $x \sim x'$, then $f(x) \sim f(x')$; and
(b) if $x \prec x'$, then $f(x) \prec f(x')$.

The function f is called a similarity mapping.

For ordered sets an analogous definition of similarity could be formulated.

Continuous Mappings on Compact Sets—Weierstrass Theorem

Compactness is a property which, if it holds for a set or space X it also holds for a set or space Y under a continuous mapping $f: X \to Y$. In other words, it is an invariant of continuous mappings.

Theorem 5.7 Let f be a continuous mapping defined on a set X. Then, if X is compact, so is its image $f(X)$.

Proof Let $f(x^1), f(x^2), \ldots$ be a sequence of points in $f(X)$. Note that every point in $f(X)$ is nothing but the image of a point in X. Hence, there must be a sequence of points $x^1, x^2, \ldots$ in X such that $f(x^n)$ is the image of x^n. By hypothesis X is compact. A subsequence $x^{i_1}, x^{i_2}, \ldots$ for $i_1 < i_2 < \cdots$

must then exist such that $\lim_{n\to\infty} x^{i_n} = x$ where $x \in X$. But, by definition f is a continuous function. Thus $\lim_{n\to\infty} f(x^{i_n}) = f(x)$, where $f(x) \in f(X)$. Hence $f(X)$ is a compact set since any sequence in $f(X)$ has a convergent subsequence.

The theorem we discuss next plays an essential role in mathematical programming and lies at the foundations of several theorems related to the notion of optimization in theoretical as well as applied economics. The properties of set compactness and function continuity are again most relevant. It is known as the generalized Weierstrass theorem.

Theorem 5.8 Let f: $X \to R$ be any continuous real-valued function and X a compact set. Then $f(X)$ is bounded and attains its glb and lub.

Proof By Theorem 5.7 if X is compact and f is continuous, then $f(X)$ is compact. By theorems 4.17 and 4.18 (see also Remark 4.5), if $f(X)$ is compact then it is closed and bounded. Call $m^0 = f(x^0)$ and $m^1 = f(x^1)$ the glb and lub of $f(X)$. We have then m^0, $m^1 \in f(X)$ since it is closed. (Refer to Remark 4.2.)

We now introduce a new terminology and restate the Weierstrass theorem using it.

Definition 5.17 Let f: $X \to R$ be the mapping of Theorem 5.8; m^0, the glb of $f(X)$, is called a minimum of $f(X)$ and m^1, the lub of $f(X)$, is called a maximum of $f(X)$. The points x^0 and $x^1 \in X$ such that $m^0 = f(x^0)$ and $m^1 = f(x^1)$, are, respectively, called a minimizer and a maximizer of $f(X)$.

The Weierstrass theorem reads then as follows:

Theorem 5.9 Let f: $X \to R$ be any continuous real-valued function, and X a compact set. Then $f(X)$ has a minimum and a maximum.

Note that to say $f(X)$ has a minimum at $f(x^0)$ is equivalent to saying for all $x \in X, f(x^0) \leq f(x)$. Analogously $f(X)$ has a maximum at $f(x^1)$ means that for all $x \in X$, $f(x) \leq f(x^1)$. The points x^0 and x^1 are, of course, the minimizer and maximizer of Definition 5.17.

It is convenient to observe, however, that the Weierstrass theorem states a sufficient though not a necessary condition for the existence of a minimum or maximum. The requirement of compactness of X in f: $X \to Y$ is a very strong one. Functions on sets that do not satisfy it may still have a minimum or a maximum. A function defined on a set X which is only bounded from below (above) might very well have a minimum (maximum) despite the fact X is not compact.

Continuous Mappings on Connected Sets—Bolzano Theorem

Like compactness, connectedness is an invariant under continuous mappings.

Theorem 5.10 Let f: $X \to Y$ be a continuous function, where X is a connected set. Then $f(X)$ is connected.

Proof Assume, contradicting our hypothesis, that $f(X)$ is not connected. We could then find two nonempty and closed sets, say P and Q, such that $P \cup Q = f(X)$ and $P \cap Q = \varnothing$. By Remark 5.2,

$$f^{-1}(P) \cup f^{-1}(Q) \subset f^{-1}(P \cup Q) = f^{-1}(f(X)) = X.$$

Certainly $f^{-1}(P)$ and $f^{-1}(Q)$ are not empty. They are also closed according to Theorem 5.6. Hence by Remark 5.3 we have

$$f^{-1}(P) \cap f^{-1}(Q) = f^{-1}(P \cap Q) = \varnothing$$

and X could be partitioned into two subsets. That would mean X is not connected, violating our assumption.

A particular case of the above theorem occurs when f is a continuous real-valued function. This property, sometimes known as the Bolzano theorem, can be formulated as follows:

Theorem 5.11 Let f: $X \to R$ be a continuous real-valued function, where X is a connected set. Then $f(X)$ is an interval.

As an interesting application of the preceding theorem let us prove the well-known intermediate-value theorem of calculus.

Theorem 5.12 Let f: $X \to R$ be a continuous real-valued function, where $X = [a, b]$. Then for each number K between $f(a)$ and $f(b)$ there is a number $k \in [a, b]$ such that $f(k) = K$.

Proof $[a, b]$ is an interval and thus connected. Since f is continuous, $f([a, b])$ is an interval by Theorem 5.11. By hypothesis K lies between $f(a)$ and $f(b)$ which are elements of $f([a, b])$. Thus $K \in f([a, b])$. Therefore, there exists $k \in [a, b]$ such that $f(k) = K$.

5.5 Convex and Concave Functions

The definition of set convexity was given in Chapter 4. Now our interest will be focused upon the properties of certain types of mappings, analyzed in terms of convexity notions. Convex and concave functions play an important role in the theory of mathematical programming. They play, subsequently, a fundamental role in economic problems which may be expressed in the formal language of mathematical programming.

Definition 5.18 Let f: $X \to Y$ be a function, where $X \subset R^n$ is a convex set. Then f is said to be convex over X (or simply convex) if for any two points x^1, $x^2 \in X$ and a scalar $\lambda \in [0, 1]$,

$$f[\lambda x^2 + (1 - \lambda)x^1] \leq \lambda f(x^2) + (1 - \lambda)f(x^1),$$

and it is said to be concave over X (or simply concave) if

$$f[\lambda x^2 + (1 - \lambda)x^1] \geq \lambda f(x^2) + (1 - \lambda)f(x^1).$$

Example 1 Figure 5.3 (a) shows a convex function in one variable while Fig. 5.3 (b) shows a concave one. Observe that a line segment joining any two points of the convex (concave) function lies on or above (below) it.

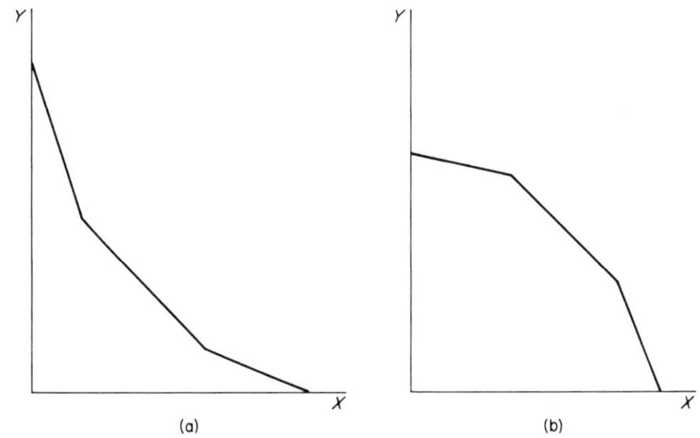

Figure 5.3

Example 2 Note that the functions $y = e^x$ and $y = \ln x$ are convex and concave, respectively.

The following properties are immediate consequences of Definition 5.16.

Remark 5.9 Let $f(x)$ define a convex function. Then $-f(x)$ defines a concave function.

Remark 5.10 A function may be convex over some subsets of its domain and concave over others.

Example The function, the graph of which is shown in Fig. 5.4, is convex for $a \leq x \leq b$ and concave for $b \leq x \leq c$.

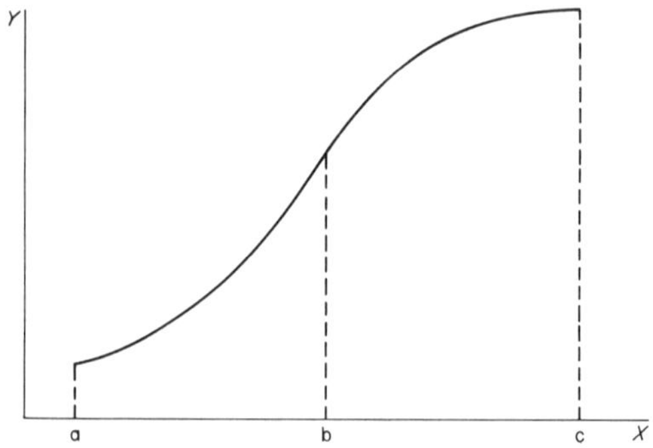

Figure 5.4

Remark 5.11 A function may be simultaneously convex and concave over its whole domain.

Example A linear function $y = \alpha + \beta x$ obviously satisfies Remark 5.11 since for any two points

$$x^1, x^2, \alpha + \beta[\lambda x^1 + (1 - \lambda)x^2] = \alpha + \lambda\beta x^1 + (1 - \lambda)\beta x^2.$$

Definition 5.19 Let f: $X \to Y$ be a function, where $X \subset R^n$ is a convex set. Then f is said to be strictly convex if for any two points x^1, $x^2 \in X$ such that $x^1 \neq x^2$, and a scalar $\lambda \in]0, 1[$,

$$f[\lambda x^2 + (1 - \lambda)x^1] < \lambda f(x^2) + (1 - \lambda)f(x^1);$$

and it is said to be strictly concave if

$$f[\lambda x^2 + (1 - \lambda)x^1] > \lambda f(x^2) + (1 - \lambda)f(x^1).$$

Example Figures 5.5(a) and (b) show the graphs of a strictly convex and strictly concave function, respectively.

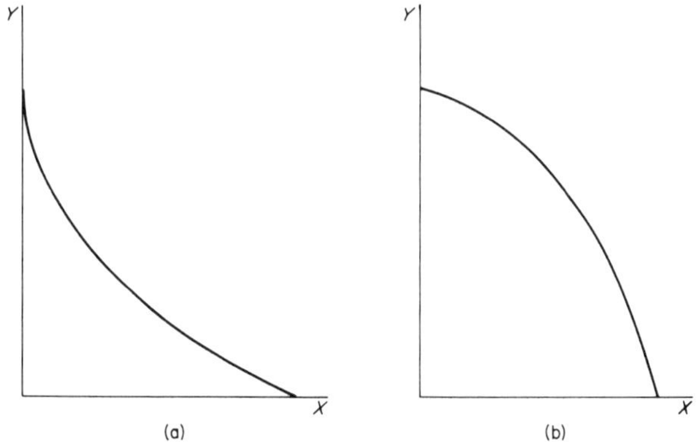

Figure 5.5

Note that not all of the previous remarks apply to strictly convex (concave) functions.

Remark 5.12 Let $f(x)$ define a strictly convex function. Then $-f(x)$ defines a strictly concave function.

Remark 5.13 A function may be strictly convex over some subsets of its domain and strictly concave over others.

Note that the graph of the example after Remark 5.10 shows a function with the mentioned characteristics. The following property is obvious.

Remark 5.14 A function cannot be simultaneously strictly convex and strictly concave over any subset.

The next two properties, with which we close for the moment our study of convex and concave functions, are particularly important. Later on we shall have the opportunity of returning to this topic, in relation to a different problem.

Theorem 5.13 Let $f(x)$ define a convex function on $X \subset R^n$. Then if the set $V = \{x \mid f(x) \leq c\}$, where c is any given point in X, is not empty, it is convex.

Proof Let us take the points x^1, $x^2 \in V$. Since $f(x)$ defines a convex function, we have

$$f[\lambda x^2 + (1 - \lambda)x^1] \leq \lambda f(x^2) + (1 - \lambda)f(x^1).$$

By definition of the set V, $f(x^1), f(x^2) \leq c$. Then

$$f[\lambda x^2 + (1 - \lambda)x^1] \leq \lambda c + (1 - \lambda)c$$

or

$$f[\lambda x^2 + (1 - \lambda)x^1] \leq c$$

so that $\lambda x^2 + (1 - \lambda)x^1 \in V$, and V is convex.

The following theorem can be proved in an analogous way, and the proof is left as an exercise.

Theorem 5.14 Let $f(x)$ define a concave function on $X \subset R^n$. Then if the set $V = \{x \mid f(x) \geq c\}$, where c is given point in X, is not empty, it is convex.

Note that the set $f(x) = c$ is not in general convex. It is usually called a contour of the function f. Since c can be any point in X, it is apparent that different contours of the function may be determined.

From the ways in which the set V has been defined in Theorems 5.13 and 5.14 as well as from the results of both theorems the following conclusion is immediate.

Remark 5.15 If $f(x)$ defines a convex function, its contour $f(x) = c$ is the upper bound of a convex set. If $f(x)$ defines a concave function, its contour is the lower bound of a convex set.

Example Figure 5.6(a) shows the first and Fig. 5.6(b) the second case stated in the preceding remark.

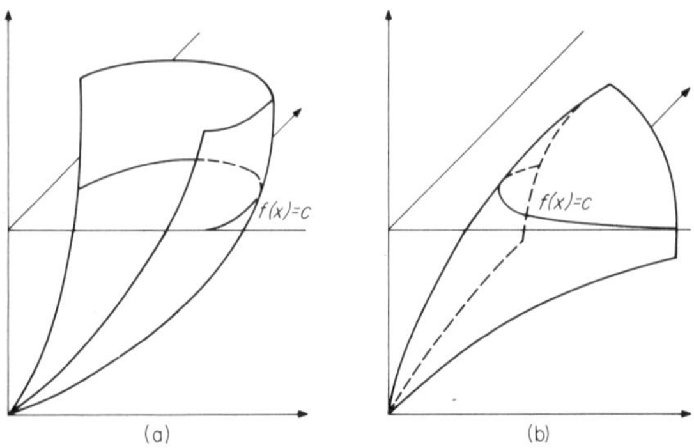

Figure 5.6

It should be realized that Theorems 5.13 and 5,14, and subsequently Remark 5.15, state a sufficient though not necessary condition. In other words, for the contour $f(x) = c$ to be the upper bound (lower bound) of a convex set, it is sufficient that the function $f(x)$ be convex (concave). However, functions do exist which without being convex (concave) have the property just mentioned for a convex (concave) function. We can summarize these results in the form of a definition.

Definition 5.20 Let f define a function. If its contour is the upper bound of a convex set, f is said to be convex contoured. It is said to be concave contoured if its contour is the lower bound of a convex set.

In terms of this quite general definition, it is obvious that convex and concave functions are special cases of convex contoured and concave contoured functions, respectively.

Example 1 Figure 5.7(a) is the graph of a function which is not convex but convex contoured. Figure 5.7(b) shows a concave contoured function which is not concave.

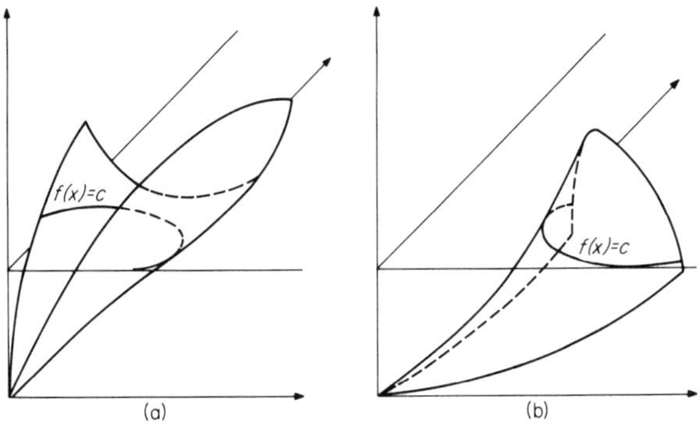

Figure 5.7

Example 2 "Neoclassical" production functions in economics constitute a typical case of concave-contoured functions. They are characterized by the fact of possessing contours, the so-called product isoquants, which are lower bounds of convex sets. Note that whereas a production function with diminishing returns to scale is typically concave, and as such concave contoured, a function with increasing returns to scale is concave contoured though not concave.

Utility functions are also concave contoured. Their contours, indifference curves in the neoclassical theory, are lower bounds of convex sets. Whether or not a utility function is concave depends on the utility index chosen. See also the example following Definition 5.15.

5.6 Linear Single-Valued Function

We close this chapter by introducing the general definition of a concept which will become our principal concern in Part II.

Definition 5.21 Let X and Y be subsets of R^m and R^n, respectively. A function f: $X \to Y$ is said to be linear if for x, $x' \in X$ and the scalar λ the following conditions are satisfied:

(1) If $y = f(x)$ and $y' = f(x')$ then $y + y' = f(x + x')$;
(2) if $y = f(x)$, then $\lambda y = f(\lambda x)$.

A fundamental property of a linear mapping is given in the following theorem:

Theorem 5.15 Let X be a convex subset of R^n and f a continuous linear function on X. Then $f(X)$ is a convex set.

Proof Take two points x, $x' \in X$ such that $y = f(x)$ and $y' = f(x')$. By Definition 5.21 we have $\lambda y = f(\lambda x)$ and $(1-\lambda)y' = f[(1-\lambda)x']$, so that

$$\lambda y + (1-\lambda)y' = f[\lambda x + (1-\lambda)x'].$$

Since $\lambda x + (1-\lambda)x' \in X$ we have $\lambda y + (1-\lambda)y' \in f(X)$ and $f(X)$ is convex.

EXERCISES

Single-Valued Functions

E.5.1 Study the following formulas, draw the corresponding graphs, and determine which of them describe a real single-valued function $f\colon R \to R$. If a function happens to be bijective, also give the formula and draw the graph of the inverse function: (a) $f(x) = (1-3x)/5$; (b) $f(x) = (4-x^2)^{1/2}$; (c) $f(x) = (1-x)/x$; (d) $f(x) = 1/x^3$; (e) $f(x) = x^2 - 2$; (f) $f(x) = (2-x)^{1/2}$.

E.5.2 Let f_1 and f_2 be two real-valued functions defined as follows: $f_1(x) = x^2 + 2x + 1$, $f_2(x) = 3x - 2$. Find (a) $f_1 f_2$; (b) $f_2 f_1$; (c) $f_1 f_1$; (d) $f_2 f_2$.

Continuity

E.5.3 Locate the discontinuities of the real-valued functions defined by the following formulas; (a) $f(x) = e^{1/x}$ if $x \neq 0$, $f(0) = 0$; (b) $f(x) = 1/(1 - e^{1/x})$ if $x \neq 0$, $f(0) = 0$; (c) $f(x) = (\sin x)/x$ if $x \neq 0$, $f(0) = 0$.

E.5.4 Using Remark 5.5 show that the real-valued function defined by $f(x) = \sin x$ is continuous.

Mappings and Set Properties

E.5.5 Show that following sets are homeomorphic: (a) R and the open interval $]-1, 1[$; (b) the unit interval $[0, 1]$ and the interval $[a, b]$.

Hint: Define $f\colon]-1, 1[\to R$ by $f(x) = \tan \frac{1}{2}x$, and $g\colon [0, 1] \to [a, b]$ by $(b-a)x + a$. Show that f, f^{-1}, g, and g^{-1} are continuous.

E.5.6 Show that the set of rationals with the natural order $(Q, \leq)$ is similar to the same set with the inverse order $(Q, \geq)$.

Hint: Construct a mapping f on Q defined by $f(x) = -x$, and show that f is a similarity mapping.

E.5.7 Let $A = \{a, b\}$. A set X is connected if and only if the only continuous mappings $f\colon X \to A$ are the constant ones $f(x) = a$ or $f(x) = b$. Sketch a proof.

Convex and Concave Functions

E.5.8 Let $X \subset R^n$ be convex and $f\colon X \to X$ continuous. Show that a necessary and sufficient condition for f to be convex is that $f((x^1 + x^2)/2) \geq (f(x^1) + f(x^2))/2$, $x^1, x^2 \in X$.

E.5.9 Let $X \subset R^n$ be convex. Show that $\sum_{i=1}^{n} \lambda_i f_i$ is convex provided $f_i : X \to X$ is convex and $\lambda_i \geq 0$, $i = 1, \ldots, n$.

E.5.10 Prove Theorem 5.14.

Hint: Study Theorem 5.13 carefully.

PROBLEMS

P.5.1 Let X and Y be two sets, and $\{X_i\}$ the collection of all subsets of X. A function $\mu\colon \{X_i\} \to Y$ which assigns to each element $X_i \in \{X_i\}$ an element (a nonempty subset) of Y is called a single-valued (multivalued) set function. A probability function p belongs to this category.

If S is the sample space, then p maps $\{S_i\}$, where $S_i \subset S$, into the interval $[0, 1]$. Explain the probabilistic meaning of each of the following axioms which p is supposed to satisfy: (a) for every i, $p(S_i) \geq 0$; (b) $p(S) = 1$; (c) for a sequence of sets $S_1, S_2, \ldots$ such that $S_i \cap S_j = \varnothing$, $i \neq j$, $p(\bigcup_i S_i) = \sum_i p(S_i)$. Is a probability function single valued?

P.5.2 Let X be sequentially compact, and f continuous on X. Prove that $f(X)$ is sequentially compact.

Hint: Show that for any convergent subsequence $x^{i_1}, x^{i_2}, \ldots$ there is a convergent subsequence $f(x^{i_1}), f(x^{i_2}), \ldots$, and that the corresponding limits are $x^0 \in X$, and $f(x^0) \in f(X)$.

P.5.3 Prove that the following two definitions are equivalent: (a) the function $f\colon X \to R$ is said to be bounded if and only if there is a number $a \in R$ such that $|f(x)| < a$ for every $x \in X$; (b) the function $f\colon X \to R$ is said to be bounded if and only if $f(X)$ is a bounded set.

P.5.4 Using previous results, prove that a continuous function from a compact set X into R is bounded.

P.5.5 Assume Y_j, the production set of the jth producer is closed, convex, and contains θ. Let $p = (p_1, \ldots, p_k)$ be a given price system. Since a production $y_j \in Y_j$ comprises positive (outputs) as well as negative components (inputs), the function $\pi_j : R^k \to R$ defined by $\pi_j(p) = \max\, pY_j$ is called the profit function of the jth producer. This means: the producer chooses an y_j so as to maximize π_j. The set $S_j(p) = \{y_j | y_j \in Y_j, py_j = \max pY_j\}$ is called the set of profit maximizers for given p.

With these elements: (a) Specify the supply function of the jth producer. (b) Specify when is it to be expected that the supply function will be single valued and when multivalued. (c) State additional conditions for Y_j, for p, or for both such that $S_j(p) = \varnothing$, $S_j(p) = Y_j$, S_j is a proper subset of Y_j, and $S_j(p)$ consists of a single point.

Hint: See Problem P.2.4, the discussion of Chapter 9, and Debreu [4].

NOTES

A similar but more complete discussion of point to point mappings is given by Kuratowski [2]. A modern calculus approach to functions is presented by Apostol [1]. A classical book on functions is that of Titchmarsh [1]. See also Berge [2] and Dunford and Schwartz [1]. The latter contains an excellent treatment of measure functions. Theorems on the relations between the properties of a space and the properties of a mapping play essential roles in the framework of theoretical economics (for examples see Debreu [2, 4], Herstein and Milnor [1], Karlin [1], and Nikaido [3, 4]).

CHAPTER 6

POINT-TO-SET MAPPINGS

The aim of this chapter is to introduce and discuss some of the properties of a more general mapping concept than that studied so far. Point-to-set mappings not only constitute an extremely interesting general category which widens considerably our views concerning the idea of function relationships, they have also proved to be a kind of analytical tool whose usefulness for theoretical economics can hardly be overestimated.

A brief comment on the terminology used seems appropriate at this point. Throughout this chapter the terms multivalued function, multivalued transformation and point-to-set mapping are considered synonyms. The less descriptive term, correspondence, may also be considered to have the same meaning. Also, some authors prefer to refer to set-valued functions.

The concept of a family of sets will be denoted, as before, by $\{Y_i\}$. To simplify notation, $\{Y\}$ will indicate the family or collection of all nonempty subsets of a set Y. Later, in a somewhat different context, we shall have to refer to the collection of all convex subsets of a certain set. If Y denotes the given set, then we shall indicate the family of all its nonempty subsets possessing the desired property by $R(Y)$.

Finally, let us say that a single-valued function will continue to be represented by f or other lowercase roman letters, indexed or not, while F or other capital roman letters, with or without indices, will be reserved for multivalued functions.

6.1 Multivalued Function

Definition 6.1 Let X and Y be two sets. An assignment which associates with each element $x \in X$ a nonempty subset Y_i of Y such that at least one Y_i contains more than one element is called a multivalued function from X into Y.

It is sometimes useful to think of a multivalued function as a mapping from a set X into a "set of sets," that is, the collection $\{Y\}$ of all nonempty subsets of Y. We would then define a point-to-set mapping as follows:

Definition 6.2 Let X be a set and $\{Y\}$ the collection of all nonempty subsets of a set Y. A mapping F: $X \to \{Y\}$ which assigns to each point $x \in X$ an element $Y_i \in \{Y\}$ such that at least one Y_i contains more than one element $y \in Y$ is called a multivalued function from X into $\{Y\}$.

Example 1 Let $X = Y$ be the closed interval [0, 100]. Suppose F assigns to each number $x \in X$ a subset $\{y \mid y \in Y, 0 \leq y \leq x\}$. Clearly F is a multivalued function. Note, for instance, that F assigns $\{0\}$ to $0 \in X$, [0, 7] to $7 \in X$, [0, 90] to $90 \in X$, [0, 100] to $100 \in X$, and so forth.

Example 2 Let $X \subset R^k$ be the consumption set of a consumer. X is supposed to be closed, convex, and preordered by a preference relation which satisfies the following convexity assumption: For any two commodity bundles $x^1, x^2 \in X$ and the scalar $\lambda \in]0, 1[$, if $x^2 \succ x^1$, then $\lambda x^2 + (1 - \lambda)x^1 \succ x^1$.

Suppose $p = (p_1, \ldots, p_k)$ is a given price system and w the total wealth the consumer possesses. Since obviously $p \in R^k$ and $w \in R$, we may consider the "price–wealth pair" (p, w) as a point of the space R^{k+1}.

The consumer's expenditure must clearly satisfy the constraint $\sum_{i=1}^{k} p_i x_i \leq w$, in short $px \leq w$. Or, in other words, his set of feasible consumptions is $X' = \{x \mid x \in X, px \leq w\}$.

Assume now the consumer behaves so as to satisfy his preferences. That is, for any given combination (p, w), he chooses that consumption which, in terms of his preference preordering, is a largest element of his set of feasible consumptions. This means he chooses, within the set of optimal consumptions, $X'' = \{x \mid x \in X', x \text{ is a largest element of } X' \text{ for the preordering } \precsim\}$.

We have thus defined a multivalued function which assigns to each (p, w) in a suitable subset of R^{k+1} a set of optimal consumptions in X. In general, such a mapping is called a demand correspondence.

That the mapping is multivalued and not merely single valued is a direct consequence of our initial assumption about the convexity of the preference preordering. Figure 6.1(a) illustrates the case for a two-commodity space.

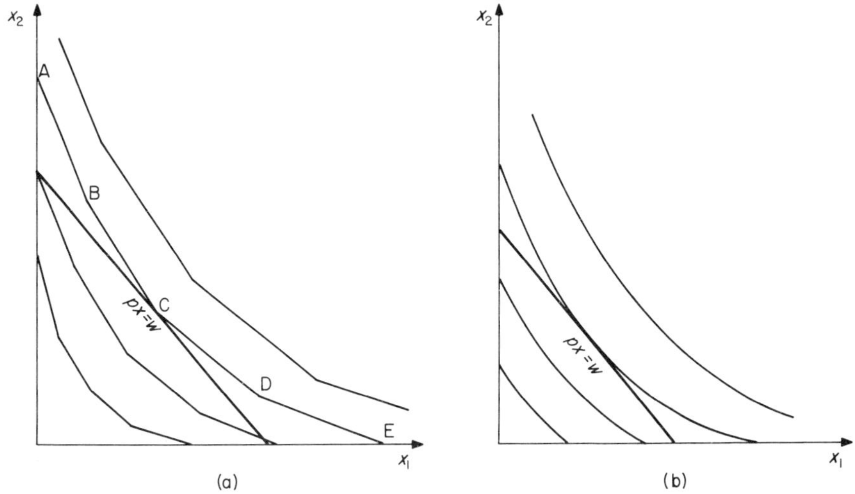

Figure 6.1

ABCDE represents an indifference class (that is, a set of consumptions which are indifferent from the point of view of the consumer) of a preordering which satisfies the postulated convexity property. The line $px = w$ is, of course, the wealth constraint of the consumer. If w is given and p is allowed to vary, notice that the demand correspondence will assign to each (p, w) either a point or a closed segment.

As a special case let us suppose the preference preordering satisfies a stronger convexity assumption, for example, for any two commodity bundles $x^1, x^2 \in X$ and $\lambda \in]0, 1[$, if $x^2 \sim x^1$, then $\lambda x^2 + (1 - \lambda)x^1 \succ x^1$. When this is the case, we can easily see that the demand mapping becomes a single-valued function. Figure 6.1(b) illustrates this for a two-commodity space.

These questions of the theory of consumption will be discussed again, in more detail, in Chapter 9.

Domain and Range

The concepts of domain, image, and range of a multivalued function are similar to those of a point-to-point mapping.

For a point-to-set mapping $F: X \to Y$, the domain is simply the set $X = \{x \mid F(x) \neq \varnothing\}$. The nonempty set $F(x) \subset Y$ is called the image or image set of the point $x \in X$. Finally, the set $\bigcup_{x \in X} F(x)$ is the range of the mapping.

Analogous to the image set of a point we can define the image set of a subset of X. Let $X_i \subset X$. Then $Y_i \subset Y$ is the image set of X_i under F if $F(X_i) = Y_i$.

In Chapter 5 we have often denoted a single-valued function by writing $x \to f(x)$ or simply $f(x)$. Analogously, we shall frequently denote a multivalued function by means of $x \to F(x)$ or $F(x)$.

Several properties which we studied for single-valued functions in Chapter 5 may be extended without difficulty to cover the case of point-to-set mappings. We give next the point-to-set equivalents of Theorems 5.1 and 5.2. The proofs are left as exercises. They can be immediately obtained by following methods similar to those of Chapter 5.

Theorem 6.1 Let $F: X \to Y$ be a point-to-set mapping and $\{X_i\}_{i \in I}$ an indexed family of subsets of X. Then

$$F\Big(\bigcup_{i \in I} X_i\Big) = \bigcup_{i \in I} F(X_i).$$

Theorem 6.2 Let $F: X \to Y$ be a point-to-set mapping and $\{X_i\}_{i \in I}$ an indexed family of subsets of X. Then

$$F\Big(\bigcap_{i \in I} X_i\Big) \subset \bigcap_{i \in I} F(X_i).$$

At this stage of our discussion it should be clear that a single-valued function is a special case of a point-to-set mapping. Precisely, it is a "point-to-set mapping" which satisfies the additional condition that the nonempty subset of Y which is assigned to each point of the domain consists of no more and no less than one point. We shall refer again to this special character of single-valued functions later, in somewhat diffetent contexts.

Graph of a Multivalued Function

The concepts of the graph of a relation and of a single-valued function have been studied in earlier chapters. The extension of the notion of a graph to a point-to-set mapping is straightforward.

Definition 6.3 Let $x \to F(x)$ represent a point-to-set mapping from a set X to a set Y. The graph of F is the set

$$\{(x, y) \mid (x, y) \in X \times Y, y \in F(x)\}.$$

Note that we are referring to a subset of the direct product $X \times Y$ such that, for any ordered pair belonging to it, the second element is in the image set of the first one.

Example Let a multivalued function from $X = [0, 100]$ to $Y = [0, 100]$ be defined by $F(x) = \{y \mid \frac{1}{3}x \leq y \leq x, x \in X\}$. The graph of $F(x)$ is obviously the closed shaded area of Fig. 6.2. Note that to each number $x \in X$ we have

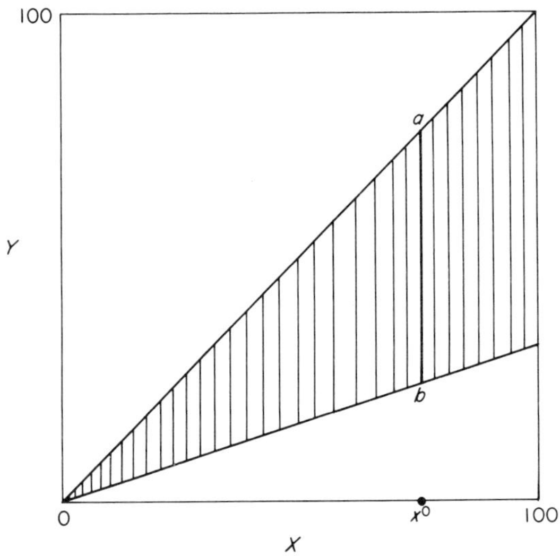

Figure 6.2

assigned a closed segment of Y. So, for instance, the segment $[a, b]$ is the image set of $x^0 \in X$.

Inverse of a Point-to-Set Mapping

To a single-valued function f: $X \to Y$ there corresponds a single-valued inverse function f^{-1}: $Y \to X$ if and only if f is bijective. This we already know from Definition 5.11.

If f is not bijective but only surjective, that is, onto, then there is no single-valued inverse function f^{-1}: $Y \to X$. However, it is clear that the inverse image of the set Y, that is, $f^{-1}(Y)$ can be regarded as a point-to-set mapping from Y into X.

The above comment serves to introduce us to the problem of the inverses of point-to-set transformations. It is clear that for a point-to-set transformation the conditions for the existence of an inverse transformation are quite less restrictive than they are for point-to-point mappings.

Definition 6.4 Let F be a point-to-set mapping with domain X and range Y. The inverse of F is the function F^{-1} with domain Y and range X defined by $F^{-1}(y) = \{x \mid x \in X, y \in F(x)\}$.

For the rest of this chapter we shall not be concerned with inverses of multivalued functions, and the above definition is all that we shall learn about the point.

6.2 Continuity

Continuity is the principal property of point-to-set mappings we shall be interested in. And it should be said at the beginning that defining the concept of continuity for a point-to-set mapping is a more complex task than it was for a single-valued function.

We analyze, first, the question of the continuity of a mapping at a point and generalize then the idea for the whole set of points which constitutes the domain. This is exactly the same path we followed when dealing with the continuity of single-valued functions in Chapter 5. However, for a multivalued function, there are three continuity concepts we ought to be able to distinguish.

For the remainder of this chapter we shall be studying multivalued functions, denoted in general by $F\colon X \to Y$, where the crucial assumption is made that *Y is a compact set*. This assumption will, of course, considerably reduce the generality of our results. However, it will facilitate obtaining these results and provide us with a set of tools with still remarkable analytical power.

Upper Semicontinuity

Definition 6.5 Let $F\colon X \to Y$ be a multivalued function and let X and Y be subsets of R^m and R^n, respectively. Then F is said to be upper semicontinuous at a point $x^0 \in X$ if $\lim_{k\to\infty} x^k = x^0$, $y^k \in F(x^k)$ and $\lim_{k\to\infty} y^k = y^0$ implies $y^0 \in F(x^0)$.

What the definition tells us is simply this. Call L the set containing the limit points of every sequence $(y^k \mid k \in N)$ where $y^k \in F(x^k)$. Call $F(x^0)$ the image set of x^0. Then, for upper semicontinuity, we must have $L \subset F(x^0)$.

It is clear that if we apply the above definition to a point-to-point rather than a point-to-set mapping, the concept of upper semicontinuity reduces to that of continuity.

Example Let the mapping $F\colon X \to Y$ be represented by the graph in Fig. 6.3. Note that the image set of x^0 is the closed interval $[a, b]$. Obviously L, the set of all limit points of sequences in the graph of F, is a subset of $F(x^0)$.

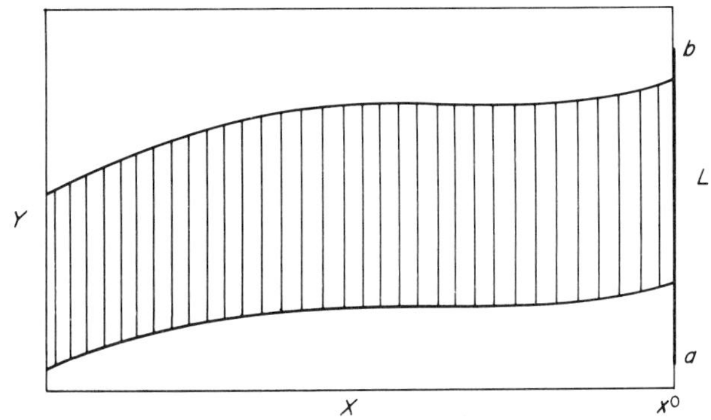

Figure 6.3

It will be useful to give here an alternative definition of upper semicontinuity in terms of neighborhoods rather than sequences. This may help us to improve our understanding of the concept, and at the same time, provide us with some tools we shall be using later. A good exercise would be to show the equivalence of both definitions.

Definition 6.6 Let $F\colon X \to Y$ be a multivalued function and let $\{S\}$ represent the collection of all open subsets of Y such that $F(x^0) \subset S$ for each $S \in \{S\}$ and some $x^0 \in X$. Then F is said to be upper semicontinuous at $x^0 \in X$ if there is a neighborhood $\delta(x^0)$ such that if $x \in \delta(x^0)$, then $F(x) \subset S$ for each $S \in \{S\}$.

We can now extend our concept of upper semicontinuity.

Definition 6.7 A multivalued function $F\colon X \to Y$ is said to be upper semicontinuous on X—briefly, upper semicontinuous—if it is upper semicontinuous at each point $x \in X$.

An immediate and important result of Definitions 6.5–6.7 is summarized in the following remark.

Remark 6.1 A multivalued function $F\colon X \to Y$ is upper semicontinuous if and only if its graph is closed in $X \times Y$.

Lower Semicontinuity

Definition 6.8 Let F: $X \to Y$ be a multivalued function and let X and Y be subsets of R^m and R^n, respectively. Then F is said to be lower semicontinuous at a point $x^0 \in X$ if $\lim_{k \to \infty} x^k = x^0$ and $y^0 \in F(x^0)$ implies that there is a sequence $(y^k | k \in N)$ such that $y^k \in F(x^k)$ and $\lim_{k \to \infty} y^k = y^0$.

Our last definition requires that the image set $F(x^0)$ of x^0 be a subset of the set of limit points of all sequences. In other terms, let x^0 be the limit to which x^k tends and let y^0 be in the image set of x^0. For lower semicontinuity, we must then have a sequence $(y^k | k \in N)$ the limit of which is y^0. This sequence belongs to $F(x^k)$ for all k.

Here we can also see that for a point-to-point mapping lower semicontinuity reduces to continuity.

Example Let the mapping F: $X \to Y$ be represented by Fig. 6.4. Note that the image set of x^0 is the closed interval $[a, b]$. It is a subset of the set of limit points L.

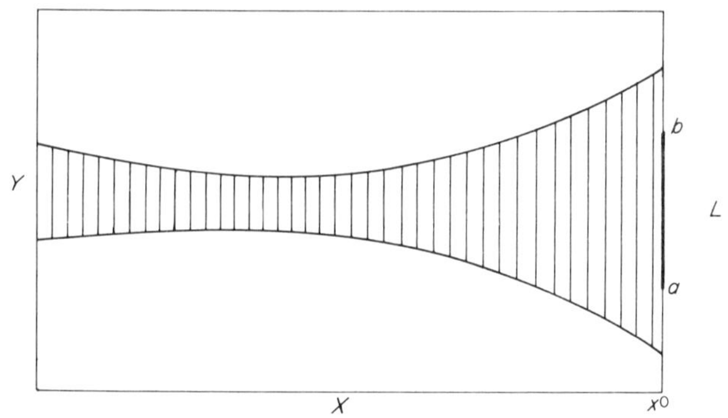

Figure 6.4

We give here an alternative definition using neighborhoods instead of sequences. Again, a careful comparison of the two definitions is strongly advisable.

Definition 6.9 Let F: $X \to Y$ be a multivalued function and let $\{S\}$ represent the collection of all open subsets of Y such that $F(x^0) \cap S \neq \emptyset$ for each $S \in \{S\}$ and some $x^0 \in X$. Then F is said to be lower semicontinuous at x^0 if there is a neighborhood $\delta(x^0)$ such that if $x \in \delta(x^0)$ then $F(x) \cap S \neq \emptyset$ for each $S \in \{S\}$.

Now for lower semicontinuity on X we can state the following definition:

Definition 6.10 A multivalued function F: $X \to Y$ is said to be lower semicontinuous on X—briefly lower semicontinuous—if it is lower semicontinuous at each point $x \in X$.

Continuity

Definition 6.11 A multivalued function F: $X \to Y$ is said to be continuous at a point $x^0 \in X$ if it is both upper and lower semicontinuous at x^0.

Example Let F: $X \to Y$ be represented by Fig. 6.5. It is easy to see that, since $L \subset F(x^0)$, the function is upper semicontinuous at x^0. Moreover, F is lower semicontinuous at x^0 for the condition $F(x^0) \subset L$ is obviously satisfied. Thus, according to Definition 6.11, F: $X \to Y$ is continuous at the point $x^0 \in X$. Note that $L = F(x^0)$ that is, $L \subset F(x^0)$ and $F(x^0) \subset L$.

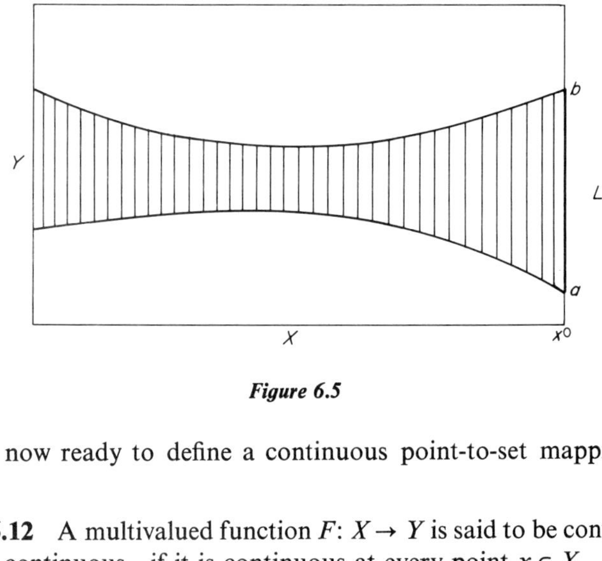

Figure 6.5

We are now ready to define a continuous point-to-set mapping over a set X.

Definition 6.12 A multivalued function F: $X \to Y$ is said to be continuous on X—briefly, continuous—if it is continuous at every point $x \in X$.

Two results which immediately follow from our definitions on continuity are given next (compare this result with Remark 5.8).

Remark 6.2 Let X be a subset of R^m and Y_j a compact subset of R^{n_j}, $j = 1, \ldots, r$, so that $Y = \prod_{j=1}^{r} Y_j$ is a compact subset of $R^{\Sigma n_j}$. Further, let $F_j : X \to Y_j, j = 1, \ldots, r$ be a multivalued function. Finally, consider the multivalued function F: $X \to Y$ defined by $F(x) = \prod_{j=1}^{r} F_j(x)$. We have the following.

(1) If every F_j is upper semicontinuous at $x \in X$, then F is upper semicontinuous at $x \in X$.

(2) If every F_j is lower semicontinuous at $x \in X$, then F is lower semicontinuous at $x \in X$.

Compactness and Upper Semicontinuity

Consider a multivalued function F: $X \to Y$. From our assumption regarding the *compactness of the set* Y together with our definitions of upper semicontinuity, one result follows immediately.

Remark 6.3 Let F: $X \to Y$ be an upper semicontinuous point-to-set mapping. Then, for each $x \in X$, $F(x)$ is a compact set. Furthermore, for any compact $S \subset X$, the set $F(S)$ is compact.

Note that the above formulation is closely in line with the result of Remark 6.1, concerning the closedness of the graph of an upper semicontinuous point-to-set mapping.

Consider now a finite collection of upper semicontinuous point-to-set mappings from X to Y. Since each of them maps into a closed set, it is clear the intersection of these is also a closed set. But the latter may also be regarded as the range of an upper semicontinuous multivalued function. In essence, we have proved the following:

Remark 6.4 Let $\{F_i\}_{i \in I}$ be a finite collection of upper semicontinuous point-to-set mappings from X to Y such that $\bigcap_{i \in I} F_i(X) \neq \emptyset$. Then $F = \bigcap_{i \in I} F_i$ is an upper semicontinuous mapping from X to Y.

Composite Mappings and Continuity

Let us enumerate some important results on composite mappings. Suppose X_1 and X_2 are subsets of R^{m_1} and R^{m_2}, respectively, and Y is a subset of R^n. From our definitions on continuity we have the following remarks:

Remark 6.5 If f: $X_1 \to X_2$ is a continuous single-valued function and F: $X_2 \to Y$ an upper semicontinuous multivalued function, then Ff: $X_1 \to Y$ is upper semicontinuous.

Remark 6.6 If f: $X_1 \to X_2$ is a continuous single-valued function and F: $X_2 \to Y$ a lower semicontinuous multivalued function, then Ff: $X_1 \to Y$ is lower semicontinuous.

Remark 6.7 If $F_1 : X_1 \to X_2$ and $F_2 : X_2 \to Y$ are upper semicontinuous multivalued functions, then $F_2 F_1 : X_1 \to Y$ is upper semicontinuous.

Remark 6.8 If $F_1: X_1 \to X_2$ and $F_2: X_2 \to Y$ are lower semicontinuous multivalued functions, then $F_2 F_1: X_1 \to Y$ is lower semicontinuous.

It would be useful to compare these results with that of Theorem 5.5. Note that in this chapter the composite mapping Ff and $F_2 F_1$ are defined, respectively, by $Ff(x) = F(f(x))$ and $F_2 F_1(x) = F_2(F_1(x))$.

6.3 Some Theorems on Maxima

In this last part of the chapter we shall apply some properties of continuous mappings to the study of certain theorems which play an important role in economics, particularly in relation to the problem of optimal solutions.

Upper and Lower Semicontinuous Point-to-Point Mappings

First of all, we need to define some weaker continuity concepts for single-valued functions than that discussed in Chapter 5. This can be done, for instance, by substituting a weak inequality for the second equation of Definition 5.13. Obviously, according to the direction of the inequality, we would obtain two different (weak) continuity concepts.

The above procedure is precisely what we shall use to define upper and lower semicontinuity for a point-to-point mapping. Before proceeding to formulate our definition, a warning with regard to terminology is necessary.

The designations upper or lower semicontinuity denote quite different ideas when applied to point-to-set mappings and to point-to-point mappings. It is unfortunate that we have the same terms to designate two distinct sets of ideas. However, there is a mathematical tradition with respect to the use of these terms and we shall follow it here. Confusion may be avoided by noting carefully in relation to which class of mappings the terms are being used in the next theorems. A comparison of Definitions 5.13, 6.5, 6.8, and 6.13 would surely help to clarify the concepts.

Definition 6.13 Let $f: X \to Y$ be a single-valued function where X and Y are subsets of R^m and R^n, respectively. Let $(x^k \mid k \in N)$ represent any sequence, where $x^k \in X$. Then

(1) f is said to be upper semicontinuous at $x^0 \in X$ if $\lim_{k\to\infty} x^k = x^0$ implies $\lim_{k\to\infty} f(x^k) \leq f(x^0)$;

(2) f is said to be lower semicontinuous at $x^0 \in X$ if $\lim_{k\to\infty} x^k = x^0$ implies $\lim_{k\to\infty} f(x^k) \geq f(x^0)$.

Take a point x close enough to x^0 such that the distance $d(x, x^0) \leq \delta$, where δ depends on a number $\varepsilon > 0$. Clearly, for the function $x \rightarrow f(x)$ upper semicontinuity at $x^0 \in X$ means $f(x) - f(x^0) \leq \varepsilon$. Note thus that lower semicontinuity of $f(x)$ is equivalent to upper semicontinuity of $-f(x)$.

Finally note that according to Definition 6.13 a function which is simultaneously upper and lower semicontinuous at $x^0 \in X$ is continuous at $x^0 \in X$.

Example The function of the graph in Fig. 6.6 is upper semicontinuous at x^0.

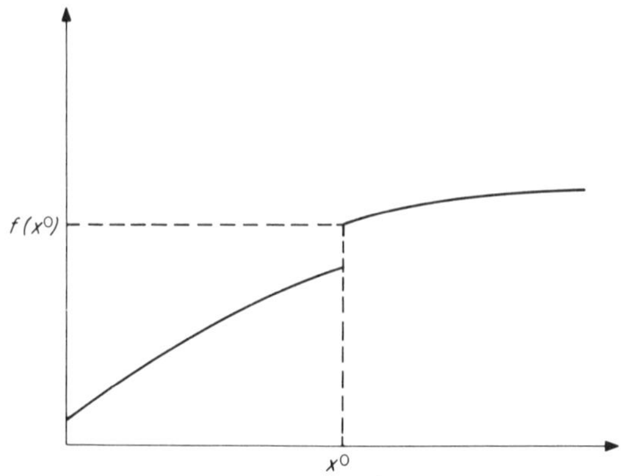

Figure 6.6

Theorems on Maxima†

Theorem 6.3 Let $f: X \times Y \rightarrow R$ be a lower semicontinuous real-valued function and $F: X \rightarrow Y$ a lower semicontinuous point-to-set mapping. Then, the real-valued function $m: X \rightarrow R$ defined by $m(x) = \max\{f(x, y) \mid y \in F(x)\}$ is lower semicontinuous.

Proof Consider the points x^0, y^0 such that the following relations are satisfied: $x^0 \in X$, $y^0 \in F(x^0)$, $f(x^0, y^0) \geq m(x^0) - \delta$, where $\delta > 0$. Consider next the pair (x, y). Clearly, it is possible to choose two neighborhoods of x^0 and y^0, say $\alpha(x^0)$ and $\beta(y^0)$ such that $(x, y) \in \alpha(x^0) \times \beta(y^0)$ implies $f(x^0, y^0) - \delta \geq m(x^0) - 2\delta$. The multivalued function F is lower semicontinuous by hypothesis. We can thus select a new neighborhood of x^0, say $\gamma(x^0)$, such that for $x, x \in \gamma(x^0)$ implies $F(x) \cap \beta(y^0) \neq \varnothing$. Now, $x \in \alpha(x^0) \cap \gamma(x^0)$ implies $m(x) \geq m(x^0) - 2\delta$. Therefore m is lower semicontinuous.

† Theorems 6.3, 6.4, and 6.5 in this subsection adapted from Berge [2].

Theorem 6.4 Let $f: X \times Y \to R$ be an upper semicontinuous real-valued function and $F: X \to Y$ an upper semicontinuous point-to-set mapping. Then, the real-valued function $m: X \to R$ defined by $m(x) = \max\{f(x, y) | y \in F(x)\}$ is upper semicontinuous.

Proof First take a point $x^0 \in X$. For any y in the image set of x^0, that is, $y \in F(x^0)$, we can choose appropriate neighborhoods $\alpha_y(x^0)$ and $\beta(y)$ such that if the pair $(x, y) \in \alpha_y(x^0) \times \beta(y)$, then necessarily $f(x, y) \leq f(x^0, y) + \delta$, where $\delta > 0$. Note that $F(x^0)$ is a compact set by Remark 6.3 and has thus a finite covering. We may construct such a covering by means of a finite number of neighborhoods $\beta(y^i)$ of the points y^i, $i = 1, \ldots, r$. Suppose $\{\beta(y^i)\} = \{\beta(y^1), \ldots, \beta(y^r)\}$ is such a covering of $F(x^0)$.

Let us define next $\beta(F(x^0)) = \bigcup_{i=1}^r \beta(y^i)$ and let $\gamma(x^0) = \bigcap_{i=1}^r \alpha_{y^i}(x^0)$. We have now that for x and y such that $x \in \gamma(x^0)$ and $y \in \beta(F(x^0))$, $f(x, y) \leq \max_i f(x^0, y^i) + \delta \leq m(x^0) + \delta$. But, because of the upper semicontinuity of the multivalued function F, a neighborhood $\alpha(x^0)$ must exist such that if $x \in \alpha(x^0)$, $F(x) \subset \beta(F(x^0))$. We have then that $x \in \alpha(x^0) \cap \gamma(x^0)$ implies $m(x) = \max_{y \in F(x)} f(x, y) \leq m(x^0) + \delta$ and m is upper semicontinuous.

As a joint result of Theorems 6.3 and 6.4 we have the following remark.

Remark 6.9 Let the real-valued function $f: X \times Y \to R$ be continuous. Furthermore, let the point-to-set mapping $F: X \to Y$ be continuous. Then the real-valued function $m: X \to R$ defined by $m(x) = \max\{f(x, y) | y \in F(x)\}$ is also continuous. We can now state the main theorem of this part.

Theorem 6.5 Let $f: Y \to R$ be a continuous real-valued function and $F: X \to Y$ a continuous point-to-set mapping. Then, the real-valued function $m: X \to R$ defined by $m(x) = \max\{f(y) | y \in F(x)\}$ is continuous, and the point-to-set mapping $G: X \to Y$ defined by $G(x) = \{y | y \in F(x), f(y) = m(x)\}$ is upper semicontinuous.

Proof Obviously f is continuous on $X \times Y$. Thus, m is a continuous real-valued function. Define now the point-to-set mapping $H: X \to Y$ by $H(x) = \{y | m(x) - f(y) \leq 0\}$. $H(x)$ is a closed set in Y so that $H: X \to Y$ is upper semicontinuous. Since $F: X \to Y$ is continuous, $F(x)$ is closed and we know that $H(x) \cap F(x)$ is closed and therefore upper semicontinuous. But, by definition of H, $H(x) \cap F(x) = G(x)$. Hence $G: X \to Y$ is an upper semicontinuous mapping.

Example Let $Y \subset R^k$, with $\theta \in Y$, be the production set of a producer. Y is supposed to be compact and convex.

In production theory it is customary to assume the producer behaves so as to maximize his profit $\pi = py$. Note here that the point $y \in Y$ can be taken as representing outputs (positive components) as well as inputs (negative components), so that py really means profit.

Based upon the postulated behavioral assumption we may define a mapping which assigns to each price system $p \in R^k$ a set of optimal productions, that is, productions which yield maximum profit. In the general case we confront here a multivalued function, customarily called a supply correspondence. However, depending on technological factors which in turn determine the properties of Y, the special case of a single-valued function may play, under certain circumstances, a most important role.

Let us remain in the general case of a multivalued function. We denote the supply correspondence as $S: R^k \to Y$ and define it by $S(p) = \{y \mid y \in Y, py = \max pY\}$.

Another useful element we need here is the so-called profit function of the producer, a real-valued point-to-point mapping $\pi = R^k \to R$ defined by $\pi(p) = \max pY$.

The two mappings S and π play a very important role both in production theory and in general equilibrium theory, see Chapter 9. However, what is essential in economic analysis is to show that for a given $p \in R^k$ optimal productions exist and the profit function is, indeed, defined. In other words, the fruitfulness of the analysis depends strongly on the possibility of guaranteeing the upper semicontinuity of S and the continuity of π at the given p.

The aim of the present example is to show the use of Theorem 6.5.

Theorem 6.5 states that if $f: Y \to R$ is a continuous real-valued function and $F: X \to Y$ continuous, then $m: X \to R$, defined by $m(x) = \max\{f(y) \mid y \in F(x)\}$, is continuous and $G: X \to Y$, defined by $G(x) = \{y \mid y \in F(x), f(y) = m(x)\}$ is upper semicontinuous. We now apply the theorem as follows:

(1) We define the function $f: Y \to R$ by $f(y) = py$ which is obviously continuous for given p.

(2) In lieu of $F: X \to Y$, we construct $F: R^k \to Y$ defined by $F(p) = Y$. As a constant mapping it is trivially continuous.

(3) In lieu of $m: X \to R$ defined by $m(x) = \max\{f(y) \mid y \in F(x)\}$, we have the profit function $\pi: R^k \to R$, already defined by $\pi(p) = \max pY$, and which can alternatively be defined by $\pi(p) = \max\{f(y) \mid y \in F(p)\}$.

(4) In lieu of $G: X \to Y$ defined by $G(x) = \{y \mid y \in F(x), f(y) = m(x)\}$, we have the supply correspondence $S: R^k \to Y$ which has been defined by $S(p) = \{y \mid y \in Y, py = \max pY\}$ and can, alternatively, also be defined by $S(p) = \{y \mid y \in F(p), f(y) = \pi(p)\}$.

The application of Theorem 6.5 leads us to recognize the upper semicontinuity of S and the continuity of π, which is what we wanted.

6.4 Linear Multivalued Function

The concept of a linear point-to-point mapping has been discussed at the end of Chapter 5. We conclude now our study of multivalued function by considering the corresponding linear case.

Definition 6.14 Let X and Y be subsets of R^m and R^n, respectively. A function F: $X \to Y$ is said to be linear if for $x, x' \in X$ and the scalar λ the following conditions are satisfied:

(1) if $y \in F(x)$ and $y' \in F(x')$, then $y + y' \in F(x + x')$;
(2) if $y \in F(x)$, then $\lambda y \in F(\lambda x)$.

A fundamental property of a linear mapping is given in the following theorem.

Theorem 6.6 Let X be a convex subset of R^n and F a continuous linear function on X. Then $F(X)$ is convex.

Proof Take two points $y \in F(X)$ and $y' \in F(X)$. We must then have x and x' in X such that $y \in F(x)$ and $y' \in F(x')$. By Definition 6.14 we have

$$\lambda y \in F(\lambda x) \qquad \text{and} \qquad (1-\lambda)y' \in F[(1-\lambda)x'],$$

so that

$$\lambda y + (1-\lambda)y' \in F[\lambda x + (1-\lambda)x'].$$

Since $\lambda x + (1-\lambda)x' \in X$, we have $\lambda y + (1-\lambda)y' \in F(X)$, and $F(X)$ is convex.

The preceding theorem is a generalization of Theorem 5.15.

EXERCISES

E.6.1 Consider the relation in R^2 defined by $\tilde{R} = \{(x, y) \mid x^2 + y^2 = 9\}$. From a preceding exercise we know that this relation does not satisfy the definition of a single-valued function. Does it, however, satisfy the definition of a multivalued function?

Hint: Note, for instance, that $\tilde{R}$ relates 0 to $\{3, -3\}$.

E.6.2 Using the appropriate definitions determine whether the following relations in R^2 might be classified as single-valued or multivalued mappings: (a) $(x^2/a^2) + (y^2/b^2) = 1$; (b) $(x^2/a^2) - (y^2/b^2) = 1$.

Hint: The given formulas are the equations of an ellipse and a hyperbola, respectively.

E.6.3 Let $A = \{a, b, c, d, e, g\}$ and $B = \{1, 2, 3\}$. Furthermore, let $f: A \to B$ be defined as follows: $f(a) = 1$; $f(b) = 2$; $f(c) = 2$; $f(d) = 2$; $f(e) = 3$; $f(g) = 1$. Clearly, $f^{-1}: B \to A$ does not exist in the sense of Definition 5.11; that is, f does not have a single-valued inverse. Would it be proper to say, however, that f possesses a multivalued inverse?

E.6.4 Give examples of single-valued functions having single-valued inverse, multivalued inverse, and no inverse.

E.6.5 Draw the graphs of the following two mappings $F: X \to R$, where $X = [0, 2]$, and classify them according to their continuity properties at $x = 1$:

(a) $F(x) = [-1 + 0.25x, x^2 - 1]$ for $x \in [0, 2]$;
(b) $F(x) = [-1 + 0.25x, x^2 - 1]$ for $x \in [0, 1[$, $x \in]1, 2]$ and $F(x) = [-1, 1]$ for $x = 1$.

E.6.6 Draw the graph and state the continuity property of $F: X \to R$, where $X = [0, 2]$, at $x = 1$, if $F(x) = [-1 + 0.25x, x^2 - 1]$ when $x \in [0, 1[$, $x \in]1, 2]$, and $F(x) = [-\frac{3}{4}, -\frac{1}{4}]$ when $x = 1$.

E.6.7 Let $X = [0, 2]$, and suppose $F: X \to R$ is defined as follows: If $x \in [0, 1[$, $F(x) = [-1, -1 + 0.25x]$; if $x \in]1, 2]$, $F(x) = [x^2 - 1, 3]$; if $x = 1$ $F(x) = A \cup B$, where $A = [-1, -\frac{3}{4}]$ and $B = [0, 3]$. Classify the function according to its continuity property at $x = 1$ and draw the graph of it.

E.6.8 Give an example to illustrate Remark 6.1: $F: X \to Y$ is upper semicontinuous if and only if its graph is closed in $X \times Y$.

E.6.9 Consider the mappings (a) and (b) of E.6.5. Show that the image $F(S)$ of any closed interval $S \subset X$ is a compact set.

E.6.10 Let $F_1: X \to R$ and $F_2: X \to R$ be two mappings with domain $X = [0, 3]$ defined by $F_1(x) = [x^2 - 2, x^2]$ and $F_2(x) = [x^2 - 3, x^2 - 1]$, respectively. Show that

(a) F_1 and F_2 are both upper semicontinuous and lower semicontinuous at $x = 3$;
(b) F_1 and F_2 have closed graphs on $X \times R$ and thus, $F_1(X)$ and $F_2(X)$ are compact sets;
(c) $F_1 \cap F_2$ is an upper semicontinuous mapping on X.

PROBLEMS

P.6.1 During most of this chapter we assumed the compactness of Y in $F: X \to Y$. Without this assumption, a mapping satisfying Definition 6.5 is called closed rather than upper semicontinuous. Prove the following two propositions: (a) the closedness of each image set $F(x)$ together with the upper semicontinuity of F, in the sense of Definition 6.6, imply the closedness of F; (b) the closedness of F together with the compactness of Y imply the upper semicontinuity of F, in the sense of Definition 6.6.

Hint: See Nikaido [4].

P.6.2 Consider the demand correspondence of Example 2, after Definition 6.2. What conditions could be postulated as to ensure the convexity of the image set of each pair (p, w)?

P.6.3 In the economic example following Theorem 6.5, conditions have been examined under which the mapping S is upper semicontinuous. Examine economic meaningful conditions under which, in addition, every image set $S(p)$ is convex.

P.6.4 Suppose an economy consists of m demand correspondences D_i (see Problem P.6.2), n supply correspondences S_j (see Problem P.6.3), and call $D = \sum_{i=1}^{m} D_i$ and $S = \sum_{j=1}^{n} S_j$ the aggregate demand and supply correspondences, respectively. Show that if each D_i and each S_j is upper semicontinuous, and if each $D_i(p, w)$ and each $S_j(p)$ is a convex set, then both D and S are upper semicontinuous, and both $D(p, w)$ and $S(p)$ are convex sets, respectively.

P.6.5 It is easy to show (see Chapter 9) that once the initial resources of each consumer have been determined, and a profit distribution mechanism has been specified, the wealth of each consumer becomes a function of prices. Demand correspondences, subsequently, become functions of prices only.

Let $r_i \in X_i$ be the initial resources of the ith consumer. For the economy as a whole, we have $\sum_{i=1}^{m} r_i = r \in X = \sum_{i=1}^{m} X_i$.

Given $r \in X$ and the other elements of problem P.6.4: (a) How would you define an aggregate excess-demand correspondence Z? (b) How would you guarantee its upper semicontinuity? (c) How would you ensure the convexity of each image set $Z(p)$?

NOTES

The discussion of this chapter is based on the work of Berge [2] and Debreu [4]. The presentation of the theorems on maxima follows Berge [2]. The economic examples are due to Debreu [4]. For a more general discussion of the question of upper semicontinuity and the relations between closed and upper semicontinuous mappings see Nikaido [4]. The term correspondence is due to Bourbaki [1] and has since found acceptance among economists.

CHAPTER 7

FIXED POINTS

7.1 Introduction

The question we propose to study in this chapter is the following: Under which conditions does a continuous mapping have a point which is equal to its image? A point with such a property is generally called a fixed point.

Let us begin our analysis by giving a formal definition.

Definition 7.1 A point x^0 is said to be a fixed point of a single-valued function $f\colon X \to X$ if $x^0 = f(x^0)$. More generally, it is said to be a fixed point of a multivalued function $F\colon X \to X$ if $x^0 \in F(x^0)$.

The concept is straightforward. For a point-to-point mapping we only require that the point and its image be equal. For a point-to-set mapping the point has to belong to its image set.

The study of fixed points constitutes an important chapter of point-set topology.

In theoretical economics the interest in fixed points is closely related to equilibrium analysis. We do not exaggerate if we say that the problem of the existence of a general equilibrium in a closed-economy model is nothing but the problem of the existence of a fixed point for some suitable mapping, defined in terms of the components of the model. Thus, general equilibrium theorems frequently are based upon fixed-point theorems or, in other words, the solution of a general equilibrium problem presupposes the solution of a fixed-point problem.

Two very simple examples will serve to illustrate the reason for our concern with fixed points.

Example 1 Suppose we have a closed economy in which n industries produce n goods. There is no fixed-capital formation, so that goods can be grouped into two categories: inputs to industries and consumption goods. The only primary factor is labor.

It is convenient to consider labor as an "industry" too (it produces the labor services needed in each industry and it uses the consumption goods produced in each of them). In this way, all goods, including consumption goods and even labor services, have the character of "intermediate goods." Obviously, the economy output is the only source of inputs.

A production (including labor services) can then be represented by a point

$$x = (x_1, \ldots, x_n) \in \Omega,$$

where Ω is the nonnegative orthant of R^n.

If we assume a fixed technology such that to any output or product point x a distinct input or requirement point y is assigned, it is clear that we have determined a point-to-point mapping $x \to f(x)$ or $y = f(x)$.

Since output is the only source of input we must have

$$y \leq x, \qquad \text{that is,} \qquad f(x) \leq x.$$

We may of course be only interested in an output $x \gg \theta$, that is, $x_i > 0$ for all i, and thus confine our study to points of the set $\Omega \backslash \{\theta\}$.

If we now define an interior equilibrium of the economy as a state for which

$$f(x^0) = x^0,$$

it is clear that we are simply asking about the existence of a production $x^0 \in \{\Omega \backslash \{\theta\}\}$ for which $f(x^0) = x^0$. Our equilibrium problem becomes a fixed-point problem.

Example 2 Consider the same model as before but drop the assumption about the fixed technology. Suppose instead that the same production can be achieved by different technologies (combination of inputs). To each output x we have then assigned a set of points. In other words, we have determined a point-to-set mapping $x \to F(x)$, where $F(x)$ is a subset of Ω or, making the same restriction as in the previous example, of $\Omega \backslash \{\theta\}$.

It is easy to see that, under similar considerations as in Example 1, the question about the existence of an interior equilibrium is identical to the question about the existence of a point $x^0 \in F(x^0)$.

An acceptable general answer to the question put at the beginning of this chapter—under which conditions does a continuous mapping have a fixed point?—requires some additional topological information. In Section 7.2 we give the necessary information in a condensed package. In Section 7.3 we study fixed points of point-to-point mappings. The results are then extended to point-to-set mappings in Section 7.4.

However, a very simple fixed-point theorem can be proved, without additional knowledge, for a single-valued function continuous over the interval [0, 1]. This may serve to give some insight into the problem, particularly to those who are willing to accept the more general results of Sections 7.3 and 7.4, without getting in trouble with the concepts presented in Section 7.2.

In Theorem 5.12 (intermediate-value theorem) we proved that a continuous function f: $[a, b] \to R$, $f(a) \neq f(b)$, always has a value $f(k) = K$ between $f(a)$ and $f(b)$ for every $k \in [a, b]$.

Consider now the special case $K = 0$. From Theorem 5.12, we then obtain the following:

Remark 7.1 Let f: $X \to R$ be a continuous real-valued function, where $X = [a, b]$. Then, if $f(a)f(b) < 0$, there is an $x \in X$ such that $f(x) = 0$.

Remark 7.1 will be used in the proof of the following fixed-point theorem.

Theorem 7.1 Let f: $X \to X$ be a continuous real-valued function, where $X = [0, 1]$. Then there is an $x^0 \in X$ such that $f(x^0) = x^0$.

Proof Assume first the case $f(0) = 0$ or $f(1) = 1$. The theorem is trivially true.

Next consider the case $f(0) > 0$ and $f(1) < 1$ (see Fig. 7.1). Let the function g: $X \to R$ be defined by $g(x) = x - f(x)$. It is obvious that if $g(x^0) = 0$, then $f(x^0) = x^0$. Now, for $g(0) = -f(0) < 0$, while $g(1) = 1 - f(1) > 0$, Remark 7.1 indicates that there must be an $x^0 \in X = [0, 1]$ such that $g(x^0) = 0$ and then $f(x^0) = x^0$.

Let us now develop the concepts we shall be using in Sections 7.3 and 7.4, when approaching the fixed-point problem in general terms.

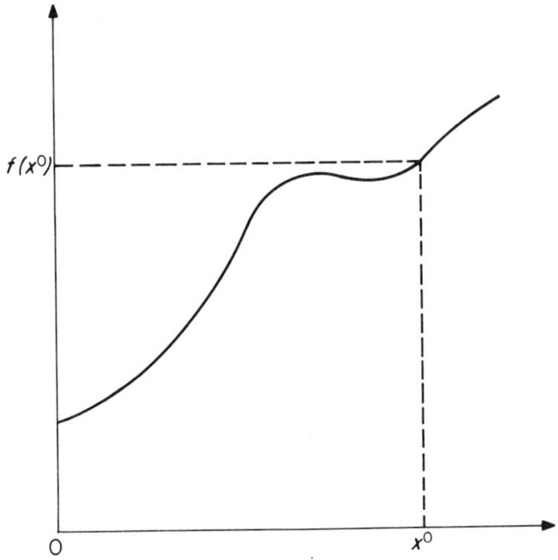

Figure 7.1

7.2 Simplicial Topology: Elementary Concepts

For the considerations which follow, our frame of reference will be the Euclidean space R^n. The point $\theta \in R^n$ will denote the origin of coordinates of the space, that is, the point with all its components equal to zero.

Simplexes and Simplicial Complexes

The first concept we need in our present context is that of $n + 1$ linearly independent points in the space R^n. For the purposes of this chapter a sum of points in R^n, denoted $\sum x^i$, and a product of a point and a scalar, written λx, has the same meaning as in Section 4.6.

Definition 7.2 Let $x^0, \ldots, x^n$ be $n + 1$ points in R^n, and $\lambda_0, \ldots, \lambda_n$ be $n + 1$ scalars. The points are said to be linearly independent if and only if $\sum_{i=0}^n \lambda_i x^i = \theta$ together with $\sum_{i=0}^n \lambda_i = 0$ imply $\lambda_i = 0$ for all i.

It is easy to deduce from the preceding definition that three linearly independent points do not lie on the same line; four do not lie on the same plane; $n + 1$ do not lie on the same $(n - 1)$-dimensional hyperplane.

The central concept of our current study is defined next.

Definition 7.3 Let $x^0, \ldots, x^n$ be $n + 1$ linearly independent points. The set of all points $x = \sum_{i=0}^n \lambda_i x^i$ is said to be a Euclidean n-dimensional simplex (briefly a simplex), denoted $S = S^n = x^0 \cdots x^n$, if and only if $\sum_{i=0}^n \lambda_i = 1$ and

$\lambda_i > 0$ for all i. The points $x^0, \ldots, x^n$ which span S are called the vertices of the simplex. The set $\{x^0, \ldots, x^n\}$ of all vertices of the simplex is called the 0-skeleton of S. The numbers λ_i are said to be the barycentric coordinates of S. Finally, the number of vertices minus one is called the dimension of $S = S^n$.

A simplex satisfying Definition 7.3 is clearly a convex, open set of points.

Note that a vertex may be regarded as a 0-dimensional simplex whereas the empty simplex is (-1) dimensional. For the latter we write S^{-1}.

It is important to recognize that, in the context of the present chapter, the expressions $x^0x^1, x^0x^1x^2, \ldots, x^0x^1 \cdots x^n$ refer to 1, 2, ..., and n-dimensional simplexes, respectively. They do not denote products of points.

In general, the simplex S^0 is a point x^0, the simplex $S^1 = x^0x^1$ a segment (without end points), the simplex $S^2 = x^0x^1x^2$ a plane triangle (without edges), the simplex $S^3 = x^0x^1x^2x^3$ the interior of a tetrahedron, and so forth.

Example The interior points of the two figures in Fig. 7.2 are examples of two- and three-dimensional simplexes. Note that edges and surface do not belong to them.

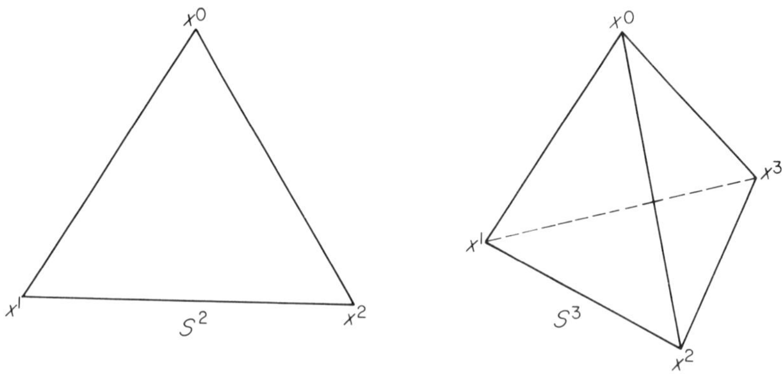

Figure 7.2

That vertices, edges, and faces of the above figures do constitute simplexes in their own right is the conclusion of the next general definition.

Definition 7.4 Let $S = S^n = x^0 \cdots x^n$ be a simplex. Then the empty simplex S^{-1} is a boundary simplex and any simplex $S^{i_k} = x^{i_0} \cdots x^{i_k}$, for $i_0 < \cdots < i_k \leq n$ is an i_k-dimensional boundary simplex of S. The simplex S itself and any boundary simplex are said to be faces of S. Strictly speaking, S^{-1} and S are called improper faces of the simplex S.

Note that the boundary simplex of $S^0 = x^0$ is the (-1)-dimensional, empty simplex S^{-1}.

Sometimes it is necessary to consider the closure of a simplex rather than the simplex of Definition 7.3.

Definition 7.5 Let $S = S^n = x^0 \cdots x^n$ be a simplex. The set of all x such that $x = \sum_{i=0}^{n} \lambda_i x^i$ for $\sum_{i=0}^{n} \lambda_i = 1$ and $\lambda_i \geq 0$ defines the closure of the simplex S, and is called the closed simplex $\bar{S} = \overline{S^n} = \overline{x^0 \cdots x^n}$. The points of S^n for $\lambda_i > 0$ are interior points of both the simplex and closed simplex. The difference $\overline{S^n} \setminus S^n = B^{n-1}$ is called the boundary of the simplex S.

The following properties are consequences of the given definitions.

Remark 7.2 Let S be an n-dimensional simplex and let $\bigcup S^{i_k}$ represent the union of all its faces. Then $\bigcup S^{i_k} = \bar{S}$.

Remark 7.3 Let S be an n-dimensional simplex. Then the simplexes $S^{i_k} = x^{i_0} \cdots x^{i_k}$ are disjoint.

Remark 7.4 A closed simplex $\overline{S^n}$ is a compact (and thus closed and bounded) set. Furthermore, it is the smallest convex set which contains its vertices.

Example The triangle (with edges) and the tetrahedron (with its surface) of the example following Definition 7.3 are closed simplexes. We write, respectively, $\overline{S^2}$ and $\overline{S^3}$. It would be very useful to verify the last three remarks using these figures.

For the sake of completeness we introduce here, briefly, the notion of an oriented simplex. However, we shall not be using this concept in our further discussions. It is by no means necessary for the understanding of the topics dealt with in Sections 7.3 and 7.4.

Definition 7.6 Let S^n be an n-dimensional simplex. By the oriented simplex $\langle S^n \rangle$ we mean the simplex $(-1)^k S^n$, where the exponent k is the number of inversions in one of the $(n + 1)!$ permutations of the elements $x^0, \ldots, x^n$.

Since the number of inversions in a permutation may be either even or odd but, obviously, not both, a simplex may be oriented in one of two possible directions. It must be either $\langle S^n \rangle = +S^n$ or $\langle S^n \rangle = -S^n$.

Example 1 Oriented one-, two-, and three-dimensional simplexes are shown in Fig. 7.3.

Example 2 Note that for a zero-dimensional simplex it does not make any difference whether or not we refer to its "orientation."

A one-dimensional simplex $S^1 = x^0 x^1$ can obviously be oriented in two different directions, "from x^0 to x^1" or "from x^1 to x^0."

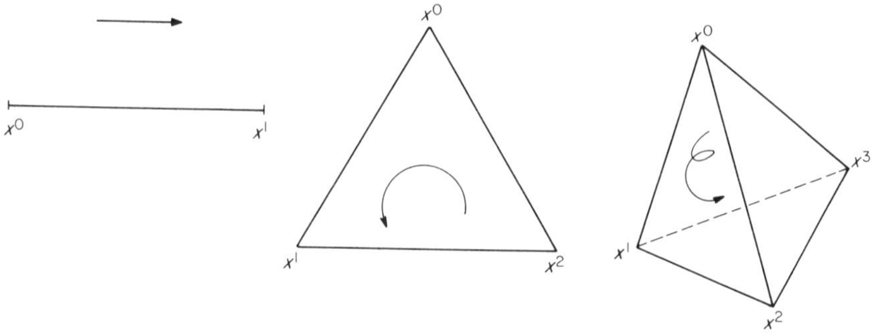

Figure 7.3

The two orientations of a two-dimensional simplex are given by

$$\langle x^0 x^1 x^2 \rangle = \langle x^1 x^2 x^0 \rangle = \langle x^2 x^0 x^1 \rangle$$

and

$$\langle x^0 x^2 x^1 \rangle = \langle x^2 x^1 x^0 \rangle = \langle x^1 x^0 x^2 \rangle$$

and shown in Fig. 7.4.

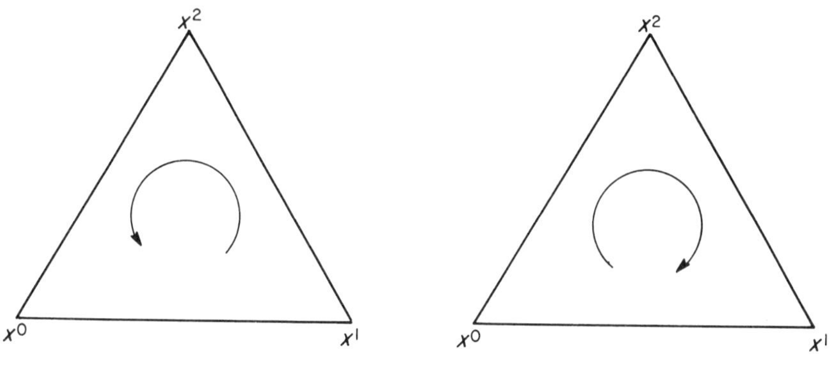

Figure 7.4

Leaving this brief introduction to oriented simplexes, we come back to our main problem.

Simplexes, in the terms of Definition 7.3, may be regarded as the building blocks of complexes, the concept which is defined next.

Definition 7.7 Let $\{S\}$ represent a finite set of simplexes in R^n. Then $\{S\}$ is said to be a finite simplicial complex K, briefly a complex, if the following conditions are satisfied: (a) If a simplex occurs in K, then each of its faces also occurs in K; (b) if S_1 and S_2 are any two simplexes of K, then $S_1 \cap S_2 = \varnothing$. The dimension of K is taken as the largest dimension of the simplexes which occur in K. A subset Λ of K is said to be a subcomplex of K if it is itself a complex.

The preceding definition is straightforward. A complex is merely a collection of simplexes (with their faces) with the particular property that the intersection of distinct elements of the collection is empty.

When necessary to avoid confusion, we shall denote the n-dimensional complex K by K^n.

Example Let $S^2 = x^0x^1x^2$ be a two-dimensional simplex. The two-dimensional complex associated with S^2 can be written

$$\begin{aligned} K^2 &= \{S^{-1}\} \cup \{S^0\} \cup \{S^1\} \cup \{S^2\} \\ &= \{S^{-1}\} \cup \{x^0, x^1, x^2\} \cup \{x^0x^1, x^0x^2, x^1x^2\} \cup \{x^0x^1x^2\} \\ &= \{S^{-1}, x^0, x^1, x^2, x^0x^1, x^0x^2, x^1x^2, x^0x^1x^2\}. \end{aligned}$$

K^2 is thus a set containing the empty simplex, all the vertices of S^2, all the edges of S^2 and, finally, the simplex S^2 itself.

The one-dimensional complex $K^1 = \{S^{-1}, x^0, x^1, x^0x^1\}$ is clearly a subcomplex of K^2. Here K^1 is the complex associated with the one-dimensional simplex x^0x^1.

Note that the intersection of any two simplexes in the complex K^2 is empty. The same is also true, of course, for the intersection of any two simplexes in K^1.

Polyhedra, Simplicial Decomposition, Subdivisions

The geometric concept of a polyhedron can be introduced into our discussion of complexes in the following way.

Definition 7.8 Let $\bar{K}$ represent the set-theoretical union of all simplexes of a simplicial complex K. Then $P = \bar{K}$ is called a polyhedron, and K is said to be a simplicial decomposition or a triangulation of $P = \bar{K}$.

It is necessary to distinguish very carefully between a polyhedron, which according to Definition 7.8 possesses the intuitive geometric properties of a body, and the concept of a complex. The former is a union of simplexes, the latter a collection of disjoint simplexes. It is in this sense that a complex comes to be a simplicial decomposition of a polyhedron, that is, a decomposition of a polyhedron into a collection of disjoint simplexes.

The idea can be extended to cover the so-called simplicial decomposition of the n-dimensional simplex S^n. By this we ought to understand the decomposition of the closed simplex $\overline{S^n}$, as it should be consistently expressed. If the complex K constitutes such a triangulation, it is clear that $\overline{K} = \overline{S^n}$. The closed simplex $\overline{S^n}$ is itself a polyhedron, though we may frequently use the equivalent expression $\overline{K}$.

Example Consider $S^2 = x^0x^1x^2$ and the corresponding closed simplex $\overline{S^2} = \overline{x^0x^1x^2}$. It is easy to see, from our previous example, that the complex

$$K^2 = \{S^{-1}, x^0, x^1, x^2, x^0x^1, x^0x^2, x^1x^2, x^0x^1x^2\}$$

is a simplicial decomposition or triangulation of the closed simplex $\overline{S^2}$.

On the other hand, the union of all simplexes contained in K^2 is the polyhedron $\overline{K^2}$. Using the notation for union of sets we may write

$$\overline{K^2} = S^{-1} \cup x^0 \cup x^1 \cup x^2 \cup x^0x^1 \cup x^0x^2 \cup x^1x^2 \cup x^0x^1x^2.$$

The notations $\overline{S^2}$ and $\overline{K^2}$ are associated with two different definitions. However, both have substantially the same idea as subject. In this sense, we may very well regard $\overline{S^2}$ and $\overline{K^2}$ as two expressions for the same concept.

One should not consider as synonymous the terms simplicial decomposition (or triangulation) and subdivision of a complex.

A subdivision of a complex is simply associated with the intuitive notion of "cutting" a complex into parts, according to a general principle and to some specific criterion. The relation between the concepts of a triangulation and a subdivision is given below.

Definition 7.9 The simplicial complex K' is said to constitute a subdivision of the simplicial complex K if the following conditions are met: (a) If $\overline{K}$ and $\overline{K'}$ are the polyhedra of K and K', respectively, then $\overline{K} = \overline{K'}$. (b) Each (open) simplex S' of K' is contained in a (open) simplex S of K.

As the polyhedra are equal, that is, $P = \overline{K} = \overline{K'}$, we may very well consider K and K' as two distinct triangulations of P. On the other hand, since each simplex of K' is a simplex of K, we deduce that each simplex of K is the union of simplexes of K' which intersect the simplex of K. This will become clear in the example given below.

Among the possible subdivisions of a triangulation K, there is one type which deserves special attention. We shall call such a type a barycentric subdivision.

Let K^n denote an n-dimensional complex and K^{n-1} an $(n-1)$-dimensional subcomplex. The boundary of the simplex $S^n \in K^n$ will be denoted, according to definition, $B^{n-1} = \overline{S^n} \setminus S^n$. By B_1^{n-1} we shall denote the subcomplex of K^{n-1} which covers the boundary B^{n-1}, that is, the collection of subsets of K^{n-1} such that each point of B^{n-1} is in at least one set of the collection. Finally, we understand the barycenter x of S^n to be the point for which all barycentric coordinates, relative to the vertices x^i are equal to $1/(n+1)$.

We thus have the following definition:

Definition 7.10 The first barycentric subdivision of the simplex S^n consists of the complex B_1^{n-1}, the simplexes $xS^k = xx^{i_0} \cdots x^{i_k}$ for $S^k \in B_1^{n-1}$, and the vertex x. The union of the first barycentric subdivision of the simplexes of K^n is called the first barycentric subdivision of K^n and denoted $K' = K_1{}^n$.

Taking further subdivisions of subdivisions we arrive, recurrently, at the following general concept.

Definition 7.11 The νth barycentric subdivision of K^n is the first barycentric subdivision of the $(\nu-1)$th barycentric subdivision of K^n and denoted $K^{(\nu)} = K_\nu{}^n$.

Note that a zero-dimensional complex K^0 has itself, K^0, as its only subdivision. On the other hand, we may regard K^n as the zeroth barycentric subdivision of K^n.

A subdivision is said to be simplicial when the elements into which a complex is subdivided are simplexes. It is easy to see that a barycentric subdivision is in this sense simplicial.

Example Let the complex $K^2 = \{S^{-1}, x^0, x^1, x^2, x^0x^1, x^0x^2, x^1x^2, x^0x^1x^2\}$ be a triangulation of the simplex $\overline{S^2} = \overline{x^0x^1x^2}$.

Consider the simplex $S^2 = x^0x^1x^2$, which is, of course, an element of K^2. S^2 is represented in Fig. 7.5(a).

The boundary of S^2 is the perimeter of the triangle in Fig. 7.5(a). We denote it $B^{2-1} = B^1 = \overline{S^2} \setminus S^2$. By $B_1^{2-1} = B_1{}^1$ we denote the subcomplex of K^2 which covers the boundary B^1 of S^2. We have thus, $B_1{}^1 = \{S^{-1}, x^0, x^1, x^2, x^0x^1, x^0x^2, x^1x^2\}$.

Finally, the point $x = x^{012}$ is the barycenter of S^2. All its barycentric coordinates are equal to $1/(n+1)$.

According to Definition 7.10 we can now define the first barycentric subdivision of $\overline{S^2}$ as the collection $\{B_1{}^1, \{xS^k \mid S^k \in B_1{}^1\}, x\}$.

To find the first barycentric subdivisions of the lower dimensional simplexes in K^2, we proceed in the same way. Of course, only the one-dimensional

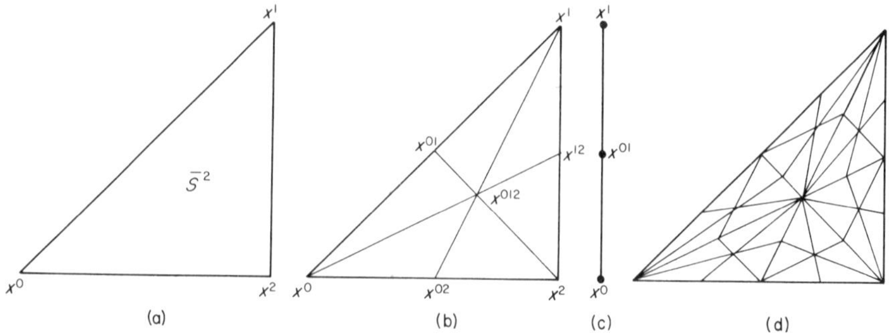

Figure 7.5 (Adapted from S. S. Cairns, "Introductory Toplogy," Copyright © 1961, The Ronald Press Company, New York.)

simplexes in K^2 are relevant in our context and their first barycentric subdivisions are straightforward.

By definition, the first barycentric subdivision of the complex K^2 is the union of the first barycentric subdivisions of all simplexes in K^2.

Figure 7.5(b) shows the first barycentric subdivision of K^2. Figure 7.5(c) represents the first barycentric subdivision of the subcomplex of K^2 associated with the simplex x^0x^1. Finally, Fig. 7.5(d) represents the second barycentric subdivision of K^2.

Simplicial Mappings and Sperner's Lemma

The next concept we shall be concerned with is that of a simplicial mapping. A special class of simplicial mapping, the Sperner's mapping as it might be called, will also play a crucial role in relation to Theorem 7.6.

We shall begin this part of our study by introducing the concept of an affine mapping. The concept of a simplicial mapping will be defined thereafter.

Definition 7.12 Let S^n and S^m be two simplexes of dimensions n and m, respectively. A continuous mapping $f: S^n \to S^m$ is said to be an affine mapping from the simplex S^n into the simplex S^m if it satisfies the following conditions:

(a) it assigns to each vertex x^i $(i = 0, 1, \ldots, n)$ of S^n, some vertex $f(x^i) = y^j$ $(j \in \{0, 1, \ldots, m\})$ of S^m;

(b) it assigns to each point $x \in \overline{S^n}$, defined such that $x = \sum \lambda_i x^i$, $\sum \lambda_i = 1$, $\lambda_i \geq 0$, a point $f(x) = y \in \overline{S^m}$, defined such that $y = \sum f(\lambda_i x^i) = \sum \lambda_i f(x^i)$.

It is interesting to note that interior points of S^n are mapped by the affine mapping f into interior points of S^m. According to the definition, points in S^n with barycentric coordinates > 0 will be carried over into points in S^m

with barycentric coordinates > 0. Such points are obviously interior points of the respective simplexes.

Knowing what an affine mapping means, we can pass on to describe a simplicial mapping. The former is included in the definition of the latter.

Definition 7.13 Let $\overline{K}$ and $\overline{L}$ be two polyhedra. A mapping f: $\overline{K} \to \overline{L}$ is said to be a simplicial mapping from the complex K into the complex L if it satisfies the following conditions:

(a) The vertices of any simplex S of K are mapped into the vertices of a simplex S' of L.

(b) The interior points of any simplex S of K are mapped according to the affine mapping determined by (a).

A simplicial mapping is continuous. For, from the fact that the simplicial mapping f is defined linearly in the coordinates of $x \in \overline{K}$ we immediately deduce that if $(x^n | n \in N)$ converges to $x \in \overline{K}$, then $\big(f(x^n) | n \in N\big)$ converges to $f(x) \in \overline{L}$.

It should be noticed that a simplicial mapping is defined from a polyhedron to a polyhedron, but that for each of them a simplicial decomposition is implied. Hence, a mapping which is simplicial for a given decomposition will not be, in general, simplicial for others.

An immediate consequence of Definition 7.13 is given in the following remark.

Remark 7.5 Let K and L be simplicial complexes. Let there be a function f which assigns to each vertex x^i of K a vertex $y^j = f(x^i)$ of L, in such a way that to vertices of a simplex S of K there always correspond vertices of a simplex S' of L. Then f is a uniquely determined simplicial mapping.

A special and important type of mapping which fits the conditions of Remark 7.5, will be our next subject of inquiry. But before doing so let us discover what is meant by a carrier simplex.

Definition 7.14 Let $\overline{K}$ be a polyhedron with simplicial decomposition K. Let x be a point of $\overline{K}$. Then, the (open) simplex of K which contains the point $x \in \overline{K}$ is called the carrier simplex (briefly the carrier) of x.

Theorem 7.2 Let K' be a subdivision of the complex K. Let the function f: $K' \to K$ assign to each vertex $x^{i'}$ of K' an arbitrary vertex $x = f(x^{i'})$ of the carrier of $x^{i'}$ in K. Then f is a simplicial mapping.

Proof Take an arbitrary r-dimensional simplex S' of K', and let $x^{i'}$ be one of its vertices. By definition of subdivision we know there is a simplex S in K such that $S' \subset S$. Thus, it is also true that $\overline{S}' \subset \overline{S}$, and therefore

$x^{i'} \in \bar{S}' \subset \bar{S}$. Since the carrier simplexes of the $x^{i'}$ are face simplexes of S, the images $f(x^{i'})$ are vertices of S. Therefore f is a simplicial mapping.

Very closely related to the question which has just been discussed is Theorem 7.3, usually known as Sperner's lemma. As we said, it will have an important role to play in connection with the proof of Brouwer's fixed-point theorem.

Theorem 7.3 Let S be an n-dimensional simplex, K the triangulation of $\bar{S}$, and K' a subdivision of K. Let the function f assign to each vertex x' of K' a vertex $f(x') = x^i$ ($i = 0$ or 1 or $\cdots$ or n) of its carrier simplex in K. Then, there is at least one n-dimensional simplex S_0' in K' the vertices of which can be mapped by f into the set of all vertices $x^0, \ldots, x^n$ of S.

Proof We shall approach our problem by proving a stronger assertion than that of the theorem, namely, that the number r of simplexes of the indicated type is odd. We shall proceed by induction on n.

Consider first the case $n = 0$. Since $S_0' = \{x^0\}$ the theorem is trivially true and $r = 1$.

Assume now our (stronger) version of the theorem holds for $0, 1, \ldots, n-1$, and consider whether it holds for n. Let the sequence $S_1', S_2', \ldots, S_w'$, or simply S_j' ($j = 1, \ldots, w$), represent all n-dimensional simplexes of the subdivision K'. An $(n-1)$-dimensional face simplex of a S_j' whose vertices are mapped by f into all vertices $x^1, \ldots, x^n$ will be called a distinguished face of S_j' and denoted Z'. Let r_j be the number of distinguished faces of S_j'.

We have the following alternatives:

(1) If S_j' satisfies the assertion of the theorem then obviously $r_j = 1$.

(2) If S_j' does not have the property of the theorem, but all vertices $x^1, \ldots, x^n$ occur among the images of the vertices of S_j', then each occurs twice, and $r_j = 2$.

(3) If not all $x^1, \ldots, x^n$ occur among the images, then simply $r_j = 0$.

By means of the congruence notation we can summarize our findings in a convenient manner. Remember that two integers a, b are said to be congruent integers modulo m [written: $a \equiv b \pmod{m}$], if and only if there exists an integer k such that $a - b \equiv km$; in other words, if m divides the difference $a - b$. In our problem we may summarize by stating that, in all cases,

$$r \equiv \sum r_j \pmod{2}$$

and confine ourselves to prove that the number $\sum r_j$ is odd.

Let Z' be an $(n-1)$-dimensional simplex mapped by the function f into $\{x^1, \ldots, x^n\}$. We now have two possibilities.

(1) If Z' happens to be an interior simplex of S, then it is a distinguished face of two of the S_j'. It will be counted twice in $\sum r_j$.

(2) If Z' happens to lie on an $(n-1)$-dimensional face of S, then this face must be $x^1 \cdots x^n$ by the carrier conditions of the theorem (note that no other face could map into $x^1 \cdots x^n$). Z' is a distinguished face of one S_j' and counted only once in $\sum r_j$.

From the $(n-1)$-dimensional simplexes Z', those of the first kind have been shown to appear twice in $\sum r_j$; those of the second kind are odd by assumption. Therefore, $\sum r_j$ is odd.

Example Let Fig. 7.6 represent the simplex $\bar{S} = \overline{x^0 x^1 x^2}$ with a subdivision. Try to follow step by step the reasoning of Theorem 7.3 by means of Fig. 7.6. Note that the shaded triangle satisfies the assertion of the theorem.

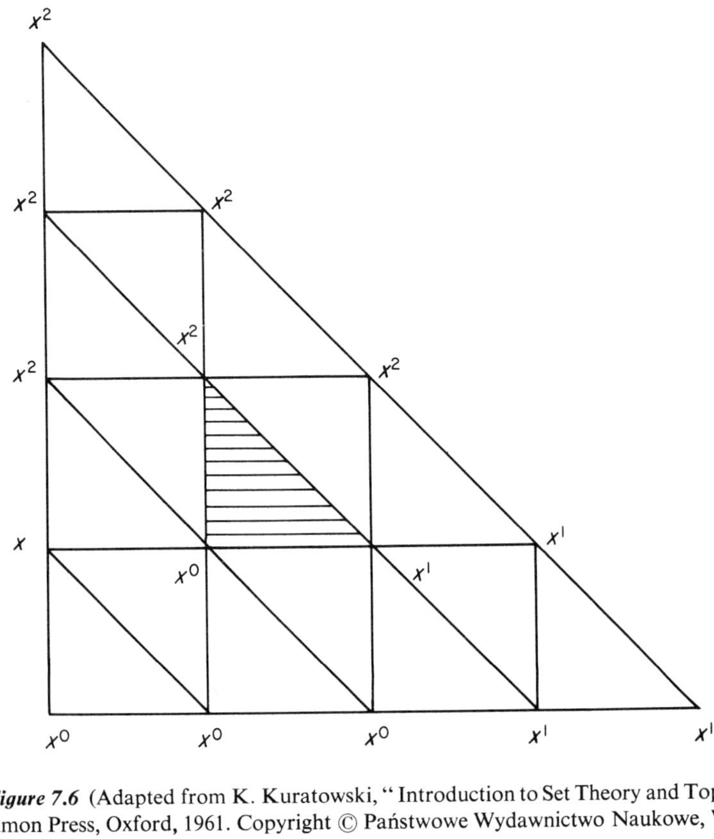

Figure 7.6 (Adapted from K. Kuratowski, " Introduction to Set Theory and Topology", Pergamon Press, Oxford, 1961. Copyright © Państwowe Wydawnictwo Naukowe, Warsaw.

7.3 Point-to-Point Mappings

Now our aim is to answer the question about the conditions under which it is certain that a single-valued function will possess at least one fixed point.

The answer we look for, at a very general level, is the content of a famous fixed-point theorem due to Brouwer [1]. The theorem is the main concern of our current study.

The topological ideas we have just examined in Section 7.2 form the necessary background for the proof of Brouwer's theorem. The two theorems which come next serve to bridge the gap between our topological considerations and the general fixed-point problem.

Lebesgue Lemma and Knaster–Kuratowski–Mazurkiewicz Theorem

The following theorem is customarily known in mathematical literature as the Lebesgue lemma.

Theorem 7.4 Let $F = \{A_1, \ldots, A_m\}$ be a finite collection of closed sets of a compact metric space S. Then there exists a number $\sigma > 0$, called the Lebesgue number of F, such that if a set L of S with diameter less than σ intersects each of the sets $A_{i_1}, \ldots, A_{i_k}$, then $A_{i_1} \cap \cdots \cap A_{i_k} \neq \emptyset$.

Proof Suppose no such number exists. Then, for each integer n ($n = 1, 2, \ldots$) there exists a nonempty set L_n with the following properties:

(1) L_n has diameter $< 1/n$, and common points with each element of a subcollection F_n of F;

(2) the intersection of the elements in F_n is empty.

F is a finite collection and has only $2^m - 1$ nonempty subcollections. Then, some of them will be repeated infinitely often in the sequence $F_1, F_2, \ldots$. Assume, for instance, $\{A_{i_1}, \ldots, A_{i_k}\} = F_{n_1} = F_{n_2} = \cdots$, and choose in each set L_{n_i} a point x^{n_i}.

We have determined in this way a sequence $(x^{n_1}, x^{n_2}, \ldots)$. Since S is compact we can select a convergent subsequence of $(x^{n_1}, x^{n_2}, \ldots)$ such that each neighborhood of its limit point x contains points of the sets $A_{i_1}, \ldots, A_{i_k}$. It is obvious thus that $x \in A_{i_1} \cap \cdots \cap A_{i_k}$ or, what is the same, $A_{i_1} \cap \cdots \cap A_{i_k} \neq \emptyset$. This contradicts, however, our original assumption.

The Lebesgue's and Sperner's lemmas will now be used to prove the following theorem due to Knaster *et al.* [1].

Theorem 7.5 Let the collection of closed sets $A_0, \ldots, A_n$ satisfy the condition $\overline{x^{i_0} \cdots x^{i_k}} \subset A_{i_0} \cup \cdots \cup A_{i_k}$ for each face of a simplex $S = x^0 \cdots x^n$. Then $A_0 \cap \cdots \cap A_n \neq \emptyset$.

Proof Call σ the Lebesgue number of $\{A_0, \ldots, A_n\}$ and let K' be a simplicial subdivision of $\bar{S}$ into simplexes of diameter $< \sigma$. Define a function f which assigns to each vertex of K' a number $f(v)$ from among the integers 0, 1, ..., n, under the following conditions:

(1) $x^{f(v)}$ is a vertex of the face of S which is the carrier simplex of v;
(2) $v \in A_{f(v)}$.

From Condition (1) and Sperner's lemma (see Theorem 7.3) we know there exists an n-dimensional simplex S' in K' whose vertices are mapped by f into all integers 0, 1, ..., n. Further, we know from Condition (2) that the simplex S' in K' has a common point with each set A_i of the collection $\{A_0, \ldots, A_n\}$. Now, since the diameter of S' is less than the Lebesgue number σ of $\{A_0, \ldots, A_n\}$, we must certainly have $A_0 \cap \cdots \cap A_n \neq \varnothing$.

Brouwer's Fixed-Point Theorem

We shall now be able to answer in quite general terms the central question of this chapter. It should be remembered, however, that for the time being we are only concerned with point-to-point mappings.

Theorem 7.6 Let $x \to f(x)$ be a continuous point-to-point mapping from a closed simplex $\bar{S}$ into itself. Then $f(x)$ has a fixed point, that is, a point x such that $x = f(x)$.

Proof Let x be defined in terms of its barycentric coordinates as follows:

(1) $x = \lambda_0 x^0 + \lambda_1 x^1 + \cdots + \lambda_n x^n$, with
(2) $\lambda_i \geq 0$ for $i = 0, 1, \ldots, n$, and
(3) $\sum_{i=0}^{n} \lambda_i = 1$.

Correspondingly, its image x' may be defined as

(4) $x' = \lambda_0' x^0 + \lambda_1' x^1 + \cdots + \lambda_n' x^n$, with
(5) $\lambda_i' > 0$ for $i = 0, 1, \ldots, n$, and
(6) $\sum_{i=0}^{n} \lambda_i' = 1$.

For $i = 0, 1, \ldots, n$ call A_i the set of all points with

(7) $\lambda_i' \leq \lambda_i$.

It is clear that each A_i is a closed set and thus satisfies the assumptions for Theorem 7.5. Since, if x is a point of the face $x^{i_0} \cdots x^{i_k}$, then

$$\lambda_{i_0} + \cdots + \lambda_{i_k} = 1 \geq \lambda_{i_0}' + \cdots + \lambda_{i_k}',$$

and the Relation (7) must hold for at least one i of $i_0, i_1, \ldots, i_k$. We thus have $x \in A_{i_0} \cup \cdots \cup A_{i_k}$. So $\overline{x^{i_0} \cdots x^{i_k}} \subset A_{i_0} \cup \cdots \cup A_{i_k}$.

Now the assertion of Theorem 7.5 holds. Consequently, there is an $x \in A_1 \cap \cdots \cap A_n$, so that $\lambda_i' \leq \lambda_i$ for all i. So there exists a point between x and its image $f(x) = x'$ for which the inequalities of (7) are satisfied for all i. From (3) and (6) it becomes clear that the latter is only possible if in (7) $\lambda_i' = \lambda_i$ for all i; this just means if $f(x) = x$.

We pointed out in Remark 7.3 that a closed simplex $\overline{S} = \overline{S^n}$ may be regarded as the smallest convex and compact (closed and bounded) set which contains its vertices. From here it is easy to see that Brouwer's theorem is applicable to a closed n-sphere, a closed n-cube and, in general, to any set homeomorphic to $\bar{S}$ (see the concept of a homeomorphism in Section 5.4).

Having the notion of a homeomorphism as background, we can formulate the results of Brouwer's theorem in the following very general terms:

Remark 7.6 Let X be a nonempty, compact, convex subset of R^n and f a continuous single-valued function from X to X. Then f has a fixed point.

Another interesting corollary of Brouwer's theorem concerns the mapping from a closed simplex $\overline{S} = \overline{S^n}$ into its boundary $B = B^{n-1}$. One can prove that such a mapping does not have a fixed point. It is particularly illustrative if we take an n-dimensional closed sphere $\bar{S}$ in the place of our simplex, and consider the mapping from $\bar{S}$ into its surface B. But let us first introduce a useful concept.

Definition 7.15 Let B be a subset of R^n. The continuous mapping $r\colon R^n \to B$ is said to be a retraction mapping of R^n if for all $x \in B$ we have $r(x) = x$. The set B is then called a retract of R^n.

We can now formulate our corollary in the following terms.

Remark 7.7 Let $\bar{S}$ be the closed unit sphere in R^n and B its surface. Then B is not a retract of $\bar{S}$, that is, there is no continuous mapping r from $\bar{S}$ into B such that $r(x) = x$ for $x \in B$.

Proof Suppose such a function r exists. Then it is clear that the function $f(x) = -r(x)$ would map $\bar{S}$ onto $f(\bar{S}) \subset \bar{S}$ without a fixed point, contradicting Brouwer's Theorem. To see this consider both the following situations:

(1) If $x \in (\bar{S} \setminus B)$, then $f(x) \neq x$ for $f(x) \in B$;

(2) if $x \in B$, then $f(x) = -x$ since $f(x) = -r(x)$ and so $f(x) \neq x$.

7.4 Point-to-Set Mappings

The last part of this chapter is dedicated to the study of the conditions under which it is guaranteed that a multivalued transformation will have a fixed point.

The main theorem on fixed points for point-to-set mappings is due to Kakutani [1], and it consists fundamentally of a generalization of the theorem of Brouwer on fixed points for point-to-point mappings. Here, we shall confine ourselves to the study of Kakutani's theorem and of some of its corollaries.

A good understanding of Brouwer's theorem together with a clear idea about sequences and barycentric subdivisions should constitute the main requirement for a successful understanding of Kakutani's theorem.

Kakutani's Fixed-Point Theorem†

The version of the fixed point theorem we give here follows, with minor notational changes, Kakutani's original paper [1].

Kakutani's own notation differs significantly from the notation we have been using so far. Unfortunately, a conversion from Kakutani's notation into ours was not possible without certain inconsistencies in regard to our system. We do not need to make a general comment about the points in which our present notation departs from that used so far. Nevertheless, we hope that a careful observation of every element used in the proof, whenever it is first introduced, may help avoid confusion.

Theorem 7.7 Let $x \to F(x)$ be an upper semicontinuous point-to-set mapping from a closed n-dimensional simplex $\bar{S}$ into $R(\bar{S})$, the family of all closed, convex subsets of $\bar{S}$. Then $F(x)$ has a fixed point; that is, there is a point $x \in \bar{S}$ such that $x \in F(x)$.

Proof Let $K^{(\nu)}$ be the νth barycentric subdivision of $\bar{S}$ and let x^ν represent a vertex of $K^{(\nu)}$ (note that here the superscript does not mean ν-vertex but rather vertex of the νth subdivision). To each vertex x^ν assign an arbitrary point y^ν of the image set $F(x^\nu)$. Now, the mapping $x^\nu \to y^\nu$, including its extension inside each simplex of $K^{(\nu)}$, forms a continuous, single-valued function from $\bar{S}$ into itself, that is, $f_\nu: \bar{S} \to \bar{S}$.

According to Brouwer's theorem $f_\nu: \bar{S} \to \bar{S}$ has a fixed point; that is, there is an $x_\nu \in \bar{S}$ such that $x_\nu = f_\nu(x_\nu)$. (Here, departing from our practice, the subscript designates a point rather than one of its components.) Consider the sequence (x_ν), where $\nu = 1, 2, \ldots$. Since $\bar{S}$ is compact it has a subsequence (x_{ν_j}), $j = 1, 2, \ldots$, which converges to a point $x \in \bar{S}$. As we shall prove, $x \in \bar{S}$ is the fixed point of our point-to-set mapping.

Let $S^{(\nu)} = S^{n,(\nu)}$ represent the n-dimensional simplex of $K^{(\nu)}$ which contains the point x_ν. We call the $n + 1$ vertices of $S^{(\nu)}$ $x_0{}^\nu, x_1{}^\nu, \ldots, x_n{}^\nu$, that is,

† Theorems 7.7 and 7.8 in this subsection adapted from Kakutani [1]. Copyright © 1941 by the Duke University Press.

x_i^{ν} where $i = 0, 1, \ldots, n$. Now take the sequences $(x_i^{\nu_j})$ where $j = 1, 2, \ldots$. It converges to x for $i = 0, 1, \ldots, n$. For some (λ_i^{ν}), where again $i = 0, 1, \ldots, n$ and $\nu = 1, 2, \ldots$, we can write

$$x_\nu = \sum_{i=0}^{n} \lambda_i^{\nu} x_i^{\nu} \qquad \text{with} \qquad \sum_{i=0}^{n} \lambda_i^{\nu} = 1 \quad \text{and} \quad \lambda_i^{\nu} \geq 0.$$

Now, for $i = 0, 1, \ldots, n$ and $\nu = 1, 2, \ldots$, write $f_\nu(x_i^{\nu}) = y_i^{\nu}$. For $\nu = 1, 2, \ldots$, we have $y_i^{\nu} \in F(x_i^{\nu})$ and thus $x_\nu = f_\nu(x_\nu) = \sum_{i=0}^{n} \lambda_i^{\nu} y_i^{\nu}$.

Consider the sequence of indices (ν_j) where $j = 1, 2, \ldots$. Select a subsequence (ν_j'), $j = 1, 2, \ldots$, such that the subsequences $(y_i^{\nu_j'})$, $j = 1, 2, \ldots$ and $(\lambda_i^{\nu_j'})$, $j = 1, 2, \ldots$ converge for $i = 0, 1, \ldots, n$. We can then write $\lim_{j\to\infty} y_i^{\nu_j'} = y_i$ and $\lim_{j\to\infty} \lambda_i^{\nu_j'} = \lambda_i$, for $i = 0, 1, \ldots, n$. We have now

$$x = \sum_{i=0}^{n} \lambda_i y_i \qquad \text{with} \quad \sum_{i=0}^{n} \lambda_i = 1 \quad \text{and} \quad \lambda_i \geq 0.$$

Furthermore, for $i = 0, 1, \ldots, n$ we have

$$\lim_{j\to\infty} x_i^{\nu_j'} = x; \qquad \lim_{j\to\infty} y_i^{\nu_j'} = y_i, \qquad y_i^{\nu_j'} \in F(x_i^{\nu_j'}).$$

By the assumed upper semicontinuity of $F(x)$ we clearly have $y_i \in F(x)$ for $i = 0, 1, \ldots, n$. Therefore, because of the convexity of $F(x)$ $x = \sum_{i=0}^{n} \lambda_i y_i \in F(x)$, and x is a fixed point of the point-to-set mapping.

The relationship of Brouwer's to Kakutani's theorem is very easy to discover. Suppose in the latter that each $F(x)$ consists of only one point, $f(x)$. In this case, the upper semicontinuity of $F(x)$ becomes simply the continuity of $f(x)$. Obviously, Brouwer's theorem is a special case of Kakutani's.

Kakutani's theorem—in analogy to what happened with Brouwer's theorem—can be shown to hold even if we take an arbitrary compact, convex set X in R^n, instead of the closed simplex $\bar{S}$.

Suppose such a set X is contained in $\bar{S}$, and define a continuous retracting single-valued function $x \to f(x)$ from $\bar{S}$ onto X such that

(1) for any $x \in X, f(x) = x$;
(2) for any $x \in \bar{S}, f(x) \in X$.

Under these conditions, it is apparent that $F(f(x))$ is an upper semicontinuous multivalued function from $\bar{S}$ into $R(X) \subset R(\bar{S})$. By Theorem 7.7 there exists a point $x \in \bar{S}$ such that $x \in F(f(x))$. But we also have $F(f(x)) \subset X$. Thus, we must have $x \in X$. From the definition of $f(x)$ we clearly obtain $x \in F(x) \subset \bar{S}$.

This result, originally formulated by Kakutani as corollary to the main theorem, can be stated as follows.

Remark 7.8 Let $x \to F(x)$ be an upper semicontinuous point-to-set mapping from a convex, compact set X into itself, such that each $F(x)$ is convex and compact. Then $x \to F(x)$ has a fixed point.

A theorem which plays a remarkable role in the theory of mathematical programming will serve to close this chapter. It is often referred to as Kakutani's theorem 2 and its proof follows immediately from the previous remark. We also take the opportunity to return to our customary notation.

Theorem 7.8 Let X and Y be two compact, convex sets in R^m and R^n, respectively. Let $Z = X \times Y$ define their direct product in R^{m+n}. Finally, let U and V be two closed subsets of $X \times Y$ such that the following two conditions are satisfied: (1) for any point $x^0 \in X$, the set U_{x^0} of all points $y \in Y$ such that $(x^0, y) \in U$, is nonempty, closed and convex; (2) for any point $y^0 \in Y$, the set V_{y^0} of all points $x \in X$ such that $(x, y^0) \in V$, is nonempty, closed and convex. Then U and V have a common point.

Proof Assume there exists a multivalued function $z \to F(z)$ from Z into $R(Z)$, defined as follows: If $z = (x, y)$ then $F(z) = V_y \times U_x$. Note first that the mapping $F(z)$ is upper semicontinuous since, by assumption, both U and V are closed sets. From this property and the previous remark $z \to F(z)$ has a fixed point. Call this point z^0. Obviously, we have $z^0 \in Z$ such that $z^0 \in F(z^0)$. But this means that there exists a pair of points $x^0 \in X$ and $y^0 \in Y$ such that $(x^0, y^0) \in V_{y^0} \times U_{x^0}$, where $x^0 \in V_{y^0}$ and $y^0 \in U_{x^0}$. Clearly $z^0 = (x^0, y^0) \in U \times V$ is the common point.

EXERCISES

E.7.1 Determine whether the following real single-valued functions have fixed points and, if the result is positive, indicate the corresponding values of x:

(a) f: $[1, 10] \to R$, defined by $f(x) = x^{1/2} + 2$;
(b) f: $R \to R$, defined by $f(x) = x^{-1}$.

Hint: Draw the graph of each function.

E.7.2 Decide whether the following multivalued functions have fixed points and, if the answer is affirmative, locate them in a graph:

(a) F: $[1, 10] \to [1, 10]$ such that

$$\begin{aligned} F(x) &= [4 + 0.5x, 9], && \text{if } x \in [0, 7[; \\ F(x) &= [4, 7[\cup]7, 9], && \text{if } x = 7; \\ F(x) &= [1 + 0.5x, 7], && \text{if } x \in [7, 10]. \end{aligned}$$

(b) All the same as in (a) above but with

$$F(x) = [4, 7[\cup [7, 9], \qquad \text{if } x = 7.$$

E.7.3 Consider the simplex S^n. Show that $\overline{S^n} \setminus S$, the boundary of S^n, is equal to the union of all the face simplexes having dimension less than n.

Hint: The face simplexes of dimension less than three are S^{-1}, the vertices, the open edges, and the open triangles which form the surface of a tetrahedron.

E.7.4 Consider the simplex $S^3 = x^0x^1x^2x^3$. (a) Find the complexes K^3, K^2, K associated with S^3; (b) show that K^3 is a simplicial decomposition of $\overline{S^3}$; (c) reconstruct the polyhedron $P = \overline{K^3}$.

E.7.5 Let $S^3 = x^0x^1x^2x^3$. (a) Find the first barycentric subdivision of S^3; (b) Find $K' = K_1{}^3$.

Hint: Use Definition 7.10 and study the example following Definition 7.11. Then find B^2 and $B_1{}^2$.

E.7.6 Illustrate Sperner's lemma with $S^3 = x^0x^1x^2x^3$.

E.7.7 Show that Theorem 7.1 does not hold for a continuous, real-valued function f: $X \to X$ when $X =]0, 1[$.

E.7.8 Let $X = \{x \mid x \in R^2, xx = 1\}$. Show that a continuous mapping from X onto a proper subset of itself has a fixed point.

Hint: X is the equation of a circumference.

E.7.9 Show that for the closed 2-sphere $\bar{S} = \{x \mid x \in R^2, xx \leq 1\}$ and its boundary $B = \{x \mid x \in R^2, xx = 1\}$, there is no continuous mapping r: $\bar{S} \to B$ such that there is $x = r(x)$ for $x \in B$.

Hint: Study, as in the proof of Remark 7.7, a function $f(x) = -r(x)$.

E.7.10 In terms of the conditions of Kakutani's fixed-point theorem determine why the following set-valued mappings do not possess fixed points:

(a) F: $X \to X$, where $X = [0, 10]$, and

$$\begin{aligned} F(x) &= [3 + 0.5x, 5 + 0.5x], && \text{if } x \in [0, 5[; \\ F(x) &= [6, 7], && \text{if } x = 5; \\ F(x) &= [3 + 0.5x, 5 + 0.5x], && \text{if } x \in]0, 5]. \end{aligned}$$

(b) F: $X \to X$, where $X = [0, 10]$, and

$$\begin{aligned} F(x) &= [6, 10], && \text{if } x \in [0, 5[; \\ F(x) &= [0, 4[\cup]6, 10], && \text{if } x = 5; \\ F(x) &= [0, 4], && \text{if } x \in]5, 10]. \end{aligned}$$

Hint: Study the upper semicontinuity of F and the convexity of the set $F(x)$ at the relevant point of X.

PROBLEMS

P.7.1 Let X and Y be two homeomorphic topological spaces. Show that a necessary and sufficient condition for each continuous mapping $f\colon X \to X$ to have a fixed point is that each continuous mapping $g\colon Y \to Y$ has a fixed point.

Hint: Study the composite mappings from X to Y and from Y to X.

P.7.2 Let $\bar{S}$ be a closed n-dimensional simplex, and $f\colon \bar{S} \to \bar{S}$ a continuous mapping. Suppose that $f(x)$ is in the boundary of S, and $x \neq f(x)$, whenever x is in the boundary of S. Prove that $\bar{S} = f(\bar{S})$.

Hint: If $\bar{S} \neq f(\bar{S})$ Brouwer's fixed point theorem is violated.

P.7.3 Let $X \subset R^m$ and $Y \subset R^n$ be compact, convex sets, and $f\colon X \times Y \to X \times Y$ a continuous mapping. Assume: (a) for every $x^0 \in X$ and any $a \in R$, the set $\{y \mid y \in Y, f(x^0, y) \leq a\}$ is convex; (b) for every $y^0 \in Y$ and any $b \in R$, the set $\{x \mid x \in X, f(x, y^0) \geq b\}$ is convex. Prove: $\max_{x \in X} \min_{y \in Y} f(x, y) = \min_{y \in Y} \max_{x \in X} f(x, y)$.

Hint: The suggested proof of this "saddle point" theorem requires the use of Theorem 7.8. It can be done by studying the sets $U = \{z^0 \mid z^0 = (x^0, y^0) \in X \times Y,\ f(x^0, y^0) = \min_{y \in Y} f(x^0, y)\}$, and $V = \{z^0 \mid z^0 = (x^0, y^0) \in X \times Y,\ f(x^0, y^0) = \max_{x \in X} f(x, y^0)\}$. See also Kakutani [1].

P.7.4 Construct a numerical example of a mapping having a "saddle point" as an illustration of Problem P.7.3.

P.7.5 Why is it necessary the convexity of $F(x)$, for every $x \in X$, for the validity of Kakutani's fixed-point theorem? Discuss.

NOTES

Chapter seven is mainly based on the work of Alexandroff and Hopf [1], Cairns [1], Franz [1, 2], and Kuratowski [2]. For an interesting discussion of Sperner's lemma see Tompkins [1]. The proof of Brouwer's theorem given here is essentially due to Knaster *et al.* [1]. It would be useful to compare the approach followed here with Berge's [2]. For a different proof of Brouwer's theorem see Franz [2]. In the proof of the fixed-point theorem for multivalued mappings we follow Kakutani [1]. Generalizations of the theorem are given by Eilenberg and Montgomery [1] and Begle [1].

CHAPTER 8

ALGEBRAIC STRUCTURES

The present chapter may be considered as a very brief account of concepts which are usually the object of abstract algebra. As such, it directly follows from Chapter 1, though pointing to a theoretical frame which is different from the frame in which we have been so far. Both general topology and the theory of abstract algebraic systems may be built up beginning with the elementary concepts of set, function, and operation.

The principal reason for which we include this chapter is, however, that it provides a simple way to bridge the gap between the mathematical theories of Parts I and II.

Part I, Chapters 4–7 in particular, was mainly concerned with topological properties of sets of points.

The link between the concepts of a topological and a metric space was discussed in Section 4.2. Example 1, immediately following Definition 4.20, mentioned the connection between a vector space and the topological space associated with it.

The concept of a vector space will underlie the whole discussion of Part II. But our basic interest will concentrate on the algebraic rather than the topological aspects related to a vector space.

8.1 Abstract Systems and Homomorphisms

The notion of an abstract algebraic system or algebraic structure is somewhat more complex than that of a set. It presupposes the existence of a set, but that is not enough. A certain set of relations is supposed to hold for the elements of the basic set. Besides, the set of elements is supposed to be closed under certain operations.

From this point of view, a system may be regarded as a kind of "structured," "organized" set. The idea was clearly present in the discussion of some of the earlier chapters, though not explicitly stated.

Definition 8.1 A set E of elements, together with a set $\{\tilde{R}\}$ of relations between the elements of E and a set O of operations defined on E, is called an abstract system or an abstract algebraic structure, and denoted $S = \{E, \{\tilde{R}\}, O\}$.

In general, it will be enough to characterize a system by indicating the set of elements under consideration together with the relevant operations. We shall write then, for instance, $S = \{E, \cdot\}$, $S = \{E, +, \cdot\}$, and so on.

E, $\{\tilde{R}\}$, and O are the fundamental components of a system. The theory of the system under question, however, is a construction of postulates, theorems, and definitions.

The aspect which distinguishes a function from a homomorphism is quite analogous to that which makes the difference between an unstructured and a structured set or system. A homomorphism maps not only a set into a set, but precisely a structured set into a structured set. Loosely speaking, it is a structure-preserving mapping. A formal definition is given next.

Definition 8.2 Let $S = \{E, \{\tilde{R}\}, O\}$ and $S' = \{E', \{\tilde{R}'\}, O'\}$ be two abstract systems where $x, y \in E$, $\tilde{R}_i \in \{\tilde{R}\}$, $\tilde{R}_i' \in \{\tilde{R}'\}$, $o_j \in O$, $o_j' \in O'$, $i = 1, \ldots, m$, and $j = 1, \ldots, n$. Let $h: E \to E'$ be a single-valued function such that the following conditions are satisfied:

(1) $x\tilde{R}_i y \in E$ implies $h(x)\tilde{R}_i'h(y) \in E'$ for $i = 1, \ldots, m$;
(2) $h(xo_j y) = h(x)o_j'h(y)$ for $j = 1, \ldots, n$.

Then h is said to be a homomorphism of S into S'.

A homomorphism h of S to S' may be classified according to an association rule, in the sense of Section 5.2.

A transformation from E onto E' defines a surjective homomorphism, a transformation from E into E' an injective one. A very important case occurs when the mapping $h: E \to E'$ is bijective, that is, one to one in the sense of Definition 5.8. In such a case the mapping is said to be an isomorphism.

When the homomorphism implies a mapping from E into itself, it is called an endomorphism. Bijective endomorphisms are referred to as automorphisms.

8.2 Groups

Let us consider first the concept of a group in general.

Definition 8.3 A group is a system $G = \{E, *\}$ having one closed binary operation and satisfying the following axioms:

G1 Associative law: For any $x, y, z \in E$

$$(x * y) * z = x * (y * z).$$

G2 Identity element: There is $e \in E$ such that

$$x * e = e * x = x \qquad \text{for any} \quad x \in E.$$

G3 Inverse element: For each $x \in E$ there is a unique $x' \in E$ such that

$$x * x' = x' * x = e.$$

Let $G^1 = \{E_1, *\}$ and $G^2 = \{E_2, \odot\}$ represent two groups. Clearly, a mapping $h: G^1 \to G^2$ (or $h = E_1 \to E_2$) is a group homomorphism if, for $x, y \in E_1$,

$$h(x * y) = h(x) \odot h(y).$$

Some particular types of groups can be characterized now, by specifying the operation of Definition 8.3. Two different types will attract our attention here.

Additive Groups

Definition 8.4 An additive group is a system $G = \{E, +\}$ which satisfies the following postulates:

G1′ For any $x, y, z \in E$

$$(x + y) + z = x + (y + z).$$

G2′ There is $0 \in E$ such that

$$x + 0 = 0 + x = x \qquad \text{for any} \quad x \in E.$$

G3′ For each $x \in E$ there is $-x \in E$ such that

$$x + (-x) = (-x) + x = 0.$$

The inverse element $-x$ is called the additive inverse or the negative of x.

Example The set of complex numbers with the operation addition may be regarded as an additive group. Consider the complex numbers u and v and denote by Re and Im their real and imaginary parts, respectively. Then

$$u + v = \operatorname{Re}(u + v) + \operatorname{Im}(u + v), \qquad u + 0 = 0 + u = u,$$
$$u + (-u) = (-u) + u = 0.$$

If G^1 and G^2 are two additive groups and h a group homomorphism, then clearly

$$h(x + y) = h(x) + h(y),$$

for $x, y \in G^1$.

Multiplicative Groups

Definition 8.5 A multiplicative group is a system $G = \{E, \cdot\}$ which satisfies the following postulates:

G1″ For any $x, y, z \in E$,

$$(x \cdot y) \cdot z = x \cdot (y \cdot z).$$

G2″ There is $1 \in E$ such that

$$x \cdot 1 = 1 \cdot x = x \qquad \text{for any} \quad x \in E.$$

G3″ For each $x \in E$ there is $1/x \in E$ such that

$$x \cdot 1/x = (1/x)x = 1 \qquad (x \neq 0).$$

The inverse element $1/x = x^{-1}$ is also called the multiplicative inverse of x.

Example The set of positive real numbers with the operation multiplication constitutes a multiplicative group.

The mapping h from the multiplicative group G^1 to the multiplicative group G^2 is a group homomorphism if

$$h(x \cdot y) = h(x) \cdot h(y)$$

for $x, y \in G^1$.

Next we define a special type of group with the special feature that it satisfies an additional postulate to those enumerated in Definition 8.3.

Commutative or Abelian Groups

Definition 8.6 A group $G = \{E, *\}$ is said to be commutative or Abelian if it satisfies axioms G1, G2, G3 of Definition 8.3 and, additionally the following:

G4 Commutative Law: For every $x, y \in E$

$$x * y = y * x.$$

From Definitions 8.6 and 8.4 it becomes clear that a commutative additive group satisfies postulates G1′, G2′, G3′, and G4′ (additive commutativity).

Similarly, from Definitions 8.6 and 8.5, a commutative multiplicative group satisfies G1″, G2″, G3″, and G4″ (multiplicative commutativity).

Example 1 The set of integers forms an Abelian group under addition.

Example 2 The set of complex numbers constitutes a commutative group under multiplication.

Isomorphic Groups

The concept of a group homomorphism h has already been discussed. An important special case occurs when the mapping h happens to be bijective that is, one to one. Notice that in such a case, to the mapping $h\colon G^1 \to G^2$ there corresponds an inverse mapping $h^{-1}\colon G^2 \to G^1$.

Definition 8.7 Let G^1 and G^2 be two groups. G^1 and G^2 are said to be isomorphic if there exists an isomorphism from G^1 to G^2. In general, we shall say that G^1 and G^2 are homomorphic groups when h is merely a homomorphism. Clearly, isomorphic groups form a subclass of the set of homomorphic groups.

8.3 Rings, Integral Domains, and Fields

A brief account of some more complex algebraic structures is given next.

Rings

Definition 8.8 A ring $L = \{E, +, \cdot\}$ is a system having two closed operations, addition and multiplication, and which satisfies the following axioms.

L1 = G1′ Additive associativity.
L2 = G2′ Additive identity element.
L3 = G3′ Additive inverse element.
L4 = G4′ Additive commutativity.
L5 = G1″ Multiplicative associativity.
L6 Left and right distributive laws: For every $x, y, z \in E$

(1) $x(y + z) = xy + xz$;
(2) $(y + z)x = yx + zx$.

Example The set of integers with the operations addition and multiplication constitutes a ring.

Note that a ring is an Abelian group under addition which additionally satisfies Axioms G1″ and L6. If it also satisfies multiplicative commutativity, that is, Axiom G4″, then we call it a commutative ring.

If in a ring L there exist elements x and y such that $x \cdot y = 0$, $x \neq 0$ and $y \neq 0$, then we say it is a ring with zero divisors. Otherwise it is called a ring without zero divisors.

A ring L which satisfies Postulate G2″, multiplicative identity, is usually called a ring with unity. For $x \in L$ we have in this case $x \cdot 1 = 1 \cdot x = x$.

Integral Domains

Definition 8.9 A commutative ring with unity L is called an integral domain if it has no zero divisors; that is, if for two elements x and y in L, $x \cdot y = 0$, then $x = 0$ or $y = 0$.

Example The set of integers with the operations of addition and multiplication forms an integral domain.

Fields

The most "complete" structure we shall be defining here is called a field. A field is a commutative ring with unity, in which every nonzero element has a multiplicative inverse.

The concepts of an Abelian group under addition and of a field are fundamental for the definition of the algebraic structure of a vector space, which we shall consider in detail in Part II. In Chapter 10 we define a "vector space over a field."

To denote a field we shall use, in the future, the letter F. Remember that so far F was the generic symbol to express a point-to-set mapping. However, for the next chapters, the meaning of F will appear clearly from the context in which it is being used.

Let us define now a field in formal terms.

Definition 8.10 A field $F = \{E, +, \cdot\}$ is a system having two closed operations, addition and multiplication, and which satisfies the following axioms:

F1 = G1′	Additive associativity,
F2 = G2′	Additive identity,
F3 = G3′	Additive inverse,
F4 = G4′	Additive commutativity,
F5 = G1″	Multiplicative associativity,
F6 = G2″	Multiplicative identity,
F7 = G3″	Multiplicative inverse (for $x \neq 0$),
F8 = G4″	Multiplicative commutativity,
F9 = L6	Left and right distributivity.

It is clear from the preceding definition that a field satisfies the properties of both a commutative group under addition and a commutative group under multiplication; in addition it satisfies the left and right distributive laws.

Example The real numbers, the rational numbers, and the complex numbers are each examples of fields.

EXERCISES

Groups

E.8.1 Let $G = \{E, *\}$ be a group. Prove: (a) G has a unique identity element; (b) for each element in G there is a unique inverse element.

Hint: (a) Suppose e and e' are identity elements. Observe that $e * e' = e' * e = e$; (b) suppose x' and x'' are two inverses of x. Note that $x * x' = x * x'' = e$.

E.8.2 Let $G = \{E, *\}$ be a group with $2n$ elements. Show that, including the identity element e, G has two elements which are their own inverses.

E.8.3 Prove that the following structures constitute groups: (a) $\{Z, +\}$, where Z is the set of integers; (b) $\{Q \setminus \{0\}, \cdot\}$, where X is the set of rationals. (c) Are $\{Z, +\}$ and $\{Q \setminus \{0\}, \cdot\}$ commutative groups?

E.8.4 Prove that $\{U, \cdot\}$ forms a group if $U = \{x \mid x^3 - 1 = 0\}$.

Hint: $U = \{1, -\frac{1}{2} + i3^{1/2}/2, -\frac{1}{2} - i3^{1/2}/2\}$. Apply Definition 8.5.

Rings, Integral Domains, Fields

E.8.5 Determine which of the following sets constitute rings: (a) N; (b) Q; (c) R; (d) C, the set of complex numbers; (e) $\{f \mid f\colon X \to R, f \text{ is continuous on } X = [-1, 1]\}$.

Hint: In (e) consider $(f_1 + f_2)(x) = f_1(x) + f_2(x)$, and $f_1 f_2(x) = f_1(f_2(x))$.

E.8.6 Prove that $X = \{x \mid x = 2p, p \in Z\}$ forms a ring but not a field.

Hint: Note that $1 \notin X$.

E.8.7 Prove that a field is always an integral domain.

E.8.8 Show that $\{Y, +, \cdot\}$ is a field, where $Y = \{y \mid y = a + b3^{1/2}; a, b \in Q\}$.

Homomorphisms

E.8.9 Let G be homomorphic to G', and G' homomorphic to G''. Prove that the relation homomorphism is transitive, that is, G is homomorphic to G''. Furthermore, prove that the relation isomorphism of groups is an equivalence relation.

Hint: For the second part of the exercise it rests only to show that G is isomorphic to itself (reflexivity), and that if G is isomorphic to G', then G' is isomorphic to G (symmetry).

E.8.10 Indicate which of the following mappings is a group homomorphism or a group isomorphism:

(a) $f\colon G \to X$, where $G = \{Z, +\}$, $X = \{x \mid x = 2p, p \in Z\}$, and $f(p) = x = 2p$;
(b) $f\colon G \to X$, where $G = \{R \setminus \{0\}, \cdot\}$, $X \subset (R \setminus \{0\})$, and $f(x) = -x$;
(c) $f\colon G \to X$, where $G = \{R \setminus \{0\}, \cdot\}$, $X \subset (R \setminus \{0\})$, and $f(x) = x^{-1}$;
(d) $f\colon G \to X$, where $G = \{R \setminus \{0\}, \cdot\}$, $X \subset (R \setminus \{0\})$, and $f(x) = x^{1/2}$.

PROBLEMS

P.8.1 Prove that if G is a $*$-group and $a, b, c \in G$, the following propositions are true: (a) If $a * b = a * c$ or $b * a = c * a$, then $b = c$; (b) $(a * b)' = b' * a'$; (c) the equations $a * x = b$ and $y * a = b$ have a unique solution in G.

P.8.2 Let $f\colon G^1 \to G^2$ be a group isomorphism. Prove: (a) f maps the identity element of G^1 into the identity element of G^2; (b) f maps the inverse element of $x \in G^1$ into the inverse element of $f(x) \in G^2$.

P.8.3 Let $f: G^1 \to G^2$ be a homomorphism. Define the kernel of f as the set $K_f = \{x \mid x \in G^1, f(x) = e^2 \in G^2, e^2$ being the identity element of $G^2\}$. Prove that f is an isomorphism if and only if $K_f = \{e^1\}$, where e^1 is the identity element of G^1.

P.8.4 Let $R^+ = \{x \mid x \in R, x > 0\}$. Show that the following structures are isomorphic groups: $\{R, +\}, \{R^+, \cdot\}$.

P.8.5 Let $f_1 : G \to G$ and $f_2 : G \to G$ be automorphisms. Prove that $f_1 f_2$ is also an automorphism.

NOTES

A most useful and comprehensive book on abstract algebra is that of Fang [1] (see also the classical book by Birkhoff and MacLane [1]). For theory of groups the book by Alexandroff [2] may be consulted. Other algebraic structures are discussed by Van der Waerden [1].

CHAPTER 9

GENERAL EQUILIBRIUM FOR ECONOMIES WITH A FINITE NUMBER OF AGENTS AND COMMODITIES

The aim of this chapter is to illustrate the use of some of the point-set topological concepts previously discussed by means of an example chosen from the field of economics.

The modern theory of general economic equilibrium makes considerable use of such concepts. Thus, it seems adequate to refer our example to this area.

General equilibrium theory is inseparably associated with the name of Walras [1]. However, the modern, rigorous analysis of the existence of equilibria for a competitive economy comprising a finite number of economic agents and commodities is chiefly the result of the works of Wald [1], McKenzie [1], Gale [2], Nikaido [1, 2], Arrow and Debreu [1], and Debreu [4]. It is Debreu's [4] theoretical exposition that we shall follow in this chapter, though with fewer technical details and in a less general version than that to be found in the original.

Some very recent developments, in particular models assuming a continuum of agents (usually represented by the points of the closed interval [0, 1]) and/or an infinite number of commodities will not be considered here. Stability theory, an area of analysis which is so close to equilibrium theory, is also excluded from our present discussion. Some "pieces" of equilibrium theory have already been introduced as examples in earlier chapters. They will reappear now, though in a wider context.

9.1 Frame of the Problem and Basic Assumptions

In our discussion, we shall be confined to the problem of finding an answer to the question about the conditions for the existence of an equilibrium for a competitive, private-ownership economy. The concepts equilibrium and competitive, private-ownership economy, will have to be clearly and adequately defined. But this will be done later, after other more elementary notions have been introduced.

Elementary Ingredients of the Theory

Our economy is supposed to consist of n production units (called producers), m consumption units (called consumers), and k commodities (goods and services of every kind). We shall use the subscripts j, i, and h, respectively, to refer to them.

A commodity is completely identified by means of three specifications: physical properties, location at which it is available, time at which it is available.

Two goods or services belong to the same commodity class, that is, have the same index h, if and only if the three mentioned specifications are identical.

The quantity of a commodity h is represented by the value of the hth coordinate of the commodity space R^k. A point of R^k thus represents a commodity bundle.

A production plan, also production or supply, for the jth producer is a specification of the quantities of all his outputs (positive numbers) and all his required inputs (negative numbers). It is a point $y_j \in R^k$. The set of all possible productions of the jth producer is his production set Y_j. For the economy as a whole, a total production is thus $y = \sum_{j=1}^{n} y_j$ and the total production set, $Y = \sum_{j=1}^{n} Y_j$. Clearly $y \in Y$.

A consumer plan, also consumption or demand, for the ith consumer is a specification of the quantities of all labor services or outputs he renders (negative numbers) and all commodities he consumes or inputs (positive numbers). It is a point $x_i \in R^k$. All possible consumptions are contained in the consumption set X_i. The point $x = \sum_{i=1}^m x_i$ and the set $X = \sum_{i=1}^m X_i$ represent, respectively, a total consumption and the total consumption set for the economy as a whole. Here too, $x \in X$.

Finally, a price system is a nonnegative point $p = (p_1, \ldots, p_k) \in R^k$. In general, we are interested in relative rather then absolute prices. For this reason we shall regard always p as a relative or normalized price system, that is, $\sum_{n=1}^k p_h = 1$.

Notice, that because of the sign convention regarding the components of y_j, it is clear that $p \cdot y_j = p_1 y_{1,j} + \cdots + p_k y_{k,j}$ will represent the profit of the jth producer from producing y_j at a given price system p.

Production and Consumption Sets: Assumptions

Not all assumptions made below are really necessary for the theory. However, although reducing the generality of the results, they greatly simplify our task. This seems justified for the merely illustrative character of the exercise.

For each $j = 1, \ldots, n$ we assume that

(i) Y_j is compact and convex;
(ii) $Y_j \cap \Omega = \{\theta\}$;
(iii) $Y_j \cap (-Y_j) = \{\theta\}$.

As a consequence of assumptions (i), (ii), and (iii), we have

(i′) Y is compact and convex;
(ii′) $Y \cap \Omega = \{\theta\}$;
(iii′) $Y \cap (-Y) = \{\theta\}$.

Assumption (ii) implies $\theta \in Y_j$, that is, the possibility of not producing anything. Together with convexity, it excludes the possibility of increasing returns to scale in production.

Assumption (iii) accounts for the irreversible character of every economic production process.

Note, finally, that the possibility of wasteful production is allowed for by our assumptions since there is a compact subset $B \subset -(\Omega)$ which is necessarily $B \subset Y$. B contains θ but also an infinite number of points of the nonpositive orthant other than θ. Had we assumed that Y is merely closed, then we would obviously have $-\Omega \subset Y$.

For each $i = 1, \ldots, m$ we assume that

(iv) X_i is closed and convex;
(v) X_i is lower bounded.

Results of these assumptions are

(iv′) X is closed and convex;
(v′) X is lower bounded.

From the closedness and lower boundedness of each X_i one can derive the closedness of X. The convexity and the lower boundedness of X are immediate.

Assumption (v) recognizes the need for a minimum consumption for each consumer and a maximum for the amount of labor services he can provide. Related to this, convexity excludes the possibility of increasing returns in the process of transforming consumption into labor services.

These are the basic assumptions in relation to the production and consumption sets. Some additional assumptions, of a more specific nature, will be made later in the course of the exposition. We shall then summarize all of them towards the end of our discussion.

9.2 Producers and Supply

Let p be a given price system. Since $p \cdot y_j$ represents the difference of all receipts and all outlays from producing y_j at given p, we have called it the profit of the jth production unit.

Profit Maximization

It is assumed now that the jth producer chooses the production which maximizes his profit relative to p. In other words, he chooses $y_j \in Y_j$ which is a maximum for the set $p \cdot Y_j$.

In general, we have thus determined a point-to-set mapping,

$$S_j : R^k \to Y_j,$$

defined by

$$S_j(p) = \{y_j \mid y_j \in Y_j, p \cdot y_j = \max p \cdot Y_j\},$$

which will be called the supply correspondence of the jth producer. For the economy as a whole, total supply will be denoted by S and defined by $S(p) = \sum_{j=1}^{n} S_j(p)$.

Another important concept which we shall introduce in this context is the profit function of the jth producer. Formally, it is a point-to-point mapping,

$$\pi_j : R^k \to R,$$

specified by

$$\pi_j(p) = \max p \cdot Y_j.$$

For the economy as a whole, the total profit function π is defined by the formula $\pi(p) = \sum_{j=1}^{n} \pi_j(p)$.

Properties of Mappings and Sets

From the assumptions regarding the compactness and convexity of each set Y_j we can derive now the following properties. (See the example after Theorem 6.5.)

Remark 9.1 If the assumptions on production sets are satisfied, then

(1) S_j is upper semicontinuous and π_j is continuous, for each j;
(2) S is upper semicontinuous and π is continuous;
(3) $S_j(p)$ and $S(p)$ are convex sets.

9.3 Consumers and Demand

A similar analysis has to be done now for the other section of the economy, the consumers.

Consumer Preferences

Each set X_i is assumed to be totally preordered by a preference preordering $\precsim_i$. For a given $x_i' \in X_i$, the set $\{x_i | x_i \in X_i,\ x_i \sim_i x_i'\}$ is called the indifference class of x_i'.

The preordering $\precsim_i$ is assumed to be continuous. This means that given $x_i' \in X_i$, the sets $\{x_i | x_i \in X_i,\ x_i \precsim_i x_i'\}$ and $\{x_i | x_i \in X_i,\ x_i \succsim_i x_i'\}$ are both closed in X_i.

Continuity of preferences is, as we shall see soon, prerequisite for the existence of a utility function. A utility function is a useful, though not at all indispensable, piece of analysis. It is an increasing, continuous function from the preordered set X_i into R. We shall denote a utility function by $u_i(x_i)$, all i, and assume here that it exists. An existence proof is given later.

Our second assumption in the present context is that, for each consumer and regardless of budget constraints, there is no satiation or saturation consumption. Formally, no X_i has a largest element for $\precsim_i$.

Finally, and like in earlier examples, the preordering satisfies the following convexity assumption: If $x_i^2 \succ_i x_i^1$, then $\lambda x_i^2 + (1-\lambda)x_i^1 \succ_i x_i^1$, for any two bundles $x_i^2, x_i^1 \in X_i$ and $\lambda \in]0, 1[$.

Wealth Distribution and Budget Constraint

A point $w \in R^m$ will denote a particular distribution of wealth in the economy. It is thus understood that the ith component of w, that is, w_i, represents the wealth of the ith consumer.

Let $p \in R^k$ and $w \in R^m$ be given. As done in an earlier example (in Example 2, Section 6.1, we considered a unique consumer, so the index i was superfluous and $w \in R$) the point $(p, w) \in R^{k+m}$ will be identified as a price–wealth pair.

We now define for each i a point-to-set mapping

$$T_i: K_i \to X_i$$

specified by

$$T_i(p, w) = \{x_i \mid x_i \in X_i, px_i \leq w_i\}.$$

Here the nonempty set $K_i \subset R^{k+m}$ is $K_i = \{(p, w) \mid$ there is $x_i \in X_i$ such that $p \cdot x_i \leq w_i\}$, that is, the set of all (p, w) for which the set of feasible consumptions is nonempty. T_i clearly assigns to each price–wealth pair a subset (of feasible consumptions) of X_i.

We are interested in the continuity of the mapping T_i, for every i. It is easy to prove that T_i is continuous at a point $(p^0, w^0) \in K_i$ if $w_i^0 \neq \min p^0 X_i$, that is, if the wealth of the ith consumer at this state is not precisely equal to the value of the minimum of his consumption set.

A proof of the latter assertion is left as an exercise. It should be anticipated, however, that the requirement $w_i^0 \neq \min p^0 X_i$ will arise the need for some additional assumption, which will be presented later.

Preference Satisfaction and Utility Maximization

The central assumption of this part of our discussion concerns the behavior of the ith consumer. It can be stated as follows:

Given (p, w) the consumer chooses in $T_i(p, w)$ a commodity bundle which is a largest element for his preordering $\precsim_i$. If a utility function $u_i(x_i)$ has been introduced into the analysis, then we may equivalently say that the consumer chooses in $T_i(p, w)$ the commodity bundle which maximizes his utility function.

When a utility function has been constructed, an "indirect utility function"

$$v_i : K_i \to R$$

may be defined by

$$v_i(p, w) = \max u_i(x_i) = \max u_i(T_i(p, w)).$$

A demand correspondence for the ith consumer can be described now, in the general case, as a point-to-set mapping

$$D_i : K_i \to X_i$$

specified either by $D_i(p, w) = \{x_i \mid x_i \in T_i(p, w), x_i \text{ is largest element for } \precsim_i\}$ or, if a utility function has been constructed, by $D_i(p, w) = \{x_i \mid x_i \in T_i(p, w), \max u_i(x_i)\}$.

For the economy as a whole, we clearly have the total demand D defined by $D(p, w) = \sum_{i=1}^{m} D_i(p, w)$.

The way in which the demand correspondences have been expressed is certainly not the most convenient for our further analysis. It would be highly desirable to have all these point-to-set mappings expressed as functions of p only. This is precisely what we shall do after introducing the initial resources into our economy.

Resources

It is assumed that, for his initial trading, the ith consumer owns an amount r_i of resources (commodities) with the property that there is always some $x_i^0 \in X_i$ such that $r_i \gg x_i^0$.

The last part of the preceding assumption is very strong, but it serves to guarantee the continuity of the mapping T_i as it was pointed out earlier. Some alternative, perhaps weaker assumptions could be made, though we shall not discuss them here.

If r_i represents the resources of the ith consumer, then the sum $\sum_{i=1}^{m} r_i$ denotes the total resources of the economy.

Our next assumption is that the ith consumer has a contractual claim over the share α_{ij} of the profit of the jth producer. In other terms, we postulate that $\alpha_{ij} \geq 0$ for all (i, j) and $\sum_{i=1}^{m} \alpha_{ij} = 1$ for all j.

Summarizing, the sources of income for the ith consumer are (1) the sale of his initial resources, $p \cdot r_i$; (2) his profit share, $\sum_{j=1}^{n} \alpha_{ij} \pi_j(p)$; (3) sale of labor services from X_i.

Because of our sign convention, we do not need to consider (3) separately. The total wealth of the ith consumer is thus

$$w_i = p \cdot r_i + \sum_{j=1}^{n} \alpha_{ij} \cdot \pi_j(p).$$

Clearly we can write

$$D_i(p, w) = D_i\big(p, (pr_i + \sum_{j=1}^{n} \alpha_{ij}\pi_j(p))\big).$$

Note, however, that once the resources r have been introduced into our model, the wealth of each consumer is completely determined. The demand correspondence of the ith consumer becomes a point-to-set mapping of p only.

We shall call $D_i'(p)$ the new demand correspondence for the ith consumer. For the economy as a whole, we immediately have $D'(p) = \sum_{i=1}^{m} D_i'(p)$.

Properties of Mappings and Sets

In our discussion here, we are only interested in relative and nonnegative price systems p. Our set of prices is, accordingly, defined as

$$P = \{p \,|\, p \in \Omega, \sum_{h=1}^{k} p_h = 1\}.$$

It is clearly compact and convex.

The sets X_i and X have been assumed to be closed, convex, and lower bounded. It is easy to see that $D_i'(p)$ and $D'(p)$ are then compact and convex. Convexity is a consequence of the convex character of the preorderings and the definition of the sets.

We assumed $x_i^0 \ll r_i$ for each i. Necessarily then, we have $px_i^0 < pr_i$ for each i. Remember that $\theta \in Y_j$ for each j. Clearly, for any relative price system $p \in P$, and every index j, $\pi_j(p) \geq 0$. We must then have, for any $p \in P$,

$$p \cdot x_i^0 < p \cdot r_i + \sum_{j=1}^{n} \alpha_{ij} \cdot \pi_j(p).$$

The single-valued function $\pi_j(p)$ is continuous. So also is the single-valued function $p \to (w_i(p))$, which assigns to each relative price system $p \in P$ the m-tuple $w = (p \cdot r_i + \sum_{j=1} \alpha_{ij} \cdot \pi_j(p))$.

Now, since the point-to-set mapping T_i is guaranteed to be continuous at the point $(p, (p \cdot r_i + \sum_{j=1}^{n} \alpha_{ij} \cdot \pi_j(p)))$ and u_i and v_i are continuous functions, we are allowed to invoke Theorem 6.5. to assert the upper semicontinuity of the demand correspondence D_i', for each i, and clearly the upper semicontinuity of the total demand correspondence D'.

We can summarize the main results of our present investigation as follows.

Remark 9.2 If the assumptions of our consumer model are satisfied, then

(1) D_i' is upper semicontinuous, and u_i and v_i are continuous for each i;
(2) D' is upper semicontinuous;
(3) $D_i'(p)$ and $D'(p)$ are convex sets.

The properties enumerated in Remarks 9.1 and 9.2 will be used later in Theorem 9.3. We shall interrupt our present discussion to analyze briefly the problem of the existence of a continuous utility function and justify so the second part of assertion (1) in Remark 9.2.

The Utility Function

A utility function—or more properly, a utility-index function—for the ith consumer can be a useful tool of analysis.

So far, we have supposed that such a function exists, when certain assumptions about the consumption set and its preordering are satisfied. Here we shall show that our supposition was justified. Once again, we follow Debreu's exposition.

We are, of course, interested in a utility function for each of the consumers of the economy. However, for the sake of convenience, *we drop the subscript i* from the notation used throughout the discussion of this part.

We recall that a utility function $u(x)$ has been defined as a continuous, increasing, point-to-point mapping from the totally preordered set X into R.

Theorem 9.1 (Existence)† Let $X \subset R^k$ be a connected set totally preordered by $\precsim$. If the preordering is continuous, that is, $\{x \mid x \in X, x \precsim x'\}$ and $\{x \mid x \in X, x \succsim x'\}$ are closed sets for any $x' \in X$, then there exists a continuous utility function $u(x)$.

Proof Let A be a countable subset of X such that A is dense in X, that is, $\bar{A} = X$. Consider two points, x', $x'' \in A$ such that $x' \prec x''$. We claim first that, if this relation is true, then there exists $x \in X$ such that

(1) $x' \prec x \prec x''$.

To implement this first step of our proof consider the two sets

$$X' = \{x \mid x \in X, x \precsim x'\} \qquad \text{and} \qquad X'' = \{x \mid x \in X, x'' \precsim x\}.$$

They are nonempty and closed, They are also disjoint by definition, so that from the fact that X is connected,

(2) $X' \cup X'' \neq X$.

† Adapted from Debreu [4].

If no $x \in A$ were to satisfy property (1),we would have

$$A \subset X' \cup X'', \qquad \bar{A} \subset \overline{X' \cup X''}.$$

But, $X' \cup X''$ being closed, we would get

$$X' \cup X'' = \overline{X' \cup X''}, \qquad \bar{A} \subset X' \cup X''$$

and, for $\bar{A} = X$, also

$$X \subset X' \cup X'',$$

thus implying $X = X' \cup X''$ which contradicts property (2). Consequently, there is $x \in X$ satisfying relation (1).

We can now define a utility function from the countable set A into the set of rational numbers Q which, as we know, is countable. Consider an interval $[a, b]$ of Q such that $a < b$. We define our function such that if x^{α} and x^{β} are, respectively, smallest and largest elements of A, then

$$u'(x^{\alpha}) = a, \qquad u'(x^{\beta}) = b.$$

Now, eliminate from A all elements indifferent to x^{α} and x^{β}, calling A' the set of remaining elements. Call Q' the open interval $]a, b[$. Since both sets A' and Q' are countable, their elements can be ranked $x^1, x^2, \ldots, x^p, \ldots$ and r^1, $r^2, \ldots, r^p, \ldots$, respectively. It is now possible to define an increasing function from A' onto Q' such that the preordering is preserved and every element of Q' is assigned. Next, we extend our function from A to X and call the extension $u(x)$. Clearly, if $x \in A$, then $u(x) = u'(x)$. We also have $Q' \subset u(X) \subset [a, b]$.

We know from Chapter 5, Remark 5.6, that a function u: $X \to R$ is continuous if and only if the inverse image of any interval $]-\infty, c]$, $[c, \infty[$, where c is a real number, is closed in X. Let us consider the interval $[c, \infty[$ only, since a similar proof can be given for $]-\infty, c]$, and take $c \in]a, b[$. The interval $[c, \infty[$ can be considered the intersection of the intervals $[r, \infty[$ for $r \in Q'$ and $r \leq c$. Consider now the inverses. Applying Remark 5.3 and defining the sets $X^r = \{x \mid x \in X, \ r \in Q', \ u(x) \geq r\}$, we have then $X^c = \bigcap_{r \in Q', \ r \leq c} X^r$, where X^c is obviously (note here X^c does not mean complement of X)

$$X^c = \{x \mid x \in X, \ c \in]a, b[, \ u(x) \geq c\}.$$

It is easy to see that for a point $x \in X$ such that $u(x) = r$ the set X^r can also be written $X^r = \{x' \mid x' \in X, \ x' \precsim x\}$. By our assumption about preference continuity, each X^r is closed in X and therefore so is X^c.

Summarizing, a continuous utility function does exist under the condition of our theorem.

9.4 General Equilibrium

The problem we are concerned with is whether an economy like that described so far has an equilibrium.

It will be convenient to have, before answering the above question, a precise definition of what we mean by an equilibrium. Moreover, since we are considering the case for only a particular type of economic system, it is necessary to summarize all the properties we attach to a private-ownership economy by means of a formal definition.

Private-Ownership Economy

Definition 9.1 A system $\{\{X_i, \precsim_i\}, \{Y_j\}, r, \{\alpha_{ij}\}\}$—all components of which have the meaning given earlier—is called a private economy if it satisfies the following conditions:

(1) For each i there is $r_i \in R^k$ such that $\sum_{i=1}^{m} r_i = r$.
(2) For each (i, j) there is $\alpha_{ij} \geq 0$ such that $\sum_{i=1}^{m} \alpha_{ij} = 1$ for every j.

The term competitive economy has here the usual economic meaning that each economic unit is simply a quantity adjuster, that is, has no influence on prices.

Equilibrium

Definition 9.2 Given the private-ownership economy of Definition 9.1, we say that an $(m + n + 1)$-tuple $((x_i^*), (y_j^*), p^*)$ of points of R^k is an equilibrium if following conditions are satisfied.

(1) The point x_i^* is a largest element of $\{x_i | x_i \in X_i,\ p^*x_i \leq p^*r_i + \sum_{j=1}^{n} \alpha_{ij}\, p^*y_j^*\}$ for each i.

(2) The point y_j^* is a profit maximizer on Y_j, relative to the price system p^*, for each j.

(3) Total demand equals total supply, that is, $x^* - y^* = r$.

Note here that condition (1) could have also been stated in terms of the utility function $u_i(x_i)$, for each i.

Excess Demand

Consider a state of the economy characterized by a pair $((x_i), (y_j))$ and total resources r.

The sum $x - y - r = z$ will be called an excess demand if $x_i \in X_i$ for every i and $y_j \in Y_j$ for every j. The excess demand point z belongs thus to the set $X - Y - \{r\} = Z$.

$S(p)$ and $D'(p)$ are the total supply and demand correspondence, respectively. The point-to-set mapping

$$Z'(p) = D'(p) - S(p) - \{r\},$$

may be called then the excess demand correspondence. Clearly, $Z'(p)$ is a subset of $Z = X - Y - \{r\}$ defined earlier.

Our interest is only in price systems which belong to the set $P = \{p \mid p \in \Omega, \sum_{h=1}^{k} p_h = 1\}$. For such a given p, the actions of an economic agent have to satisfy the wealth constraint. It must be $p \cdot x_i \leq p \cdot r_i + \sum_{j=1}^{n} \alpha_{ij}\, p \cdot y_j$, for each i. Summing over all consumers, that is, over the index i, we have

$$p \cdot x \leq p \cdot r + p \cdot y, \qquad pz \leq 0$$

for the point $z \in Z'(p)$. So for $Z'(p)$ we have

$$pZ'(p) \leq 0.$$

An equilibrium may be analyzed by means of the wealth constraint idea for a condition $p \cdot z = 0$. However, it may also be studied by means of the weaker condition $p \cdot z \leq 0$.

There is no advantage in choosing the strong condition for equilibrium in our further discussion. On the contrary, mathematically there is a distinct benefit in dealing with the form $pz \leq 0$. If an excess demand exists which satisfies the weak budget condition, then the existence of a strong equilibrium is also guaranteed.

The existence problem is now split into two subproblems.

The first subproblem is of a quite general nature. We have to show that under suitable conditions, there exists a nonpositive excess demand point z such that $p \cdot Z'(p) \leq 0$. This amounts to proving that there is a price system $p \in P$ for which $Z(p) \cap (-\Omega) \neq \varnothing$. This is done by means of the excess-demand theorem.

The second subproblem is less general, or at least more specific. We have to show that our economic model, the attention focused on its mathematical properties, is fit to the application of the excess-demand theorem. This is done in terms of an equilibrium theorem. It seems to be intuitive that, the weaker the basic assumptions of the model are, the greater the generality of the corresponding equilibrium theorem will be.

For the sake of simplicity, we have chosen assumptions which would guarantee an almost immediate proof of our equilibrium theorem. A lower generality of our results will be, of course, the price to be paid for our choice.

Theorem 9.2 (Excess Demand)† Let $Z \subset R^k$ be a compact and convex set. Let $Z'\colon P \to Z$, or alternatively $p \to Z'(p)$, be an upper semicontinuous point-to-set mapping (excess-demand correspondence) such that for every $p \in P$ the set $Z'(p)$ is nonempty and convex and $p \cdot Z'(p) \leq 0$. Then, there exists some $p^* \in P$ such that $z \leq \theta$ for some $z \in Z'(p^*)$.

Proof By definition P is nonempty, compact and convex. By assumption Z is nonempty, compact and convex. Take $z \in Z$. Call $U(z)$ the set of all $p \in P$ which maximize pz on P. Now the point-to-set mapping $z \to U(z)$ from Z to P is upper semicontinuous, for $U(z)$ is nonempty. $U(z)$ is also a convex set. First, P is convex. Thereafter we have that either $z = \theta$ and so $U(z) = P$, or $z \neq \theta$ and so $U(z) = P \cap \{p \mid p \in R^k, pz = \max P \cdot z\}$.

Define further the point-to-set mapping,

$$V\colon P \times Z \to P \times Z,$$

which could also be written

$$(p, z) \to V(p, z)$$

and where

$$V(p, z) = U(z) \times Z'(p).$$

Clearly $P \times Z$ is nonempty, compact, and convex for so are P and Z.

We have $V(p, z)$ representing an upper semicontinuous mapping, for $U(z)$ and $Z'(p)$ are upper semicontinuous. Moreover, $V(p, z)$ is a nonempty, convex set because $U(z)$ and $Z'(p)$ are so. Now, from our results of Chapter 7 (Kakutani's theorem), $V(p, z)$ has a fixed point (p^*, z^*), that is, there is a

$$(p^*, z^*) \in U(z^*) \times Z'(p^*)$$

or $p^* \in U(z^*)$ and $z^* \in Z'(p^*)$. From our definition of $U(z)$ it follows that

$$p^* \in U(z^*) \qquad \text{implies} \qquad pz^* \leq p^*z^*.$$

But $z^* \in Z'(p)$. From the budget constraint,

$$p^* \cdot z^* \leq 0.$$

We have thus, for all $p \in P$, $pz^* \leq 0$. Therefore, it must be

$$z^* \leq \theta.$$

It is clear also that the existence of the special case $z^* = \theta$ is implied by our result. Let us return now to our economy.

† Adapted from Debreu [4].

Theorem 9.3 (Equilibrium) A competitive, private-ownership economy,

$$\{\{X_i, \precsim_i\}, \{Y_j\}, r, \{\alpha_{ij}\}\},$$

has an equilibrium if it satisfies following assumptions:

(i) Y_j is compact and convex for each j.
(ii) $Y_j \cap \Omega = \{\theta\}$.
(iii) $Y_j \cap (-Y_j) = \{\theta\}$.
(iv) X_j is closed and convex for each i.
(v) X_i is lower bounded for each i.
(vi) X_i has no satiation consumption for $\precsim_i$.
(vii) The preference preordering $\precsim_i$ on X_i is continuous for each i and it satisfies following convexity condition: If $x_i^2 \succ_i x_i^1$, then $\lambda x_i^2 + (1-\lambda)x_i^1 \succ_i x_i^1$ for $x_i^2, x_i^1 \in X_i$ and $\lambda \in]0, 1[$.
(viii) There is some $x_i^0 \in X_i$ such that $x_i^0 \ll r_i$ for each i.

Proof The total supply correspondence S is upper semicontinuous by Remark 9.1. By the same remark the set $S(p)$ is convex (and obviously compact). The total demand correspondence D' is upper semicontinuous by Remark 9.2. By the same remark the set $D'(p)$ is convex (and obviously compact). Since $Z'(p) = D'(p) - S(p) - \{r\}$, it is thus immediate that the excess-demand correspondence Z' is upper semicontinuous. Furthermore, the nonempty set $Z'(p)$ is convex (and obviously compact). Therefore the excess-demand theorem is applicable.

EXERCISES

E.9.1 Let Y be the aggregate production set of an economy. Show that the condition " Y is convex" together with "$\theta \in Y$" imply the prevalence of nonincreasing returns to scale for the whole economy.

E.9.2 If we assume that each Y_j is closed instead of compact, what additional difficulties would we encounter in our analysis of Chapter 9?

Hint: Refer specially to Theorem 9.3.

E.9.3 Suppose that if y_j^1 and y_j^2 are two feasible productions for the jth producer so is $y_j^1 + y_j^2$ also. This condition, called additivity, can also be stated as $(Y_j + Y_j) \subset Y_j$. Prove that additivity, together with $\theta \in Y_j$ and convexity imply the prevalence of constant returns to scale for the jth producer.

E.9.4 Show that if Y_j is a cone with $\theta \in Y_j$, constant returns to scale prevail for the jth producer. Under such conditions, show that his maximum (equilibrium) profit is zero.

Hint: Study Definition 18.9 and the example following Theorem 18.16 in Chapter 18.

E.9.5 State the conditions under which the supply correspondence of the jth producer reduces to a single-valued function.

Hint: Study the technology of the production set reflected in the shape of Y_j.

E.9.6 Study the following three possible convexity assumptions regarding the preordering on X_i. For any $x_i^1, \; x_i^2 \in X_i, \; \lambda \in]0, 1[$

(1) if $x_i^2 \succsim_i x_i^1$, then $\lambda x_i^2 + (1 - \lambda)x_i^1 \succsim_i x_i^1$ (weak convexity);
(2) if $x_i^2 \succ_i x_i^1$, then $\lambda x_i^2 + (1 - \lambda)x_i^1 \succ_i x_i^1$ (convexity);
(3) if $x_i^2 \sim_i x_i^1$, then $\lambda x_i^2 + (1 - \lambda)x_i^1 \succ_i x_i^1$ (strong convexity).

Prove that (a) assumption (1) allows for "thick" indifference classes, that is, indifference classes with nonvoid interior; assumption (2) allows for "broken" indifference classes, that is, indifference classes with nondegenerate closed segments; assumption (3) allows for neither "thick" nor "broken" indifference classes; (b) if the preordering is continuous, then assumption (2) implies assumption (1), and assumption (3) implies assumption (2).

Hint: Debreu [4].

E.9.7 Show that the statement "the preordering on X_i is convex" is equivalent to the statement "for every $x_i' \in X_i$, the set $\{x_i | x_i \in X_i, x_i \succsim_i x_i'\}$ is convex."

E.9.8 Let $\precsim_i$ on X_i be continuous, and satisfy convexity as in assumption (2) of exercise E.9.6. Give the sketch of a proof that if x_i' is a largest element of $T_i(p, w)$, then it is also an expenditure minimizer on the set $\{x_i | x_i \in X_i, x_i \succsim_i x_i'\}$.

E.9.9 Consider the mappings D_i, T_i, u_i, and v_i as defined in this chapter. Applying Theorem 6.5 show the upper semicontinuity of D_i, and the continuity of v_i at point (p, w).

Hint: Refer to the example following Theorem 6.5 in Chapter 6; to simplify matters, assume that X_i is compact and T_i is continuous at (p, w).

E.9.10 Using the results of E.9.5 (and with equivalent, suitable assumptions for consumers), replace all supply and demand correspondences in the economy by single-valued supply and demand functions. Thereafter, follow step by step the analysis developed in the chapter, and point out the extent in which the same is affected by the new, more restrictive assumptions.

PROBLEMS

P.9.1 The notion "private ownership" economy or "private economy" has been given in Definition 9.1. The more general concept "economy" is defined as follows: An economy is a system $\{\{X_i, \precsim_i\}, \{Y_j\}, r\}$—all components of which have the meaning given earlier in this chapter.

Discuss the sense in which Definition 9.1 is a special case of the preceding definition, and give the proper economic meaning to its restrictive conditions.

P.9.2 Let $((x_i), (y_j))$ be a state of the economy defined in P.9.1. The set of attainable states is still $A = \{((x_i), (y_j)) \mid x_i \in X_i$, for every i, $y_j \in Y_j$, for every j, $x - y = r\}$ (see also P.2.4, P.3.1, P.4.5, and the example after Remark 18.3). The set A is partially preordered by $\precsim$. A maximal element of A for $\precsim$ is called an optimum (Pareto optimum) of the economy. Finally, the state or allocation $((x_i^*), (y_j^*))$ corresponding to the equilibrium $((x_i^*), (y_j^*), p^*)$ is called an equilibrium allocation or an equilibrium relative to a price system p^*.

Suppose our economy satisfies the relevant assumptions referred to a private-ownership economy in this chapter, and assume no x_i is a satiation consumption. Then, prove the following theorem: For an economy $\{\{X_i, \precsim_i\}, \{Y_j\}, r\}$ an equilibrium relative to a price system is an optimum.

Hint: Debreu [4].

P.9.3 The following is a special case of the definition given in Problem P.9.1: A pure-exchange economy is a system $\{\{X_i, \precsim_i\}, r\}$, all components of which have the meaning given earlier in this chapter, but with each X_i only in the nonnegative orthant of R^k.

Note that for such an economy an equilibrium is an $(m+1)$-tuple $((x_i^*), p^*)$ of points of R^k such that (1) the point x_i^* is a largest element of the set $\{x_i \mid x_i \in X_i, p^*x_i \leq p^*r_i\}$ for each i; (2) total demands equals total supply, that is, $x^* - r = \theta$. Sketch a proof of the existence of such an equilibrium in the framework of the discussion of this chapter.

P.9.4 Call M the set of all consumers of a pure-exchange economy. A nonempty set $S \subset M$ is called a coalition. A coalition S is said to block an allocation or state (x_i) if there exist bundles x_i' for all $i \in S$ such that (1) $\sum_{i \in S} (x_i' - r_i) = \theta$; (2) $x_i' \succsim_i x_i$ for all $i \in S$ with at least one $x_i' \succ_i x_i$, $i \in S$.

The set of all allocations which cannot be blocked by any coalition is called the core of the economy. Show that (a) the set of all optima is the set of all allocations which cannot be blocked by the coalition M, of all consumers; (b) the core of an economy is a subset of the set of all Pareto optimal allocations.

P.9.5 Consider the pure exchange economy of P.9.3, and suppose that every preordering $\precsim_i$ is strongly convex. (a) Show that any equilibrium allocation is a core allocation. (b) Generalize the notion of core and the result of (a) to the case of an economy $\{\{X_i, \precsim_i\}, \{Y_j\}, r\}$.

Hint: See Nikaido [4].

NOTES

The point of departure of modern general equilibrium theory is by Walras [1]. The first rigorous existence proof is due to Wald [1] though, in a somewhat different context, von Neumann [1] may also have to be regarded as a pioneer in the field.

The solution of the existence problem for finite economies is the result of the work of Arrow and Debreu [1], Debreu [4, 5], Gale [2], McKenzie [1], Morishima [2], Nikaido [1, 2], and Uzawa [1].

A comparison of different general equilibrium models may be found in the work of Quirk and Saposnik [1]. This is a very useful textbook which does not require advanced mathematics. At a more advanced level Karlin [1] and Nikaido [3, 4] should be read. A masterly written new book is by Arrow and Hahn [1]. The latter contains both a comprehensive and rigorous discussion of most subjects in the field. See also Rader [2].

In historical perspective, the notion of the core of an economy is associated with the research of Edgeworth [1]. The concept of the core has game-theoretical origin. An excellent textbook on modern game theory is by Owen [1]. Studies on the core have been the subject of many important papers, in particular by Shubik [1], Scarf [1], Debreu [6], Debreu and Scarf [1], and Rader [1]. The books by Arrow and Hahn [1] and Nikaido [4] should be consulted. For an excellent discussion see Rader [2].

The study of economies with measure spaces of agents has been initiated by Aumann [1, 2]. Among the rapidly increasing literature related to this new approach we shall only mention the contributions of Hildenbrand [1] and Vind [1]. The study of economies with continuum of agents requires new mathematical techniques, especially Banach space methods and measure-theoretical tools. In the latter point the book by Dunford and Schwartz [1] appears as an almost obligatory reference.

PART II

VECTOR SPACES AND VECTOR-SPACE HOMOMORPHISMS

CHAPTER 10

VECTOR SPACES AND SUBSPACES

Part II will be concerned, essentially, with three basic mathematical concepts: that of a vector space, a linear mapping from a vector space into a vector space, and a matrix.

A vector space, as was suggested before, may be regarded as a special type of metric space. Thus, it falls within the subject of Definition 2.15 of Chapter 2. But it is also a special type of algebraic structure, in the sense of Definition 8.1 of Chapter 8, and precisely for this latter reason it will attract our attention here.

Linear single-valued functions were discussed, in general, at the end of Chapter 5. In this part we shall confine our interest to linear single-valued functions from one vector space into another. We shall learn later that mappings of this type, very often referred to as linear transformations, are homomorphisms with the meaning given in Definition 8.2 of Chapter 8.

A matrix is an operator possessing the most important property that it provides a means for describing completely, uniquely, and in a condensed manner a given linear transformation. Its significance in our present context is thus obvious.

The present chapter, together with Chapters 11 and 12 form the "core" of Part II. They contain the basic theory and the links between the three mentioned concepts.

The remainder of Part II deals with some special topics, though within the framework of the first three chapters. Most of these topics are usually covered in standard linear algebra books. A few of them, for instance, indecomposability and nonnegative matrices, are less common. They have been included here for their considerable importance in theoretical economics.

10.1 Vector Spaces over a Field

The concept of an algebraic system is known to us from Chapter 8. There, we agreed that the simplest way of denoting a system S is to indicate the set E of its elements together with the set O of operations defined on E.

The definition of a vector space involves both the concepts of an additive Abelian group and of a field. For the sake of clarity, it is thus convenient to distinguish between E, the set underlying the group, and E', the set underlying the field.

Furthermore, it seems appropriate to have different notations to indicate those operations under which E is defined to be closed and those operations under which E' is defined to be closed.

The sum of two elements of E will be called "vector addition" and denoted by $\oplus$. The product of an element of E times an element of E' will be called "vector multiplication by a scalar" and denoted $\odot$. Under these two operations, as we shall see later, the set E is closed.

On the other hand, the sum and the product of two elements of E' will be referred to as "sum of scalars" and "multiplication of scalars," respectively. We shall denote them, accordingly, by the customary symbols $+$ and $\cdot$.

We repeat, the use of two different sets of symbols to indicate operations is only intended as a means to emphasize the actual presence of two different structures. Once this aspect of the problem has been clearly understood, insistence on the distinction will become unnecessary and we shall be able to adopt a much simpler notation.

Definition 10.1 An algebraic structure $V = \{E, E', \oplus, \odot, +, \cdot\}$ is called a vector space over a field F if and only if it satisfies the following properties:

1 $G = \{E, \oplus\}$ is an Abelian group with identity element θ.
2 $F = \{E', +, \cdot\}$ is a field with identity elements 0 (additive identity) and 1 (multiplicative identity).

3 For all elements $u, v \in E$, called vectors, and all elements $h, k \in E'$, called scalars,

(a) $h \odot u \in E$,
(b) $h \odot (u \oplus v) = (h \odot u) \oplus (h \odot v)$,
(c) $(h + k) \odot u = (h \odot u) \oplus (k \odot u)$,
(d) $(h \cdot k) \odot u = h \odot (k \odot u)$,
(e) $1 \odot u = u$.

In other words, a vector space satisfies all four basic properties of an additive commutative group (see Definition 8.6) and, besides, is closed under the operation vector multiplication by a scalar.

Example 1 Let F be an arbitrary field. The set of all n-tuples of elements of F is a vector space. For any two vectors $(x_1, \ldots, x_n)$ and $(y_1, \ldots, y_n)$, vector addition is defined by $(x_1, \ldots, x_n) + (y_1, \ldots, y_n) = (x_1 + y_1, \ldots, x_n + y_n)$ and, for $k \in F$, vector multiplication by a scalar is defined by $k(x_1, \ldots, x_n) = (kx_1, \ldots, kx_n)$.

Example 2 Let F be the field of real numbers and V the set of all polynomials of degree n with coefficients $a_i \in F$: $a_0 + a_1 x + a_2 x^2 + \cdots + a_n x^n$. V forms a vector space over F in regard to addition of polynomials and multiplication of polynomials by a constant.

Example 3 Let F be the field of real numbers and V the set of all continuous, real-valued functions f defined over the interval $[a, b]$. V forms a vector space over F.

Theorem 10.1 Call θ the zero vector in the vector space V and let $-v$ be the additive inverse of $v \in V$. Then, for $0, 1, k \in F$ and any $v \in V$ we have (1) $0v = \theta$ (2) $(-1)v = -v$; (3) $k\theta = \theta$.

Proof Note that here we have dropped the notational distinction between $\oplus$ and $+$, and between $\odot$ and $\cdot$. The character of the operation should be clear from the context.

(1) From Definition 10.1 we know $v = 1v = (1 + 0)v = 1v + 0v = v + 0v$. Adding $-v$ to both sides, we get $v - v = v + 0v - v$, that is, $\theta = 0v$.

(2) For the second statement we may write $v + (-1)v = 1v - 1v = (1 - 1)v = 0v = \theta$.

(3) Finally, we have $k\theta = k[v + (-v)] = kv + k(-v) = kv + (-kv) = \theta$.

10.2 Vectors and Operations with Vectors

In Example 1 following Definition 10.1 we referred to the set of all n-tuples of elements of an arbitrary field F as a vector space. Each n-tuple represents, in our new terminology, a vector; to be precise, a finite-component vector. The corresponding vector space is said to be finite dimensional.

Two typical classes of finite-dimensional vector spaces are obtained when F is the field of complex numbers or, more restrictedly, the set of real numbers. The concept of vector (complex or real) is identical here to that of an n-component point or sequence in the terminology of Part I. In our present study we shall be concerned almost exclusively with these types of spaces, in particular with the n-dimensional real space.

A different type of finite-dimensional vector space is suggested in Example 2. There, a vector does not have the intuitive appeal of a point, in a geometric sense, but consists of a polynomial.

Vector spaces need not be finite-dimensional. Functions like the one mentioned in Example 3 after Definition 10.1 serve very well to illustrate this point. The space of functions to which f belongs is an infinite-dimensional vector space. It is suggested to check that such a space satisfies all the axioms of Definition 10.1. A formal definition of the term "dimension" will be given in Section 10.6.

Vector Addition

By definition, a vector space satisfies all the properties of a commutative group under addition. Let u, v, w be any vectors in the vector space V, and let us again use $\oplus$ to symbolize vector addition. It is immediate that the following laws are satisfied:

(1) $u \oplus v \oplus w \in V$;

(2) **G1′** $(u \oplus v) \oplus w = u \oplus (v \oplus w)$,

G2′ $u \oplus \theta = \theta \oplus u = u$,

G3′ $u \oplus (-u) = (-u) \oplus u = \theta$,

G4′ $u \oplus v = v \oplus u$.

Example 1 In the n-dimensional space V, note that the sum of $u = (u_1, \ldots, u_n)$ and $v = (v_1, \ldots, v_n)$ is the vector $(u_1 + v_1, \ldots, u_n + v_n)$.

Example 2 In the space of polynomials of degree n, the sum of $a_0 + a_1 x + a_2 x^2 + \cdots + a_n x^n$ and $b_0 + b_1 x + b_2 x^2 + \cdots + b_n x^n$ is the vector $(a_0 + b_0) + (a_1 + b_1)x + (a_2 + b_2)x^2 + \cdots + (a_n + b_n)x^n$.

Example 3 In the space of continuous, real-valued functions defined over the interval $[a, b]$, the sum of f_1 and f_2 is defined by $(f_1 + f_2)(x) = f_1(x) + f_2(x)$.

Vector Multiplication by a Scalar

From Definition 10.1 we know that a vector space is closed under the operation vector multiplication by a scalar, which we denote here once more by $\odot$.

For any two vectors u, v of a vector space V and the field elements 1, h, and k, we have

(1) $h \odot u \in V$,
(2) $h \odot (u \oplus v) = h \odot u \oplus h \odot v$,
(3) $(h + k) \odot u = h \odot u \oplus k \odot u$.
(4) $(h \cdot k) \odot u = h \odot (k \odot u)$,
(5) $1 \odot u = u \odot 1 = u$.

In the following examples we return to the simplified notation.

Example 1 For $u = (u_1, \ldots, u_n)$ in the n-dimensional space V and the field element k, $ku = (ku_1, \ldots, ku_n)$.

Example 2 For $u = a_0 + a_1 x + a_2 x^2 + \cdots + a_n x^n$ in the space of polynomials of degree n and the field element k, we have $ku = (ka_0) + (ka_1)x + (ka_2)x^2 + \cdots + (ka_n)x^n$.

Example 3 For the function f, in the space of continuous, real-valued functions defined over the interval $[a, b]$ and the field element k, we define $(kf)(x) = kf(x)$.

Dot Product

At this point we shall confine our discussion to finite-dimensional vector spaces, namely vector spaces of n-tuples of elements of the complex field F. The conclusions regarding a real n-space will follow immediately from our results.

We remind the reader that if $z = a + bi$ represents a complex number, then $\bar{z} = \overline{a + bi} = a - bi$ is also a complex number called the conjugate of z.

For two complex numbers $z_1 = a + bi$ and $z_2 = c + di$, if we define the products $x = z_1 \bar{z}_2$ and $y = z_2 \bar{z}_1$, it is immediate that $x = \bar{y}$. This property is often called conjugate symmetry. In particular, it should be noted that $z_1 \bar{z}_1$ and $z_2 \bar{z}_2$ are real numbers.

We may thus regard the simple symmetry property of the product of two real numbers a and c ($ac = ca$) as a special case of conjugate symmetry.

We can now define the dot product of two vectors.

Definition 10.2 Let V be an n-space over the complex field F and consider the vectors $u, v \in V$. The expression $u \cdot v = \sum_{i=1}^{n} u_i \bar{v}_i$, where $\bar{v}_i$ is the conjugate of the ith component of v, is called the dot product of u and v.

The idea of a product of two points was already introduced in Chapter 4. In this place we merely formalize and generalize the concept.

An important fact that should be noted is that the dot product of two vectors is a scalar (element of the field) and not a vector. In other words, a vector space is not closed under this operation.

The following remark is a direct consequence of the property of conjugate symmetry mentioned earlier.

Remark 10.1 For any pair of vectors u, v of the n-space V over the complex field F, $u \cdot v = \overline{v \cdot u}$.

Consider now the case of the n-space V over the real field F. The situation constitutes here a special case of that discussed so far, and we may state the following:

Remark 10.2 For any pair of vectors u, v of the n-vector space V over the real field F,

$$u \cdot v = \sum_{i=1}^{n} u_i v_i = \sum_{i=1}^{n} v_i u_i = v \cdot u.$$

Example 1 If $u = (0, 1, 2)$ and $v = (3, 2, 1)$, then

$$uv = (0, 1, 2)(3, 2, 1) = (3, 2, 1)(0, 1, 2) = vu = 4.$$

Example 2 If $u = (1 + i, 2 + 3i)$, and $v = (4 - 2i, 3 + 5i)$ then

$$u \cdot v = (1 + i, 2 + 3i)\overline{(4 - 2i, 3 + 5i)} = (1 + i, 2 + 3i)(4 + 2i, 3 - 5i) = 23 + 5i$$

and

$$vu = (4 - 2i, 3 + 5i)\overline{(1 + i, 2 + 3i)} = (4 - 2i, 3 + 5i)(1 - i, 2 - 3i) = 23 - 5i.$$

Note that $uv = \overline{vu}$, that is, $23 + 5i = \overline{23 - 5i}$.

10.3 Metrics

A vector space, we said, may be regarded as a special kind of metric space satisfying specific algebraic properties.

We shall now analyze how metric concepts are related to a vector space. Our treatment is again confined to the case of n-spaces over the complex or real field F. The basic ideas, however, could be generalized so as to cover other spaces as well.

Inner Product Function

The idea of a dot product (complex or real) is present throughout the considerations which follow.

Definition 10.3 Let V be a vector space over the complex field F. Consider the vectors u, v, w and the scalars h, k. Then, a complex-valued function $p : V \times V \to F$ is said to be an inner product on V if it satisfies the following conditions:

(1) $p(hu + kv, w) = hp(u, v) + kp(v, w)$;
(2) $p(u, v) = \overline{p(v, u)}$;
(3) if $u \neq \theta$, then $p(u, u) > 0$;
(4) if $u = \theta$, then $p(u, u) = 0$.

The function just defined plays a fundamental role because it is the basis of the metric concepts we shall discuss very soon. Note that it is a generalization of the idea of dot product of Definition 10.2.

The domain of p is clearly the direct product $V \times V$; its range the complex field. The image of each pair $(u, v) \in V \times V$ is an element of the set of complex numbers C.

Mappings which satisfy condition (1) of Definition 10.3 are said to be bilinear. Condition (2) requires the conjugate symmetry of the image of each pair of vectors. In standard terminology, functions which possess both properties are said to be conjugate bilinear. Finally, taking properties (3) and (4) together, a function for which the image $p(u, u)$ is positive if $u \neq \theta$, and zero if $u = \theta$, is said to be positive definite.

It is easy to see that when V is the real space the mapping is p: $V \times V \to R$. Condition (2), as it is known to us from our previous discussion, reduces to the simple symmetry property $p(u, v) = p(v, u)$.

Spaces on which an inner product function has been defined are usually called inner product spaces. They are metric spaces since, as we shall see very soon, the definition of an inner product leads to the introduction of a metric. In particular, we state the following definitions:

Definition 10.4 An n-space V over the complex field with a complex inner product function p: $V \times V \to C$ is called a unitary space.

Definition 10.5 An n-space V over the real field with a real inner product function $p : V \times V \to R$ is called a Euclidean space.

Using the notion of an inner product as a background we are now in condition to define concretely some useful metric concepts. But we remind the reader once more that in our current discussion we are confining ourselves to the consideration of spaces the elements of which are real or complex n-tuples. And, in this context, the case of the real space is of particular interest to us.

Norm of a Vector

Definition 10.6 Consider any $u \in V$, where V is a Euclidean or unitary space. Let p be an inner product function on V. The Euclidean norm or length of u is the number $\|u\| = [p(u, u)]^{1/2}$.

Note that the norm $\|u\|$ is a positive number, unless $u = \theta$.

Example 1 Consider the real vector $u = (1, 2, 2)$. According to Definition 10.6 $\|u\| = (u \cdot u)^{1/2} = (1^2 + 2^2 + 2^2)^{1/2} = 3$.

From our definition of dot product, we know that the length of a complex vector is also a real number as is shown in the following example.

Example 2 Consider now the complex vector $u = (1 + i, 2 + 3i)$. By definition

$$\begin{aligned}\|u\| &= (u \cdot u)^{1/2} = [(1 + i, 2 + 3i)\overline{(1 + i, 2 + 3i)}]^{1/2} \\ &= [(1 + i, 2 + 3i)(1 - i, 2 - 3i)]^{1/2} = (15)^{1/2}.\end{aligned}$$

Distance between Two Vectors

Definition 10.7 Let V be a Euclidean or unitary space with the inner product function p. Consider any two vectors $u, v \in V$ such that $u - v = w$. The Euclidean distance between u and v, denoted $d(u, v)$, is the number

$$d(u, v) = \|u - v\| = \|w\| = [p(w, w)]^{1/2}.$$

What the definition tells us, is that the distance between two vectors u and v is the norm or length of the vector which is the difference of u and v. If w is the difference vector, then its length is given by the positive square root of $p(w, w)$. Thus, even if u and v are complex vectors, the distance $d(u, v)$ between them is a real number.

Clearly enough $d(u, v)$ is only the image of the pair $(u, v) \in V \times V$ under the mapping $d : V \times V \to R$. The mapping d is a metric or distance function, and it is suggested as an exercise to prove that d satisfies the four axioms of Definition 2.14 in Chapter 2.

Example The distance between the vectors $u = (2, 4)$ and $v = (6, 1)$ is given by

$$d(u, v) = \|u - v\| = [(2 - 6)^2 + (4 - 1)^2]^{1/2} = (25)^{1/2} = 5.$$

Orthogonal and Normal Vectors

Before changing the subject of our discussion, we shall define some concepts for which the notion of an inner product function plays also an important role.

Definition 10.8 Let V be an inner product space (real or complex). Then the vectors $u, v \in V$ are said to be orthogonal if and only if $p(u, v) = 0$.

Note that our definition includes the interesting case when $u \cdot v = 0$ but $u, v \neq \theta$. Obviously, if any one of the vectors is the zero vector, say $u = \theta$, then it is trivially orthogonal to every $v \in V$.

Example The vectors $u = (2, -3)$ and $v = (3, 2)$ are orthogonal because $p(u, v) = uv = 0$. The concept of orthogonality and the concept of perpendicularity are equivalent.

Definition 10.9 Let V be an inner product space (real or complex). Then a vector $u \in V$ is said to be normal if and only if $\|u\| = 1$.

Consider a nonzero vector $u \in V$. Since $\|u\| > 0$, it is clear that $\|u\|^{-1} = 1/\|u\|$ is also a scalar and that we shall be able to "normalize" the vector u by multiplying each of its components by $\|u\|^{-1}$. Then the vector $\|u\|^{-1} \cdot u$ is normal since it has unit length, that is $\| \|u\|^{-1} \cdot u\| = 1$.

The concepts of orthogonality and normality appear side by side in the following definition.

Definition 10.10 Let $A = \{v^1, \ldots, v^m\}$ be a set of vectors of an inner product space V. A is said to be an orthonormal set if and only if following properties are satisfied: (1) $\|v^i\| = 1$ for each i; (2) $p(v^i, v^j) = 0$ for all $i \neq j$.

In an orthonormal set every vector is normal and orthogonal to the other vectors.

10.4 Subspaces

In the definition which follows, for the sake of consistency, we distinguish once more between the symbols $\oplus$ and $\odot$ on the one hand and $+$ and $\cdot$ on the other. However, in the examples, $+$ will represent addition of vectors as well as addition of scalars. The operator $\cdot$ will be used, when necessary, to indicate a dot product, a multiplication of a vector by a scalar or a product of two scalars.

Definition 10.11 Let $V = \{E, E', \oplus, \odot, +, \cdot\}$ be a vector space over the field F and consider a subset $B \subset E$. Then B is said to form a subspace S of V if and only if $S = \{B, E', \oplus, \odot, +, \cdot\}$ is itself a vector space over F.

The interpretation of the definition is straightforward. In particular, it means that if S is a subspace of V and $B \neq \varnothing$: (1) $\theta \in B$, that is, for every $u \in B$ we must have $u + (-u) = \theta \in B$; (2) B is closed under vector addition and vector multiplication by a scalar.

In Section 2.2 we had stated that if R^n is a set of points, then the sets $R^{n-1}, \ldots, R^2, R$ might be regarded as special kinds of subsets of it. Now, using the proper terminology, it is apparent that $R^{n-1}, \ldots, R^2, R$ constitute subspaces of the space R^n. The following examples will clarify and extend this idea.

Example 1 Let V be the space of all n-tuples of elements of the complex or real field F. The set B of all n-tuples of elements of F with zero nth component gives rise to a subspace S of V. Vector addition is now defined, for any two $(x_1, \ldots, x_{n-1}, 0)$ and $(y_1, \ldots, y_{n-1}, 0)$ in B by $(x_1 + y_1, \ldots, x_{n-1} + y_{n-1}, 0)$. Vector multiplication by a scalar is given, for any $(x_1, \ldots, x_{n-1}, 0)$ in B and k in F, by $k(x_1, \ldots, x_{n-1}, 0) = (kx_1, \ldots, kx_{n-1}, 0)$.

Example 2 Let V be the space of all nth degree polynomials $a_0 + a_1 x + a_2 x^2 + \cdots + a_n x^n$. Then the subset of all second-degree polynomials $a_0 + a_1 x + a_2 x^2$ forms a subspace of V.

Example 3 If V is the space of all continuous, real-valued functions defined over the interval $[a, b]$, we may consider the subset of all constant functions defined over the same interval as forming a subspace S of V.

Linear Combinations

Definition 10.12 Let V be a vector space over F and consider the vectors $v^1, \ldots, v^n \in V$. Then the vector $u \in V$ defined by $u = \sum_{i=1}^{n} k_i v^i$, where $k_i \in F$, is called a linear combination of the vectors $v^1, \ldots, v^n$.

Note that when we require, in addition, that $k_i \in [0, 1]$ and $\sum_{i=1}^{n} k_i = 1$ we have as a special case the convex combination introduced in Section 4.6.

Example 1 $u = (10, 8, 8)$ is a linear combination of the vectors $v^1 = (3, 1, 2)$, $v^2 = (0, 2, 0)$, $v^3 = (1, 1, 1)$ since, for the scalars 2, 1, and 4, we can write $(10, 8, 8) = 2(3, 1, 2) + (0, 2, 0) + 4(1, 1, 1)$.

Example 2 Consider the vectors $v^1 = a_0 + a_1 x + a_2 x^2$, $v^2 = b_0 + b_1 x + b_2 x^2$, and $v^3 = c_0 + c_1 x + c_2 x^2$. For the constants k_1, k_2, k_3 we have the linear combination

$$\begin{aligned} u &= k_1 v^1 + k_2 v^2 + k_3 v^3 \\ &= (k_1 a_0 + k_2 b_0 + k_3 c_0) + (k_1 a_1 + k_2 b_1 + k_3 c_1)x \\ &\quad + (k_1 a_2 + k_2 b_2 + k_3 c_2)x^2. \end{aligned}$$

Generation of Subspaces

Definition 10.13 Let V be a vector space over F and consider a set A of vectors in V. Then, the set of all linear combinations of vectors of A, denoted $[A]$, is a subspace of V and is called the "subspace spanned or generated" by A.

Example 1 Consider the space $V = R^4$ and the sets of vectors

$$\begin{aligned} A_4 &= \{(1, 0, 0, 0), (0, 1, 0, 0), (0, 0, 1, 0), (0, 0, 0, 1)\}, \\ A_3 &= \{(2, 3, 0, 0), (1, 1, 2, 0), (0, 2, 3, 0)\}, \\ A_2 &= \{(1, 1, 0, 0), (0, 1, 0, 0)\}, \\ A_1 &= \{(5, 0, 0, 0)\}. \end{aligned}$$

Observe that $[A_4] = R^4$ whereas A_3, A_2, A_1 generate, respectively, the spaces $[A_3]$, $[A_2]$, and $[A_1]$. Allowing for the corresponding zero coordinates, we may write $[A_3] = R^3$, $[A_2] = R^2$, $[A_1] = R$.

Example 2 Note that the set $A = \{1, x, x^2, x^3, \ldots, x^n\}$ generates the vector space of all polynomials of degree less than or equal to n. For the scalars $a_0, a_1, a_2, \ldots, a_n \in R$ the expression $a_0 + a_1 x + a_2 x^2 + \cdots + a_n x^n$ represents such an nth degree polynomial.

Sum, Intersection, and Direct Sum of Subspaces

Definition 10.14 Let U and W be subspaces of the vector space V. Then (1) $U + W$, the sum of U and W, is the set of all sums $u + w$, where $u \in U$ and $w \in W$; (2) $U \cap W$, the intersection of U and W, is the set of all vectors v such that $v \in U$ and $v \in W$; (3) $U \boxplus W$ is called the direct sum of U and W if and only if $U + W = V$ and $U \cap W = \{\theta\}$.

The concepts of sum and intersection of spaces are straightforward. The connection between the "sum of spaces" and the "sum of sets" as defined in Chapter 4 is also apparent. Some important properties are shown in the following two theorems.

Theorem 10.2 Let U and W be subspaces of V over F. Then $U + W$ and $U \cap W$ are also subspaces of V.

Proof First we show $U + W$ is a subspace of V. The union $U \cup W$ contains all vectors which belong to U, to W or to both. The set of all linear combinations in $U \cup W$ is the subspace $[U \cup W]$. (Note that, by definition, $[U \cup W]$ is a subspace even if $U \cup W$ is not.) Now, any vector in $[U \cup W]$ can be represented by a sum $u + w$. On the other hand, any vector in V which can be represented by a sum $u + w$ belongs to $[U \cup W]$. Hence $U + W = [U \cup W]$. Second we prove that $U \cap W$ is a subspace of V. Consider the vectors u, $w \in U \cap W$ and the scalar $k \in F$. Since U and W are subspaces of V, we have

$u + w \in U, u + w \in W, ku \in U, ku \in W$. Thus $u + w \in U \cap W$ and $ku \in U \cap W$. Therefore $U \cap W$ is a subspace of V.

Theorem 10.3 Let U and W be subspaces of V over F. Then V is a direct sum of U and V if and only if every $v \in V$ has a unique representation $v = u+w$ for some $u \in U$ and some $w \in W$.

Proof Assume first that $V = U \boxplus W$. Then each $v \in V$ can be represented by $v = u + w$. But suppose $v \in V$ has a second representation $v = u' + w'$. It must then be $u' + w' = u + w$ and hence $u' - u = w - w'$, where $u' - u \in U$, $w - w' \in W$. However, by hypothesis, U and W have only the vector θ in common, that is, $u' = u$ and $w = w'$. Therefore the representation of v is unique. Conversely, assume $v \in V$ has a unique representation $v = u + w$, where $u \in U$ and $w \in W$. Hence $U + W = V$. Furthermore, assume the existence of $z \in U \cap W$. We can then write $z = z + \theta$, where $z \in U$, $\theta \in W$ and $z = \theta + z$, where $\theta \in U$, $z \in W$. Since any such representation of z is unique we have $z = \theta$. Thus, $U \cap W = \{\theta\}$ and therefore $V = U \boxplus W$.

Example Thc notions of sum, intersection and direct sum of subspaces can easily be illustrated by means of the three-dimensional space of geometry. Consider V, the three-coordinate space of Fig. 10.1. The planes U and W are subspaces of V.

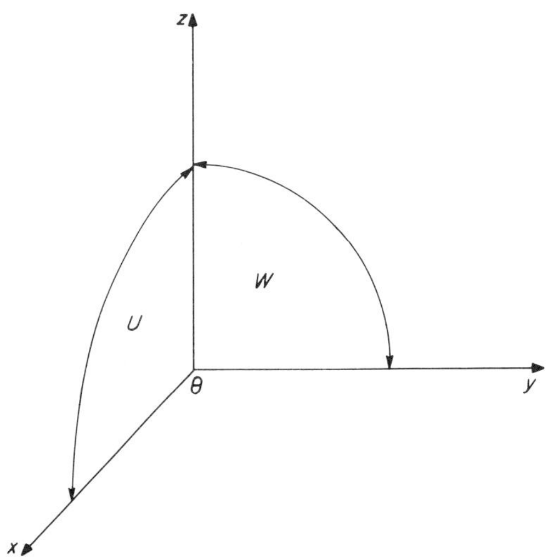

Figure 10.1

Note that while $U \cup W$ is not a subspace, $U + W$ is a subspace of V, since $U + W = [U \cup W] = V$. Clearly $U \cap W$ is the axis z, which is a subspace of V. The sum of W and the subspace represented by the axis x is a direct sum.

On the one hand, their sum is exactly the space V. On the other hand, their intersection is $\{\theta\}$. For the sake of completeness we may consider $\{\theta\}$, or the space generated by $\{\theta\}$, $[\theta]$, as a subspace of every vector space V.

10.5 Linearly Independent Set of Vectors

In Definition 7.2. of Chapter 7 we introduced the idea of $n + 1$ linearly independent points in R^n. The concept is also valid, of course, when we refer to a finite vector space V of n-tuples of F.

However, it is most important not to confuse the concept of a finite set of linearly independent *vectors*, as mentioned above, with that of a linearly independent finite *set* of vectors, as defined below.

Definition 10.15 Let V be a n-space over F and $S = \{u^1, \ldots, u^n\}$ a finite subset of V. Then S is said to be a linearly independent set of vectors if and only if every equation $\sum_{i=1}^{n} k_i u^i = 0$, $k_i \in F$, implies $k_i = 0$ for every i.

The definition thus suggests that in a linearly dependent set of vectors any vector in the set may be expressed as a linear combination of the other vectors in the set. The concepts of independence involved in Definitions 7.2. and 10.15 are thus different. However, a careful comparision of the two definitions will show that the following property, the proof of which is left as an exercise, holds true.

Remark 10.3 The $n + 1$ points $x^0, x^1, \ldots, x^n$ are linearly independent in R^n if, for $u^1 = x^1 - x^0, \ldots, u^n = x^n - x^0$, the set of vectors $S = \{u^1, \ldots, u^n\}$ is linearly independent.

In Part II we shall make considerable use of the concept of a linearly independent finite set of vectors rather than that of Definition 7.2. An extension of the idea to an infinite set of vectors is given in Section 10.6.

Example 1 Consider the two-dimensional simplex $S = x^0x^1x^2$ shown in Fig. 10.2. Suppose $x^0 = (1, 2)$, $x^1 = (9, 2)$, and $x^2 = (3, 8)$. From Fig. 10.2 it is obvious that the three points are linearly independent in the sense of Definition 7.2. Moreover, defining the vectors $u^1 = x^1 - x^0 = (8, 0)$ and $u^2 = x^2 - x^0 = (2, 6)$, according to Remark 10.3 we have that the set $A = \{u^1, u^2\}$ is linearly independent with the meaning of Definition 10.15.

Example 2 Suppose $W = \{w^1, \ldots, w^m\}$ is a linearly dependent set. We may then write $\sum_{i=1}^{m} k_i w^i = \theta$ with at least one $k_i \neq 0$. Assume $k_h \neq 0$. We have then

$$-k_h w^h = k_1 w^1 + \cdots + k_{h-1} w^{h-1} + k_{h+1} w^{h+1} + \cdots + k_m w^m.$$

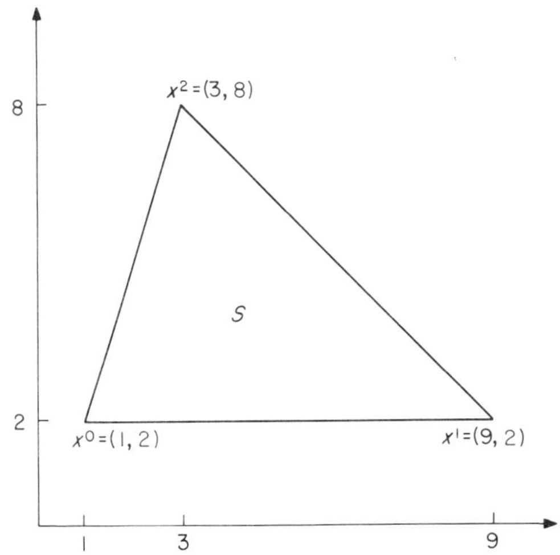

Figure 10.2

Or, what is the same, we may write, w^h has a linear combination of the remaining vectors in W; that is,

$$w^h = -k_h^{-1} \sum_{\substack{i=1\\ i\neq h}}^{m} k_i w^i.$$

Example 3 Suppose $k_i \in R$. Note the sets $A_2 = \{k_1(1,0), k_2(0,1)\}$, $A_3 = \{k_1(1,0,0), k_2(0,1,0), k_3(0,0,1)\}, \ldots,$ and $A_n = \{k_1(1,0,\ldots,0), \ldots, k_n(0,\ldots,0,1)\}$ are linearly independent in $R^2, R^3, \ldots, R^n$, respectively.

Note from Fig. 10.3 that the vectors of a linearly dependent set in R^2 lie on the same line through the origin; those of a linearly dependent set in R^3 lie on the same plane through the origin; finally, those of a linearly dependent set in R^n lie on the same $(n-1)$-hyperplane through θ.

As we can see a set of two vectors can only be linearly dependent if one vector is a multiple of the other.

For the sake of completeness we shall agree to regarding the empty set $\varnothing$ as linearly independent.

Let us study now certain properties of sets of vectors using our knowledge about linear independence.

Theorem 10.4 Let S be any finite set of vectors with $\theta \in S$. Then S is linearly dependent.

Proof Suppose $S = \{v^1, \ldots, v^m\}$ and $v^h = \theta$. With $k_h \neq 0$ and $k_i = 0$ for $i \neq h$, we may then write $\theta = \sum_{i=1}^m k_i v^i$ with at least $k_h \neq 0$. Therefore S is linearly dependent.

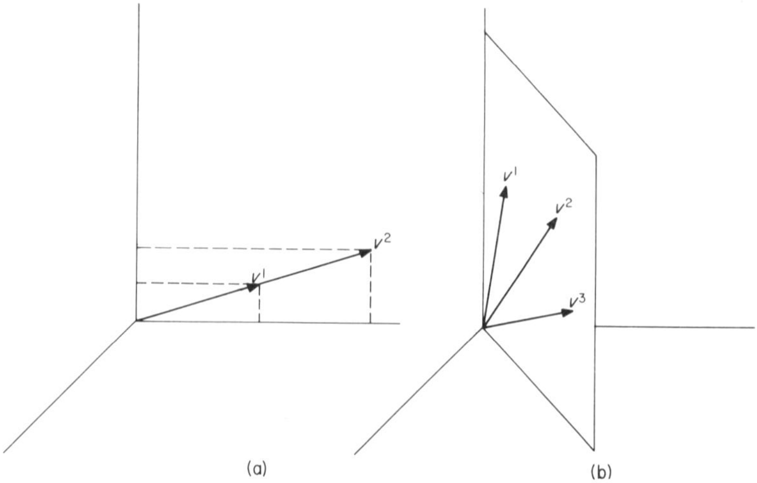

Figure 10.3 (a) $S = \{v^1, v^2\}$. (b) $S = \{v^1, v^2, v^3\}$.

Theorem 10.5 Let $S = \{v^1, \ldots, v^n\}$ be a set of vectors and consider any $A \subset S$. Then, if S is linearly independent A is linearly independent and if A is linearly dependent S is linearly dependent.

Proof Assume $S = \{v^1, \ldots, v^n\}$ is linearly independent and suppose, for $h \leq n$, $\{v^1, \ldots, v^h\} = A \subset S$ is linearly dependent. Hence, we can write $\sum_{i=1}^{h} k_i v^i = \theta$ with at least one $k_i \neq 0$. If this is true, however, we would also be able to write $\sum_{i=1}^{n} k_i v^i = \theta$ with, at least, the same $k_i \neq 0$. But then S would not be independent thus violating our assumption. The remainder of the theorem is proved similiarly.

Theorem 10.6 Let $S = \{v^1, \ldots, v^k\}$ span the space $U \neq [\theta]$. Then, there exists a linearly independent subset of S which also spans the space U.

Proof If S is a linearly independent set then the theorem is trivially true and nothing is to be proved. Suppose then that this is not the case and begin by taking one $v^j \in S$. If $[v^j] = U$, the theorem is proved. If not, take a second $v^h \in S$ so that $\{v^j, v^h\}$ is linearly independent. If $[v^j, v^h] = U$ the proof is complete. If not, then we take a third vector of S and proceed in the same way until we get a linearly independent set which spans U.

A corollary of the preceding theorem is presented in the following remark.

Remark 10.4 Any finite set of nonzero vectors contains a maximal independent subset.

Note that a set of only one vector is linearly independent.

Theorem 10.7 Let V be a vector space over F and let $S = \{v^1, \ldots, v^n\}$ be a linearly independent set of vectors which generates V. Then, any set of $n + 1$ vectors in V is linearly dependent.

Proof $S = \{v^1, \ldots, v^n\}$ is linearly independent by assumption. Moreover, $[v^1, \ldots, v^n] = V$. By Theorem 10.6 and Remark 10.4 S is a maximal independent subset. Hence, any $v^{n+1} \in V$, $v^{n+1} \notin S$ can only be a linear combination of the vectors in S. Therefore, any set containing $n + 1$ vectors must be linearly dependent.

Example The vectors $u^1 = (3, 0)$ and $u^2 = (0, 2)$ obviously span R^2. The set $\{u^1, u^2\}$ is linearly independent. It is easy to show that for any vector $u^3 \in R^2$, the set $\{u^1, u^2, u^3\}$ is dependent. For instance, consider $u^3 = (a, b)$. We have $u^3 = (a/3)\, u^1 + (b/2)u^2 = (a/3)(3, 0) + (b/2)(0, 2) = (a, b)$.

10.6 Basis and Dimension

The idea of a linearly independent set of vectors is fundamental for the definition of a basis of a vector space. The concept of dimension, closely related to it, has already been mentioned in a rather intuitive manner earlier in this chapter.

Basis

Definition 10.16 Let V be a n-space over F and B a maximal linearly independent subset of V. Then B is said to be a basis of V.

A vector space having a finite basis is called finite dimensional. If the basis is an infinite set, on the contrary, the corresponding vector space is said to be infinite dimensional. Our main interest here, as it has been pointed out earlier, is on finite-dimensional spaces. However, for the sake of completeness, we may extend Definition 10.16 so as to include infinite-dimensional spaces. The crucial point here is the definition of linear independence. An infinite set of vectors S is said to be linearly independent if every subset of S is linearly independent.

Henceforth we shall deal exclusively with finite-dimensional spaces but, to simplify matters, we shall not use the adjective "finite-dimensional" to qualify the term space.

Theorem 10.8 Let $V \neq [\theta]$ be a vector space over F. Then any vector $v \in V$ has a unique representation as a linear combination of the vectors $b^1, \ldots, b^n$ of a basis of V.

Proof Assume that, for suitable $k_i, v \in V$ can be represented by $v = \sum_{i=1}^{n} k_i b^i$. Suppose, furthermore, a second representation is possible by using the scalars h_i, $i = 1, \ldots, n$. Thus, we have $v = \sum_{i=1}^{n} h_i b^i$. Since $v - v = \theta$ we have

$$\theta = \sum_{i=1}^{n} k_i b^i - \sum_{i=1}^{n} h_i b^i = \sum_{i=1}^{n} (k_i - h_i) b^i.$$

Therefore $k_i = h_i$ for all i and the representation is unique.

Theorem 10.9 Let V be a vector space over F. Then every basis of V contains the same number of vectors.

Proof Suppose we have the two bases $B = \{b^1, \ldots, b^m\}$ and $C = \{c^1, \ldots, c^n\}$. From Theorem 10.7 we know that, since B generates V, C must have $n \leq m$ vectors or it is linearly dependent. By the same reasoning, since C generates V, B must have $m \leq n$ vectors or it is linearly dependent. Thus we have $n \leq m$ and $m \leq n$. The statement is true if and only if $n = m$, that is, if B and C contain the same number of vectors.

We know now that a real or complex n-space V has an infinite number of bases. In particular, a basis formed by the n "unit vectors" $e^1 = (1, 0, \ldots, 0)$, $e^2 = (0, 1, \ldots, 0)$, $\ldots$, $e^n = (0, 0, \ldots, 0, 1)$ is called the "usual basis" of V. What is important, however, is that all bases of V have the same number of vectors. We shall deal with this question very soon, and thus pass from an intuitive to a formal concept of dimension. However, before doing so, let us introduce the following concepts.

Definition 10.17 Let V be a vector space over F and let $B = \{b^1, \ldots, b^n\}$ be a linearly independent set of vectors which spans V. Then, (1) B is called an orthogonal basis of V if $b^i \cdot b^j = 0$ for all i, j such that $i \neq j$; (2) B is called an orthonormal basis of V if it is orthogonal and, additionally, $b^i \cdot b^i = 1$.

Definition 10.17 follows directly from Definitions 10.8–10.10, and 10.16. Suppose $B = \{b^1, \ldots, b^n\}$ forms an orthogonal basis of V. If $\|b^i\|$ represents the norm of the ith vector of B, it is clear that the set of vectors $\{\|b^1\|^{-1} \cdot b^1, \ldots, \|b^n\|^{-1} \cdot b^n\} = \{a^1, \ldots, a^n\}$ is an orthonormal basis of V. We would have $a^i \cdot a^i = 1$ and $a^i \cdot a^j = 0$ $(i \neq j)$.

Dimension

It is clear that if $B = \{b^1, \ldots, b^n\}$ is a basis of a vector space V the following two conditions are satisfied: (1) $\theta = \sum_{i=1}^{n} k_i b^i$ implies that $k_i = 0$ for $i = 1, \ldots, n$ (linear independence); (2) $[b^1, \ldots, b^n] = V$ (the set B generates V).

The following definition is now straightforward.

Definition 10.18 Let V be the vector space of n-tuples of elements of the field F and consider a basis $B = \{b^1, \ldots, b^n\}$. Then n, the number of vectors in B, is called the dimension of V and denoted $d(V) = n$.

Note that the vector space $[\theta]$ is said to be of dimension zero, that is, $d([\theta]) = 0$.

Example 1 The sets $B_1 = \{5\}$, $B_2 = \{(2, 0), (0, 3)\}$ and $B_3 = \{(3, 0, 0), (0, 1, 0), (0, 0, \frac{1}{2})\}$ are bases of one-, two-, and three-dimensional spaces, respectively.

Example 2 The set $B = \{1, x, x^2, \ldots, x^n\}$ is a basis for the vector space V of polynomials of degree n. Note that in this case $d(V) = n + 1$.

Theorem 10.10 Let V be a vector space and S any linearly independent set of vectors in V. Then S can be extended to a basis of V.

Proof Assume $B = \{b^1, \ldots, b^n\}$ is a basis of V. Suppose $S = \{s^1, \ldots, s^k\}$. Since S is linearly independent it is obvious that $k \leq n$. The space spanned by S is $[s^1, \ldots, s^k]$. If $b^i \in [s^1, \ldots, s^k]$ for all i, then clearly $[s^1, \ldots, s^k] = V$ and S is a basis of V. But, if $[s^1, \ldots, s^k] \neq V$, we can find for some j a vector $b^j \notin [s^1, \ldots, s^k]$ such that, by Theorem 10.6, $\{s^1, \ldots, s^k, b^j\}$ is linearly independent. Now, if $[s^1, \ldots, s^k, b^j] = V$, then obviously S has been extended to a basis and the theorem is proved. If $[s^1, \ldots, s^k, b^j] \neq V$, then we repeat the argument as many times as necessary.

Example Consider the space $V = R^3$ and the linearly independent set $S = \{s^1, s^2\}$, where $s^1 = (2, 0, 0)$ and $s^2 = (0, 3, 0)$. It follows immediately that $[s^1, s^2]$ is a two-dimensional subspace of V with third coordinate zero. Clearly $[s^1, s^2] \neq V$.

Now, taking some $s^3 \notin [s^1, s^2]$, for instance, $s^3 = (0, 0, 5)$, the set S can be extended to a basis of V. Note that $\{s^1, s^2, s^3\}$ is linearly independent and $[s^1, s^2, s^3] = V$.

The proof of the following theorem is left as an exercise.

Theorem 10.11 Let U and W be subspaces of a finite-dimensional vector space V. Then $d(U + W) = d(U) + d(W) - d(U \cap W)$. In particular, if $V = U \boxplus W$, then $d(V) = d(U) + d(W)$.

Example Consider the space $V = R^3$ and the subspaces U and W (see Fig. 10.4). Note that $\{(x_1, 0, 0), (0, x_2, 0)\}$ and $\{(0, x_2, 0), (0, 0, x_3)\}$ span the spaces U and W, respectively. The set $\{(x_1, 0, 0), (0, x_2, 0), (0, 0, x_3)\}$ is a basis of V.

On the one hand $U + W = V$. On the other hand $U \cap W$ is the line $k(0, x_2, 0)$ for k varying over R. Since $V = U + W$ we have $d(V) = d(U + W) = 3$. Now $d(U) = 2$, $d(W) = 2$, $d(U \cap W) = 1$. According to Theorem 10.11, $d(V) = d(U + W) = d(U) + d(W) - d(U \cap W) = 2 + 2 - 1 = 3$.

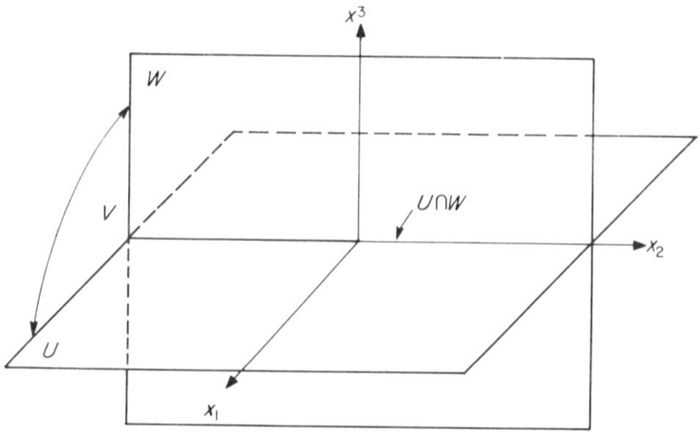

Figure 10.4

10.7 Vector-Space Homomorphisms

It is known to us from Definition 8.2 that a homomorphism is a structure-preserving mapping. Thus the definition of a vector-space homomorphism does not present special difficulties.

Definition 10.19 Let $V = \{E, F, \oplus, \odot, +, \cdot\}$ and $V' = \{E', F, \oplus, \odot, +, \cdot\}$ be two vector spaces over the field F. A function $h : E \to E'$ is called a vector-space homomorphism if for any vectors $v^1, v^2 \in V$ and any scalar $k \in F$ we have

(1) $h(v^1 \oplus v^2) = h(v^1) \oplus h(v^2) \in V'$;
(2) $h(k \odot v^1) = k \odot h(v^1) \in V'$.

Note that in the equalities (1) and (2) of Definition 10.19 the left-hand sides refer to the images of operations performed on elements of V, whereas the right-hand sides express operations performed on images, that is, on elements of V'.

A vector-space homomorphism may of course be bijective. We say then the spaces are isomorphic and call the mapping an isomorphism.

A most interesting case is that of an isomorphism from V to F_n, the space of all n-tuples of elements of the field F. We can always think of V and F_n as two different spaces, as will become apparent from the next theorem and the example following thereafter.

Theorem 10.12 Let V be an n-dimensional vector space over the field F. Call F_n the space of all n-tuples of elements of F. Then V is isomorphic to F_n.

Proof Let $B = \{b^1, \ldots, b^n\}$ be a basis of V. As we know, any vector $v \in V$ has a unique representation $v = \sum_{i=1}^n k_i b^i$ for $k_i \in F$. That is to say, for each $v \in V$, given B, there is a unique n-tuple $(k_1, \ldots, k_n) \in F_n$. There exists thus a one-to-one mapping f from V to F_n defined by $f(v) = (k_1, \ldots, k_n)$. We show next the mapping f is a structure-preserving one.

Consider the two vectors in V represented by $v = \sum_{i=1}^n k_i b^i$ and $w = \sum_{i=1}^n h_i b^i$ for the scalars $k_i, h_i \in F$. Adding the vectors we get

$$v + w = \sum_{i=1}^n k_i b^i + \sum_{i=1}^n h_i b^i = \sum_{i=1}^n (k_i + h_i) b^i.$$

Thus, for any vector $v + w \in V$ we have

$$f(v + w) = (k_1 + h_1, \ldots, k_n + h_n) = (k_1, \ldots, k_n) + (h_1, \ldots, h_n) = f(v) + f(w).$$

Finally, for any $w = \sum_{i=1}^n h_i b^i$, and $h_i, k \in F$ we have $kw = \sum_{i=1}^n kh_i b^i$. For the vector kw in V this means that

$$f(kw) = (kh_1, \ldots, kh_n) = k(h_1, \ldots, h_n) - kf(w).$$

The two operations are preserved by the mapping f. Therefore V and F_n are isomorphic.

Example V, the vector space of all polynomials of degree less than or equal to n and the space F_{n+1} of all $(n + 1)$-tuples of real numbers are isomorphic. The set $\{1, x, x^2, \ldots, x^n\}$ is a basis of V since, for $a_i \in F$, each polynomial in V has a unique representation by means of a linear combination $a_0 + a_1 x + a_2 x^2 + \cdots + a_n x^n$. There is, thus, a one-to-one correspondence between every polynomial in V and a different $(n + 1)$-tuple $(a_0, a_1, a_2, \ldots, a_n)$ in F_{n+1}.

For the operations vector addition and vector multiplication by a scalar we have similar correspondences. To the sum

$$\begin{aligned}(a_0 + a_1 x + a_2 x^2 + \cdots + a_n x^n) &+ (b_0 + b_1 x + b_2 x^2 + \cdots + b_n x^n) \\ &= (a_0 + b_0) + (a_1 + b_1)x + \cdots + (a_n + b_n)x^n \qquad \text{in } V, \qquad a_i, b_i \in F,\end{aligned}$$

there corresponds the sum

$$\begin{aligned}(a_0, a_1, a_2, \ldots, a_n) &+ (b_0, b_1, b_2, \ldots, b_n) \\ &= (a_0 + b_0, a_1 + b_1, a_2 + b_2, \ldots, a_n + b_n) \qquad \text{in } F_{n+1}.\end{aligned}$$

Similarly, to the product

$$\begin{aligned}k(a_0 &+ a_1 x + a_2 x^2 + \cdots + a_n x^n) \\ &= ka_0 + ka_1 x + ka_2 x^2 + \cdots + ka_n x^n \qquad \text{in } V, \qquad k \in F,\end{aligned}$$

corresponds the product

$$k(a_0, a_1, a_2, \ldots, a_n) = (ka_0, ka_1, ka_2, \ldots, ka_n) \qquad \text{in } F_{n+1}.$$

Clearly, V and F_{n+1} are isomorphic spaces.

EXERCISES

E.10.1 Let F be a field, X a nonempty set, and V the set of all single-valued functions, that is, $V = \{f \mid f: X \to F\}$. If for every $f_1, f_2 \in V$, $k \in F$, we have $(f_1 + f_2)(x) = f_1(x) + f_2(x)$, and $(kf)(x) = kf(x)$, for every $x \in X$, show that V is a vector space.

E.10.2 Consider the space V of the preceding exercise, and let $U \subset V$ be the set of all bounded functions $f: X \to F$. Show that U forms a subspace of V.

Hint: See the definition of a real bounded function in P.5.3.

E.10.3 An $n \times n$ real matrix is a square array of real numbers in n rows and n columns. The sum of two matrices $A = [a_{ij}]$ and $B = [b_{ij}]$, $i, j = 1, 2, \ldots, n$ is defined to be the matrix $A + B = [a_{ij} + b_{ij}]$. The product of an $n \times n$ matrix A by a scalar $k \in R$ is an $n \times n$ matrix $kA = [ka_{ij}]$. Show that the set of all $n \times n$ real matrices is a vector space.

Hint: See Chapter 12, Definition 12.7.

E.10.4 An $n \times n$ real matrix $A = [a_{ij}]$ is called symmetric if $a_{ij} = a_{ji}$, $i, j = 1, 2, \ldots, n$. Show that the set of all $n \times n$ real symmetric matrices is a subspace of the space of E.10.3.

E.10.5 Show that the field of real numbers may be regarded as a vector space over itself.

E.10.6 Let V be a vector space. Prove that the intersection of any number of subspaces of V is a subspace of V.

E.10.7 Prove Theorem 10.11.

Hint: Take a basis of the subspace $U \cap W$, and extend it to form bases of the subspaces U and W. Study the set of all vectors contained in the union of the constructed bases of U and W. Show that this is a basis of $U + W$.

E.10.8 Let $\{u^1, \ldots, u^n\}$ form a linearly independent set of vectors. Show that a vector v is a linear combination of the vectors $u^1, \ldots, u^n$ if and only if the set $\{u^1, \ldots, u^n, v\}$ is linearly dependent.

E.10.9 Call V the Euclidean n-space, and let U be a subspace of it. Prove that, if U' is the set of all vectors orthogonal to every vector in U, then $V = U \boxplus U'$.

Hint: Show that $U + U' = V$, and $U \cap U' = \{\theta\}$.

E.10.10 Let U, W be two subspaces of V. Prove that, if $d(U) = d(W)$, and $U \subset W$, then $U = W$.

PROBLEMS

P.10.1 Apply the results of E.10.1 and E.10.2 and show that (a) the set of all complex-valued functions is a vector space; (b) the set of all real-valued functions is a subspace of the space of (a); (c) the set of all real-valued bounded functions is a subspace of the spaces of (a) and (b).

P.10.2 Let $\sum_{j=1}^{n} a_{1j} x_j = 0, \ldots, \sum_{j=1}^{n} a_{mj} x_j = 0$ represent a system of m linear equations in n unknowns over R. Show that (a) the set of solutions of the system is a vector space V; (b) V is a subspace of R^n.

P.10.3 Let $\sum_{j=1}^{n} a_{1j} x_j = b_1, \ldots, \sum_{j=1}^{n} a_{mj} x_j = b_m$ be a system of m linear equations in n unknowns over R. Is its solution set, in general, a vector space? Is it a subspace of R^n?

P.10.4 Show that if $V_1, V_2, \ldots$ are all subspaces of V such that $V_1 \subset V_2 \subset \cdots$, then $\bigcup_i V_i$ is a subspace of V.

P.10.5 Show that if the set of vectors S_i generates the subspace V_i of V, and $V_1 \subset V_2 \subset \cdots$, then the set of vectors $\bigcup_i S_i$ generates the subspace $\bigcup_i V_i$ of V.

NOTES

The works of Finkbeiner [1], Halmos [2], Thrall and Tornheim [1] are all excellent books on vector space theory (see also Gantmacher [1]). For a book more oriented toward applications see Hadley [1]. Berge [2] also gives an interesting treatment of vector spaces and linear mappings in the context of metric spaces. A comprehensive treatment of linear economic theory is presented by Gale [3] (see also Nikaido [4], Lancaster [1], and Allen [1]).

CHAPTER 11

VECTOR-SPACE HOMOMORPHISMS

In Part I we were mostly interested in some topological properties of spaces and in a class of mappings (homeomorphisms) under which these properties remain invariant.

In Part II we study the properties of a space with a special algebraic structure defined on it (vector space) and a class of mappings (homomorphisms) under which this structure is preserved.

The idea of a linear single-valued mapping from a set $X \subset R^m$ to a set $Y \subset R^n$ was discussed at the end of Chapter 5. The corresponding concept which will occupy our attention here is that of a linear single-valued function from a vector space U to a vector space V.

11.1 Linear Transformations

In this and in the following chapters the terms vector-space homomorphism and linear transformation are regarded as meaning the same thing, that is, a linear mapping from a vector space to a vector space.

Definition 11.1 Let U and V be vector spaces over a field F. A mapping $t\colon U \to V$ is called a homomorphism or linear transformation of U to V if for all $x, y \in U$ and $a, b \in F$, $t(ax + by) = at(x) + bt(y)$.

It is obvious that a linear transformation may have the same space as its domain and codomain, that is, $t\colon U \to U$. The equality at the end of the preceding definition is called the linearity principle.

Example 1 Let V be the space of n-tuples of real numbers $v = (x_1, \ldots, x_n)$. Then the mapping $t\colon V \to V$ defined by $u = t(v) = kv$, for a constant k, is a linear transformation.

Example 2 Let V be the vector space of all polynomials $v = a_0 + a_1x + \cdots + a_nx^n$ over the field of real numbers. The mapping $t\colon V \to V$ which assigns to each $v \in V$ its first derivative Dv is a linear transformation. Note, in particular, that $D(u + v) = Du + Dv$ and $D(kv) = kDv$.

Sum, Scalar Multiple, and Product Transformation

A direct consequence of Definition 5.4 is the following:

Definition 11.2 The two transformations t_1 and t_2 from U to V are said to be equal if for every vector $u \in U$, $t_1(u) = t_2(u)$.

Consider any two linear transformations from a space U to a space V, and a scalar k. It is very easy to show that the sum of the two transformations and the product of either of them by k are also linear transformations from U to V.

Definition 11.3 Let t_1 and t_2 be linear mappings from U to V and $k \in F$. Then the sum $t_1 + t_2$ is the linear mapping $(t_1 + t_2)(u) = t_1(u) + t_2(u)$, for every $u \in U$. The scalar multiple kt_1 is the linear mapping $(kt_1)(u) = kt_1(u)$, for every $u \in U$.

Another important property, the proof of which is very simple and thus left as an exercise, concerns the linearity of the composite transformation of two linear transformations. The concept is carried over in the following definition.

Definition 11.4 Let U, V, and W be vector spaces over a field F. Let t_1 and t_2 be linear mappings from U to V and from V to W, respectively. Then the product or composite mapping t_2t_1 is the linear mapping from U to W defined by $(t_2t_1)(u) = t_2(t_1(u))$, for every $u \in U$.

Suppose now T is the set of all linear transformations from U to V. From our previous discussion we see that, given t_1, $t_2 \in T$ and $k \in F$, we also have $t_1 + t_2 \in T$ and $kt_1 \in T$. Thus we have proved the following remark:

Remark 11.1 Let U and V be vector spaces over the field F and T the set of all linear mappings form U to V. Then T forms a vector space over F.

Example Consider the linear transformations $t_1: R^3 \to R^2$, $t_2: R^3 \to R^2$, and $t_3: R^2 \to R^2$ defined by $t_1(x, y, z) = (x, y + z)$, $t_2(x, y, z) = (x, y + 3z)$, and $t_3(x, y) = (y, x)$. Find $t_1 + 3t_2$ and $t_3 t_1$.

First, for $t_1 + 3t_2$ we have

$$\begin{aligned}(t_1 + 3t_2)(x, y, z) &= t_1(x, y, z) + 3t_2(x, y, z)\\ &= (x, y + z) + 3(x, y + 3z) = (4x, 4y + 10z).\end{aligned}$$

Second, for $t_3 t_1$ we get

$$(t_3 t_1)(x, y, z) = t_3(t_1(x, y, z)) = t_3(x, y + z) = (y + z, x).$$

Special Transformations

Among the linear transformations there are four which deserve to be mentioned in particular for they will play important roles in our future discussions.

Definition 11.5 A linear mapping $z: U \to V$ is called a zero transformation if for any vector $u \in U$ we have $z(u) = \theta$.

Definition 11.6 A linear mapping $i: U \to V$ is called an identity transformation if for any vector $u \in U$ we have $i(u) = u$.

Definition 11.7 Let $t: U \to V$ be any linear mapping. Then a linear mapping $-t$ exists, called the negative of t, such that for any $u \in U$, $-t(u) = t(-u)$.

Definition 11.8 A linear mapping t^{-1} from a subspace of V to U is said to be the inverse of the linear mapping $t: U \to V$ if and only if $tt^{-1} = i: V \to V$ and $t^{-1}t = i: U \to U$.

From the preceding definitions and the linearity principle, the following properties follow immediately.

Remark 11.2 Let $t: U \to V$ be a linear mapping and $z, i, -t$ and t^{-1} the special mappings of Definitions 11.5–11.8. Then (1) $t + z = z + t = t$; (2) $ti = it = t$; (3) $t + (-t) = (-t) + t = z$; (4) $tt^{-1} = t^{-1}t = i$.

It should be clear to us (refer to Chapter 5) that not every transformation t has an inverse transformation t^{-1}. We shall return to this point later in our study.

11.2 Analysis of Linear Transformations

We focus our attention now on some particular aspects of linear mappings.

Range Space and Rank

Definition 11.9 Let t: $U \to V$ be a linear transformation. Then the set $R_t = \{v \mid v \in V, v = t(u)$ for some $u \in U\}$ is called the range space of t.

Since R_t is the set of all images under the mapping t, it is obviously a subset of V, the codomain of t. Furthermore, as Theorem 11.1 will show, it is a subspace of V.

Theorem 11.1 Let t: $U \to V$ be a linear mapping with range space R_t. Then R_t is a subspace of V.

Proof From Definition 11.1 we know that $t(\theta) = \theta$; so $\theta \in R_t$. Suppose we have now $v^1, v^2 \in R_t$ and any scalars $a, b \in F$. Since $v^1, v^2 \in R_t$ there must be some $u^1, u^2 \in U$ such that $t(u^1) = v^1$ and $t(u^2) = v^2$. Thus, $t(au^1 + bu^2) = av^1 + bv^2 \in R_t$. Therefore R_t is a subspace of V.

We can now speak of the dimension of R_t.

Definition 11.10 Let R_t be the range space of the linear mapping t: $U \to V$. Then the dimension of R_t is called the rank of t and denoted $r(t) = d(R_t)$.

Kernel and Nullity

Definition 11.11 Let t: $U \to V$ be a linear transformation. Then the set $K_t = \{u \mid u \in U, t(u) = \theta\}$ is called the kernel or null space of t.

K_t is clearly a subset of U, the domain of t. It follows from the definition of homomorphism that $\theta \in K_t$. Note, however, that K_t may contain vectors other than θ, for it is possible that some $u \neq \theta$ in U be mapped by t into $\theta \in V$. That K_t is a subspace of U is proved in the following theorem.

Theorem 11.2 Let t: $U \to V$ be a linear mapping with kernel K_t. Then K_t is a subspace of U.

Proof From Definition 11.1 we know that $t(\theta) = \theta$; so $\theta \in K_t$. Suppose we have now $u^1, u^2 \in K_t$ and any scalars $a, b \in F$. Since $u^1, u^2 \in K_t$, we must have $t(u^1) = t(u^2) = \theta$. Thus $t(au^1 + bu^2) = at(u^1) + bt(u^2) = \theta$. Therefore K_t is a subspace of U.

K_t being a subspace, we can speak of its dimension.

Definition 11.12 Let K_t be the kernel or null space of the linear mapping $t\colon U \to V$. Then the dimension of K_t is called the nullity of t and denoted $k(t) = d(K_t)$.

An important property relating rank and nullity of a vector-space homomorphism is given in the following theorem.

Theorem 11.3 Let $t\colon U \to V$ be a linear mapping. Then $d(U) = r(t) + k(t)$.

Proof Assume $d(U) = n$ and $k(t) = k$. A basis of K_t is thus $\{u^1, \ldots, u^k\}$ which, by Theorem 10.10 can be extended to a basis $\{u^1, \ldots, u^k, u^{k+1}, \ldots, u^n\}$ of U. Next we need to prove that the set of images of the vectors $u^{k+1}, \ldots, u^n$, that is, $\{t(u^{k+1}), \ldots, t(u^n)\}$ is a basis of R_t. Suppose first the set $\{t(u^{k+1}), \ldots, t(u^n)\}$ is linearly dependent, that is, it is not a basis of R_t. For some scalars $a_{k+1}, \ldots, a_n \in F$ we would have $\sum_{i=k+1}^n a_i t(u^i) = \theta \in V$; therefore

$$t\left(\sum_{i=k+1}^{n} a_i u^i\right) = \theta \in V.$$

This would mean that $\sum_{i=k+1}^n a_i u^i$ lies on K_t and, for suitable scalars $a_1, \ldots, a_k \in F$, we could represent it as a linear combination of basis vectors of K_t, that is, $\sum_{i=1}^k a_i u^i = \sum_{i=k+1}^n a_i u^i$. Were this true, then the set $\{u^1, \ldots, u^k, u^{k+1}, \ldots, u^n\}$ would be linearly dependent violating our assumption that it is a basis of U. Hence, the set $\{t(u^{k+1}), \ldots, t(u^n)\}$ is linearly independent.

We show next that the linearly independent set $\{t(u^{k+1}), \ldots, t(u^n)\}$ generates R_t. Consider any $v = t(u)$, for some $u \in U$. For the scalars $b_1, \ldots, b_n \in F$ we can write $u = \sum_{i=1}^n b_i u^i$ and thus

$$\begin{aligned} v = t(u) = t\left(\sum_{i=1}^{n} b_i u^i\right) &= \sum_{i=1}^{n} b_i t(u^i) \\ &= \sum_{i=1}^{k} b_i t(u^i) + \sum_{i=k+1}^{n} b_i t(u^i) = \theta + \sum_{i=k+1}^{n} b_i t(u^i), \end{aligned}$$

since $u^1, \ldots, u^k$ all lie in K_t. Hence every vector in R_t is a linear combination of the vectors $t(u^{k+1}), \ldots, t(u^n)$ and this set generates R_t. Now we assumed $d(U) = n$ and $k(t) = k$. Since $\{t(u^{k+1}), \ldots, t(u^n)\}$ generates R_t we know, applying Definition 11.10, that $r(t) = n - k$. Therefore $d(U) = r(t) + k(t)$.

Example Consider the linear transformation $t\colon R^3 \to R^3$ defined by $t(x_1, x_2, x_3) = (x_1, x_2, 0)$. Note that the null space K_t is the set of all vectors $(0, 0, x_3)$ while the range space R_t contains all the vectors $(x_1, x_2, 0)$. We have then $d(U) = 3$, $r(t) = 2$ and $k(t) = 1$.

11.3 Nonsingular and Inverse Transformations

We have already met the concept of an inverse transformation in Definition 11.8. Given $t: U \to V$, its inverse was described as "a linear mapping from a subspace of V to U." At that point, it might have seemed strange to refer to a "subspace of V" instead of V itself. But it should be clear now that the subspace referred to was none other than the range space R_t. While it is true that the spaces V and R_t can have the same dimension, this is not necessarily so.

We know that not every mapping has an inverse. On the contrary, to be precise, this is a particular property of one-to-one functions. Linear mappings having this important property are called nonsingular transformations.

Definition 11.13 Let $t: U \to V$ be a linear transformation with range space R_t. Then t is said to be nonsingular if and only if a transformation $t^{-1}: R_t \to U$ exists such that $t^{-1}t = tt^{-1} = i$ where $t^{-1}t$ represents the identity transformation on U and tt^{-1} the identity transformation on V.

This definition is very close to Definition 11.8. Obviously the mapping $t^{-1}: R_t \to U$ of the former is none other than the inverse transformation defined in the latter. That $t^{-1}t$ and tt^{-1} represent identity mappings in two different spaces—provided, of course, that U and V are not one and the same space—can be easily shown.

Consider the vectors $u^1, u^2 \in U$ and $v^1, v^2 \in R_t$, chosen so that $t(u^1) = v^1$ and $t(u^2) = v^2$. We have $(t^{-1}t)(u^1) = t^{-1}(t(u^1)) = t^{-1}(v^1) = u^1$. Analogously, $(t^{-1}t)(u^2) = u^2$. On the other hand, $(tt^{-1})(v^1) = t(t^{-1}(v^1)) = t(u^1) = v^1$. Analogously, $(tt^{-1})(v^2) = v^2$.

Furthermore, for the scalars $k_i \in F$, we have

$$(t^{-1}t)(k_1 u^1 + k_2 u^2) = t^{-1}(k_1 v^1 + k_2 v^2) = k_1 u^1 + k_2 u^2 \in U$$

and

$$(tt^{-1})(k_1 v^1 + k_2 v^2) = t(k_1 u^1 + k_2 u^2) = k_1 v^1 + k_2 v^2 \in V.$$

Thus $t^{-1}t = i: U \to U$ and $tt^{-1} = i: V \to V$.

Nonsingular transformations have some interesting properties and we shall study next some of these.

Theorem 11.4 Let $t: U \to V$ be nonsingular. Then the kernel of t is $K_t = [\theta]$.

Proof Consider $u^1, u^2 \in U$ and assume $t(u^1) = t(u^2)$. Since t is nonsingular, this implies $(t^{-1}t)(u^1) = (t^{-1}t)(u^2) = i(u^1) = i(u^2) = u^1 = u^2$. Consider next $u^3 \in K_t$. By definition, then, $t(u^3) = \theta$. But also $t(\theta) = \theta$. By our previous argument we must then have $(t^{-1}t)(u^3) = (t^{-1}t)(\theta) = u^3 = \theta$. Therefore θ is the only vector in K_t or, in other terms, $K_t = [\theta]$.

Three corollaries follow from Theorem 11.4.

Remark 11.3 Let $t\colon U \to V$ be nonsingular. Then $k(t) = 0$.

The result is obvious since the nullity of t is by definition the dimension of the null space.

Remark 11.4 Let $t\colon U \to V$ be nonsingular. Then $d(U) = r(t)$.

We know from Theorem 11.3 that $d(U) = r(t) + k(t)$. But according to Remark 11.3, we have here $k(t) = 0$. Domain and range space of nonsingular transformations thus have the same dimension.

Remark 11.5 A linear transformation $t\colon U \to V$ is nonsingular if for any $u \neq \theta$, $t(u) \neq \theta$; it is singular if there is some $u \neq \theta$ with $t(u) = \theta$.

The result is obvious since we know that, for a nonsingular linear mapping $K_t = [\theta]$.

Theorem 11.5 Let $t\colon U \to V$ be a linear transformation. Then t is nonsingular if and only if it maps a basis of U onto a basis of R_t.

Proof Assume t is nonsingular and let $\{u^1, \ldots, u^n\}$ be a basis of U. We have to show that $\{t(u^1), \ldots, t(u^n)\}$ is a basis of R_t. Suppose $\sum_{i=1}^n k_i t(u^i) = \theta$ for $k_i \in F$. Hence $t(\sum_{i=1}^n k_i u^i) = \theta$ and so $\sum_{i=1}^n k_i u^i \in K_t$. The assumption that t is nonsingular together with $\sum_{i=1}^n k_i u^i \in K_t$ imply $\sum_{i=1}^n k_i u^i = \theta$. But since $\{u^1, \ldots, u^n\}$ is linearly independent, it follows that $k_i = 0$ for all i. Hence $\{t(u^1), \ldots, t(u^n)\}$ is linearly independent. Besides, $d(U) = r(t) = n$, which is the number of vectors in the set. Therefore $\{t(u^1), \ldots, t(u^n)\}$ is a basis of R_t. Suppose now t maps the basis $\{u^1, \ldots, u^n\}$ of U onto the basis $\{t(u^1), \ldots, t(u^n)\}$ of R_t. Obviously $u^i \neq \theta$ for all i and $t(u^i) \neq \theta$ for all i. Therefore t is nonsingular.

The proof of the following property is very easy and is left as an exercise.

Remark 11.6 If the linear mappings $t_1\colon U \to V$ and $t_2\colon V \to W$ are both nonsingular, then the composite linear mapping $t_2 t_1\colon U \to W$ is also nonsingular.

Remark 11.7 Let $t_1\colon U \to V$ and $t_2\colon V \to W$ be nonsingular linear mappings and $t_2 t_1\colon U \to W$ their composition. Then $(t_2 t_1)^{-1} = t_1^{-1} t_2^{-1}\colon W \to U$ (see Fig. 11.1).

Consider the vectors $u \in U$, $v \in V$, and $w \in W$. We have

$$v = t_1(u), \qquad w = t_2(v) = t_2(t_1(u)) = (t_2 t_1)(u).$$

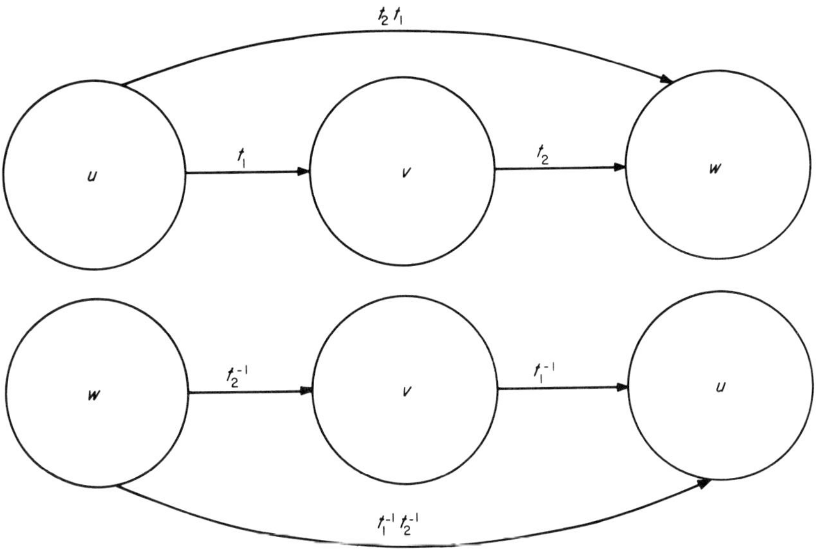

Figure 11.1

On the other hand,

$$v = t_2^{-1}(w), \qquad u = t_1^{-1}(v) = t_1^{-1}(t_2^{-1}(w)) = (t_1^{-1}t_2^{-1})(w).$$

We can see thus that $t_2 t_1 \colon U \to W$ and $(t_2 t_1)^{-1} = t_1^{-1}t_2^{-1} \colon W \to U$.

Remark 11.8 Let $t \colon U \to W$ be nonsingular. Then $(t^{-1})^{-1} = t$. Suppose $v = t(u)$. We have then $t^{-1}(v) = u$ and $(t^{-1})^{-1}(u) = v$.

Remark 11.9 Let $t \colon U \to V$ be nonsingular. Then $(kt)^{-1} = k^{-1}t^{-1}$, for $k \neq 0$.

11.4 Linear Functionals and Dual Spaces

A field F, as we know from Chapter 8, satisfies all the axioms of an additive Abelian group. Besides, if $a, b \in F$, then the product $a \cdot b \in F$. This amounts to saying that a field F also satisfies all the axioms of a vector space. We may thus regard F as a vector space over itself.

If a field F can be taken as a vector space, then it is clear that we may also define a linear mapping from a vector space V to F. We shall study this question in this section.

Definition 11.14 A mapping $q: V \to F$ is said to be a linear functional if for every $u, v \in V$ and $a, b \in F$, $q(au + bv) = aq(u) + bq(v) \in F$.

Example 1 Let V be the space of all convergent sequences of numbers $(x^n | n \in N)$. Then a functional $q: V \to F$ assigns to each sequence in V its limit $\lim_{n\to\infty} x^n = x \in F$.

Example 2 Let V be the space of all real-valued functions f integrable for $x \in [a, b]$ with integral $\int_a^b f(x)\, dx$. Note that the integral of each $f \in V$ is a real number in F. A mapping $q = V \to F$ defined by $q(f) = \int_a^b f(x)\, dx$ is a linear functional.

In Remark 11.1 it was pointed out that the set of all linear mappings from a vector space U to a vector space V is itself a vector space. The same applies, of course, to the set of all linear functionals from a space V into F.

Definition 11.15 The vector space of all linear functionals from V to F is called the dual space of V and denoted V^*.

The following theorem establishes a relationship between a basis of a space V and a basis of its dual V^*. The expression δ_{ij} appearing in the theorem is called the Kronecker delta, and defined by $\delta_{ij} = 1$ if $i = j$, and $\delta_{ij} = 0$ if $i \neq j$ $(i, j = 1, \ldots, n)$.

Theorem 11.6 Let $\{b^1, \ldots, b^n\}$ be a basis of the n-dimensional space V. Let the linear functionals $q_i : V \to F$ be $q_i(b^j) = \delta_{ij}$ $(i, j = 1, \ldots, n)$, where δ_{ij} is the Kronecker delta. Then the set $\{q_1, \ldots, q_n\}$, called the dual basis of $\{b^1, \ldots, b^n\}$, is a basis of the dual space V^*.

Proof The zero of V^* is the functional q_θ. For every $v \in V$, this means $q_\theta(v) = 0$. On the other hand, for the functionals q_i and vectors of $\{b^1, \ldots, b^n\}$ we have either $q_i(b^j) = 1$ or $q_i(b^j) = 0$. To prove $\{q_1, \ldots, q_n\}$ is a basis of V^* we show, first, that it is a linearly independent set and, second, that it spans the space V^*.

(1) Suppose $\sum_{i=1}^n k_i q_i = q_\theta = 0$ for some $k_i \in F$. Then, for the vectors b^j $(j = 1, \ldots, n)$ we will have

$$0 = q_\theta(b^1) = \left(\sum_{i=1}^n k_i q_i\right)(b^1) = \sum_{i=1}^n k_i q_i(b^1) = k_1 \delta_{11} = k_1$$

$$\vdots$$

$$0 = q_\theta(b^n) = \left(\sum_{i=1}^n k_i q_i\right)(b^n) = \sum_{i=1}^n k_i q_i(b^n) = k_n \delta_{nn} = k_n.$$

Thus, $\{q_1, \ldots, q_n\}$ is linearly independent.

(2) Consider a functional $q \in V^*$ such that, for $h_i \in F$, $h_1 = q(b^1), \ldots, h_n = q(b^n)$. We are then allowed to write

$$q(b^1) = \sum_{i=1}^{n} h_i \delta_{1i} = \sum_{i=1}^{n} h_i q_i(b^1)$$
$$\vdots$$
$$q(b^n) = \sum_{i=1}^{n} h_i \delta_{ni} = \sum_{i=1}^{n} h_i q_i(b^n).$$

In general, then, any $q \in V^*$ can be expressed as $q = \sum_{i=1}^{n} h_i q_i$, for some $h_i \in F$. Hence the set $\{q_1, \ldots, q_n\}$ spans V^*.

An interesting corollary follows from the preceding theorem.

Remark 11.10 Let V and V^* be an n-dimensional vector space and its dual space, respectively. Then the dimension of V^* is $d(V^*) = n$.

The result is immediate since, as it was shown, if $\{b^1, \ldots, b^n\}$ is a basis of V and q_i: $V \to F$ are the functionals defined in Theorem 11.6, then $\{q_1, \ldots, q_n\}$ is a basis of V^*. This set contains precisely n vectors so V^* is clearly n-dimensional.

Theorem 11.7 Let V be a vector space and V^* its dual space. Then for any nonzero vector $v \in V$ there exists some functional $q \in V^*$ such that $q(v) \neq 0$.

Proof Suppose $\{b^1, \ldots, b^n\}$ is a basis of V and $\{q_1, \ldots, q_n\}$ the dual basis for V^*. For the scalars $k_i \in F$ we can write $v = \sum_{i=1}^{n} k_i b^i$. Now, by hypothesis $v \neq \theta$. Thus, there is at least one k_i which is different from zero in $\sum_{i=1}^{n} k_i b^i$. For this nonzero k_i we have $q_i(v) = q_i(k_i b^i) = k_i q_i(b^i) \neq 0$.

Example Let $\{e^1 = (1, 0, 0), e^2 = (0, 1, 0), e^3 = (0, 0, 1)\}$ be the usual basis of R^3. Find its dual basis $\{q_1, q_2, q_3\}$. Suppose

$$q_1(x, y, z) = a_1 x + a_2 y + a_3 z,$$
$$q_2(x, y, z) = b_1 x + b_2 y + b_3 z,$$
$$q_3(x, y, z) = c_1 x + c_2 y + c_3 z,$$

such that

$$q_1(v_1) = 1, \quad q_1(v_2) = 0, \quad q_1(v_3) = 0,$$
$$q_2(v_1) = 0, \quad q_2(v_2) = 1, \quad q_2(v_3) = 0,$$
$$q_3(v_1) = 0, \quad q_3(v_2) = 0, \quad q_3(v_3) = 1.$$

Now, since

$$q_1(v_1) = q_1(1, 0, 0) = a_1 = 1,$$
$$q_1(v_2) = q_2(0, 1, 0) = b_1 = 0,$$
$$q_1(v_3) = q_1(0, 0, 1) = c_1 = 0,$$

we get

$$q_1(x, y, x) = x.$$

Following an analogous method, we obtain $q_2(x, y, z) = y$ and $q_3(x, y, z) = z$.

Therefore, the dual basis of $\{e^1, e^2, e^3\}$ is the set $\{q_1(x, y, z) = x, q_2(x, y, z) = y, q_3(x, y, z) = z\}$.

We have said that the set of all linear functionals $q: V \to F$ forms a vector space, the dual space V^*. Now, since V^* is itself a vector space, it also has a dual. We call this the second dual space or the bidual space of V, and denote it V^{**}. By definition V^{**} contains all linear functionals on V^*.

Notice that every vector $v \in V$ determines a vector (functional) $q \in V^*$. Thus, using the notation q' for the functionals contained in V^{**}, every $q \in V^*$ determines a $q' \in V^{**}$ such that $q(v) = q'(q)$.

That the functionals in V^{**} are linear too can be seen from the fact that, for every $q_1, q_2 \in V^*$ and $v \in V$,

$$q'(aq_1 + bq_2) = (aq_1 + bq_2)(v) = aq_1(v) + bq_2(v) = aq'(q_1) + bq'(q_2).$$

Continuing from this result, it is not difficult to prove that the composite linear mapping from a finite-dimensional vector space V into its bidual space V^{**} is an isomorphism.

To conclude this part of the chapter we introduce the following concept:

Definition 11.16 Let V be a vector space and V^* its dual space. Furthermore, let the set $A \subset V$ be nonempty. Then the annihilator A^0 of A is the set of all linear functionals in V^* which map each vector of A into 0, that is, $A^0 = \{q \mid q \in V^*, q(v) = 0 \text{ for every } v \in A\}$.

That the annihilator A^0 is not merely a subset but a subspace of V^* is easy to show. First, it obviously contains the functional q_θ, that is, the zero of V^*. Second, for two functionals $q_1, q_2 \in A^0$, the scalars $k_1, k_2 \in F$ and the vector $v \in A \subset V$, we have

$$(k_1q_1 + k_2q_2)(v) = k_1q_1(v) + k_2q_2(v) = k_10 + k_20 = 0.$$

This means that A^0 is always a subspace of V^*, even if the set A is not a subspace of V.

11.5 Transpose of a Linear Transformation

Let V and W be two vector spaces and, once again, let us consider the field F as a vector space over itself (see Fig. 11.2). On these spaces we define two linear mappings, $t: V \to W$ and the functional $q: W \to F$.

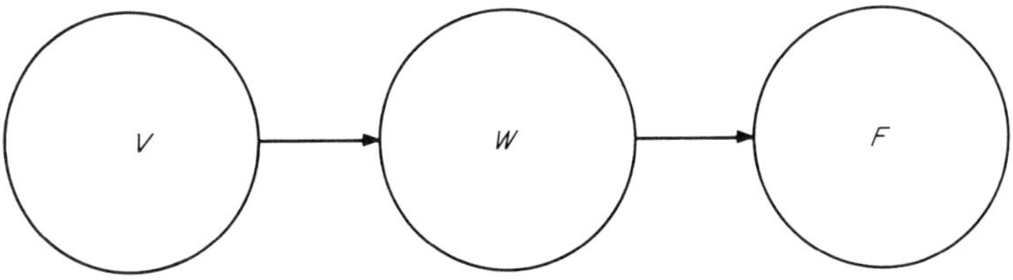

Figure 11.2

Since q is a linear mapping so also is the composite transformation $qt\colon V \to F$. Now, clearly $q \in W^*$ (the dual space of W) whereas $qt \in V^*$ (the dual space of V).

With the elements we have, we may construct a mapping from W^* to V^*, which assigns to each functional $q \in W^*$ a composite transformation $qt \in V^*$. Such a mapping will be of great importance in the context of our study and to it we dedicate most of the remainder of this section.

Definition 11.17 Let V and W be vector spaces over the field F, and consider the linear transformations $t\colon V \to W$ and $q\colon W \to F$. Then t^{T}, called the transpose of t, is a mapping from the dual space W^* to the dual space V^* defined, for every $q \in W^*$, by $t^{\mathrm{T}}(q) = qt$.

Example Suppose we have $t\colon R^2 \to R^2$ defined by $t(x, y) = (y, x + y)$ and the functional $q\colon R^2 \to R$ defined by $q(x, y) = x - 3y$. The transpose of t is then the mapping defined by

$$t^{\mathrm{T}}(q(x, y)) = q(t(x, y)) = q(y, x + y) = -3x - 2y.$$

Theorem 11.8 Let $t\colon V \to W$ be a linear mapping. Then t^{T}, the transpose of t, is linear.

Proof Consider any $q_1, q_2 \in W^*$ and $k_1, k_2 \in F$. By definition we have

$$t^{\mathrm{T}}(k_1q_1 + k_2q_2) = (k_1q_1 + k_2q_2)t = k_1(q_1t) + k_2(q_2t) = k_1t^{\mathrm{T}}(q_1) + k_2t^{\mathrm{T}}(q_2).$$

Therefore t^{T} is a linear mapping.

Theorem 11.9 Let $s\colon U \to V$ and $t\colon V \to W$ be linear transformations. Then $(ts)^{\mathrm{T}} = s^{\mathrm{T}}t^{\mathrm{T}}$.

Proof Observe first that t^{T} is a mapping from W^* to V^*, s^{T} a mapping from V^* to U^*, and $s^{\mathrm{T}}t^{\mathrm{T}}$ a mapping from W^* to U^*. Consider now any functional $q \in W^*$. By Definition 11.17 we have

$$(ts)^{\mathrm{T}}(q) = qts = (qt)s = s^{\mathrm{T}}(qt) = s^{\mathrm{T}}t^{\mathrm{T}}(q) = (s^{\mathrm{T}}t^{\mathrm{T}})(q).$$

That is, $(ts)^{\mathrm{T}} = s^{\mathrm{T}}t^{\mathrm{T}}$.

The properties of the following remark are important. Proofs are simple and thus left as exercises.

Remark 11.11 Let both s and t be linear transformations from U to V. Then

(1) $(t^{\mathrm{T}})^{-1} = (t^{-1})^{\mathrm{T}}$;
(2) $(s + t)^{\mathrm{T}} = s^{\mathrm{T}} + t^{\mathrm{T}}$;
(3) $(ct)^{\mathrm{T}} = ct^{\mathrm{T}}$;
(4) $i^{\mathrm{T}} = i$, where i and i^{T} represent the identity transformations in U and U^*, respectively;
(5) $r(t) = r(t^{\mathrm{T}})$.

11.6 Linear Algebras

The theory of vector spaces and vector space homomorphisms constitutes the foundation of that branch of mathematics usually called linear algebra. In this section we shall see how this is so.

First of all we have to make clear what we understand by a linear algebra.

A linear algebra is an algebraic structure or system in the sense used before, that is, a set of elements with a set of operations defined on it. It is, though, a more complicated system than those previously described for, on the one hand, its elements are more sophisticated mathematical entities and, on the other hand, it includes one additional operation to those mentioned in the definition of a vector space. At this time, the new operation we will refer to (it is not so new, after all, for we met it several times before) is a product of mappings or mapping composition.

Only in the next definition and there for the last time we shall return to the operational notation of Chapter 10. For the sake of "preserving the form" we distinguish between $\oplus$ and $+$ and between $\odot$ and $\cdot$. In addition, we indicate with the symbol $\otimes$ the composition of mappings.

Definition 11.18 An algebraic system $L_{\mathrm{A}} = \{T, F, \oplus, \odot, \otimes, +, \cdot\}$ is called a linear algebra over a field F if it satisfies following axioms:

1 $\{T, F, \oplus, \odot, +, \cdot\}$ is a vector space over the field F.
2 For any three elements $t_1, t_2, t_3 \in T$ and scalars $h, k \in F$,

(a) if $t_1, t_2 \in T$ then $t_2 \otimes t_1 \in T$;
(b) $t_1 \otimes (t_2 \otimes t_3) = (t_1 \otimes t_2) \otimes t_3$;
(c) $t_1 \otimes (h \odot t_2 \oplus k \odot t_3) = h \odot t_1 \otimes t_2 \oplus k \odot t_1 \otimes t_3$ and $(h \odot t_2 \oplus k \odot t_3) \otimes t_1 = h \odot t_2 \otimes t_1 \oplus k \odot t_3 \otimes t_1$.

The first axiom does not require special comment. All properties enumerated under 2 refer to the product operation $\otimes$. The operation is required to be closed [2(a)], associative [2(b)], and bilinear [2(c)].

In the context of Definition 11.18 we now analyze the set of all linear mappings from a vector space into itself. That such a set constitutes a linear algebra is proved in the following theorem. We return here to our standard, simplified system of symbols.

Theorem 11.10 Let T be the set of all linear transformations $t\colon V \to V$. Then T forms a linear algebra.

Proof We have to show that, in effect, T gives rise to a structure which satisfies postulates 1, 2(a), 2(b), and 2(c) of Definition 11.18. First, that T forms a vector space follows immediately from Remark 11.1. Second, that T is closed under the operation of composition follows from what has been stated after Definition 11.4. In particular, for $t_1\colon V \to V$, $t_2\colon V \to V$; $v^1, v^2 \in V$; and $k_1, k_2 \in F$ we have

$$\begin{aligned}(t_2 t_1)(k_1 v^1 + k_2 v^2) &= t_2\big(t_1(k_1 v^1 + k_2 v^2)\big)\\ &= t_2\big(k_1 t_1(v^1) + k_2 t_1(v^2)\big)\\ &= k_1 t_2\big(t_1(v^1)\big) + k_2 t_2\big(t_1(v^2)\big).\end{aligned}$$

Thus if t_1 and t_2 are linear mappings, so is $t_2 t_1$. Third, the associativity of the product of mappings was proved in Theorem 5.3. Finally, let us show that the composition of mappings is bilinear. Consider any transformations $t_1, t_2, t_3 \in T$ and a vector $v \in V$. Then, for any scalars $h, k \in F$ we have

$$\big(t_1(ht_2 + kt_3)\big)(v) = h(t_1 t_2)(v) + k(t_1 t_3)(v) = (ht_1 t_2 + kt_1 t_3)(v),$$

and analogously,

$$\begin{aligned}\big((ht_2 + kt_3)t_1\big)(v) &= (ht_2 + kt_3)\big(t_1(v)\big)\\ &= h(t_2 t_1)(v) + k(t_3 t_1)(v)\\ &= (ht_2 t_1 + kt_3 t_1)(v).\end{aligned}$$

Therefore, satisfying all the postulates of Definition 11.18, the set T of all linear transformations $t\colon V \to V$ forms a linear algebra.

At this stage we have acquired most of the basic elements of the theory of linear mappings we shall be using later. Our preoccupation is now to find a convenient operational device by means of which a linear transformation can be described. The answer to this question will occupy our attention in Chapter 12.

EXERCISES

E.11.1 Show that the following mappings are linear: (a) $t\colon R^3 \to R$, such that $t(x_1, x_2, x_3) = 3x_1 + 2x_2 - x_3$; (b) $t\colon R^3 \to R^3$, such that $t(x_1, x_2, x_3) = (x_1 + x_2, 2x_2, x_3)$; (c) $t\colon R^2 \to R^2$, such that $t(x_1, x_2) = (x_1 \cos\alpha - x_2 \sin\alpha, x_1 \sin\alpha + x_2 \cos\alpha)$.

E.11.2 Show that the set S of all mappings from a vector space U into a vector space V is itself a vector space, and that the subset of all linear mappings from U into V is a subspace of S.

E.11.3 Show that a linear transformation t is an isomorphism if and only if $K_t = [\theta]$.

E.11.4 Show that if t is an isomorphism from U onto V, then t^{-1} is an isomorphism from V onto U.

E.11.5 Show that if $t_1\colon U \to V$ is an isomorphism and $t_2\colon V \to W$ is an isomorphism, then also $t_2 t_1\colon U \to W$ is an isomorphism.

E.11.6 Prove the kernel of a linear transformation $t\colon U \to V$ is a convex set.

Hint: Show that if u^1, u^2 map into θ, so does $\lambda u^1 + (1 - \lambda)u^2$ for $\lambda \in [0, 1]$; or is this trivially true? Explain.

E.11.7 Let $t\colon R^3 \to R^3$ be defined by $t(x_1, x_2, x_3) = (5x_1, x_1 - x_2, x_1 + x_2 + x_3)$. Show that t is nonsingular.

Hint: Since we must have $t(x_1, x_2, x_3) = (0, 0, 0)$, K_t is the solution space of $5x_1 = 0$, $x_1 - x_2 = 0$, $x_1 + x_2 + x_3 = 0$; show thus that $K_t = [\theta]$.

E.11.8 Let $t\colon R^3 \to R^3$ be defined by $t(x_1, x_2, x_3) = (3x_1 + 6x_3, x_1 + 2x_2, x_1 + 2x_3)$. Is this mapping nonsingular?

Hint: Refer to the dimension of K_t.

E.11.9 Let V be the space of all polynomials with real coefficients, and degree equal to or less than one, $t(x) = a + bx$. Furthermore, let q_1 and q_2 be two functionals defined by $q_1(t(x)) = \int_0^1 t(x)\,dx$ and $q_2(t(x)) = \int_0^2 t(x)\,dx$, respectively. Which is the basis $\{t_1(x), t_2(x)\}$ of the space V, dual to $\{q_1, q_2\}$ of V^*?

Hint: The solution is $\{2 - 2x, -\frac{1}{2} + x\}$.

E.11.10 Let $q\colon R^2 \to R$ be the linear functional defined by $q(x_1, x_2) = 2x_1 - x_2$, and find the transpose of the following linear transformations on R^2: (a) $t(x_1, x_2) = (0, x_2)$; (b) $t(x_1, x_2) = (x_1 - x_2, 2x_1 - x_2)$; (c) $t(x_1, x_2) = (x_2, x_1)$.

PROBLEMS

P.11.1 Let S be the set of all nonsingular linear mappings on an n-dimensional vector space V. Prove that S is a group.

P.11.2 Prove Remark 11.11.

P.11.3 In the "Keynesian" system shown in Fig. 11.3, c means consumption, i investment, g public expenditure, y income, and the operator D stands for the time derivative of the succeeding variable. The constant coefficients

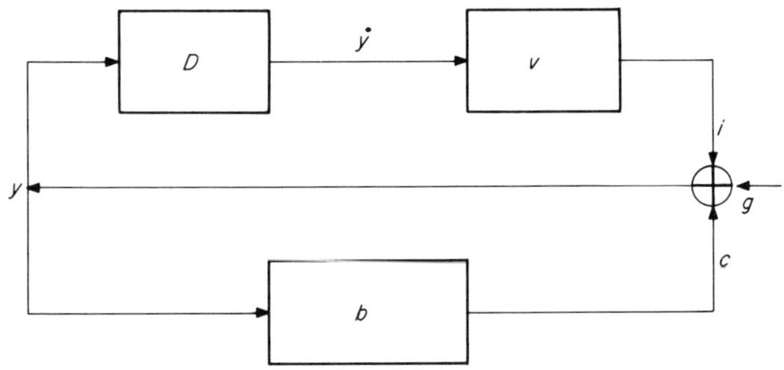

Figure 11.3

v and b are the customary accelerator and marginal propensity to consume, respectively. (a) Show that the consumption and investment functions are linear mappings and give the explicit formulas for $c = t_1(y)$ and $i = t_2(y)$, respectively. (b) Show that the overall transformation of the system is a linear mapping too, and find the formula for $y = t(g)$.

P.11.4 Show that the set of all linear mappings from the n-dimensional space U into the m-dimensional space V is a space of dimension mn.

P.11.5 Let U and V be two vector spaces and t: $U \to V$ a linear mapping. Show that $(t^T)^T = t$, and interpret this equation.

NOTES

See Notes at the end of Chapter 10.

CHAPTER 12

MATRICES

In this chapter we shall pursue two different aims. First, we shall attempt to discover what a matrix means when the concept is scrutinized within the framework of the theory of linear transformations. The results obtained in this way will greatly simplify achieving our second objective. Second, we shall survey the more traditional fields of matrix operations and matrix classification. On one hand, it is assumed the reader is acquainted with both the elementary rules of matrix algebra and the main properties of the more common matrix types. On the other hand, the elucidation of the relationship between linear mappings and matrices makes it possible to shorten considerably this part of the discussion, by simply extending to matrices the results already obtained for linear transformations in Chapter 11.

12.1 Operational Description of Linear Mappings

In Chapter 11, a linear transformation $t : U \to V$ was often accompanied by some specific algebraic rule which described to us the image $t(u) \in V$ of a vector $u \in U$. This method of describing t was general enough. However, it neither provided us with a complete and detailed description of t nor would it guarantee us the uniqueness of the representation.

It is a fact, however, that having been given a pair of bases, one for the space U, the other for the space V, it is possible to provide a complete description of t. And, what is most important, such a description is unique for any chosen pair of bases.

Theorem 12.1 Let $t : U \to V$ be a linear transformation and assume $d(U) = n$ and $d(V) = m$. Then, for a fixed pair of bases $\{u^1, \ldots, u^n\}$ and $\{v^1, \ldots, v^m\}$, t determines a unique set of mn scalars a_{ij} $(i = 1, \ldots, m; j = 1, \ldots, n)$. Conversely, such a set of scalars determines a unique linear transformation $t : U \to V$.

Proof The set $\{u^1, \ldots, u^n\}$ is the chosen basis of U. For some $x_j \in F$ a given $u \in U$ has a unique representation $u = \sum_{j=1}^n x_j u^j$. Hence, we have also $t(u) = \sum_{j=1}^n x_j t(u^j)$. The set $\{v^1, \ldots, v^m\}$ is the chosen basis of V. Each of the n vectors $t(u^j)$ is uniquely represented by a linear combination of the elements of the basis of V. For mn scalars $a_{11}, \ldots, a_{1n}, \ldots, a_{m1}, \ldots, a_{mn}$ we can write

$$t(u^1) = \sum_{i=1}^m a_{i1} v^i, \ldots, t(u^n) = \sum_{i=1}^m a_{in} v^i.$$

The mapping t is now completely and, with respect to the chosen pair of bases, uniquely described. For the image of u we have

$$t(u) = \sum_{j=1}^n x_j t(u^j) = x_1 t(u^1) + \cdots + x_n t(u^n).$$

So,

$$t(u) = x_1 \sum_{i=1}^m a_{i1} v^i + \cdots + x_n \sum_{i=1}^m a_{in} v^i = \sum_{j=1}^n a_{1j} x_j v^1 + \cdots + \sum_{j=1}^n a_{mj} x_j v^m$$

$$= \sum_{i=1}^m \left(\sum_{j=1}^n a_{ij} x_j \right) v^i.$$

Thus, the image of u under t is described by the n x_i and the mn a_{ij}. The former set of scalars expresses the vector u in terms of the basis $\{u^1, \ldots, u^n\}$. The latter sets of scalars expresses the image of u in terms of the basis $\{v^1, \ldots, v^m\}$. Again let $\{u^1, \ldots, u^n\}$ and $\{v^1, \ldots, v^m\}$ be the chosen bases for U and V, respectively. Then, conversely, any set of mn scalars $\{a_{11}, \ldots, a_{1n}, \ldots, a_{m1}, \ldots, a_{mn}\}$ determines a linear transformation $t : U \to V$ specified by $t(u^1) = \sum_{i=1}^m a_{i1} v^i, \ldots, t(u^n) = \sum_{i=1}^m a_{in} v^i$. Now since any $u \in U$ is uniquely representable as a linear combination of the elements in the basis $\{u^1, \ldots, u^n\}$, the transformation t is defined for every vector in U.

For operational reasons it is convenient to put some order in the mn scalars a_{ij} by means of which the transformation $t: U \to V$ has been described. The usual way of doing so is to order the mn scalars in a rectangular array of m rows and n columns,

$$\begin{pmatrix} a_{11} & \cdots & a_{1n} \\ \vdots & & \vdots \\ a_{m1} & \cdots & a_{mn} \end{pmatrix}.$$

Now, suppose the vector $u \in U$ is given by its coordinates $x_1, \ldots, x_n$ relative to the basis $\{u^1, \ldots, u^n\}$ while the vector $t(u) \in V$ is specified by its coordinates $y_1, \ldots, y_m$ relative to the basis $\{v^1, \ldots, v^m\}$. The mapping $u \to t(u)$ is then completely determined by describing the way in which the mn scalars a_{ij} transform the coordinates of u into the coordinates of $t(u)$. The vectors u and $t(u)$ can be written, relative to the corresponding basis, as

$$u = \begin{pmatrix} x_1 \\ \vdots \\ x_n \end{pmatrix}, \qquad t(u) = \begin{pmatrix} y_1 \\ \vdots \\ y_m \end{pmatrix},$$

and the mapping $u \to t(u)$ may be expressed by

$$\begin{pmatrix} a_{11} & \cdots & a_{1n} \\ \vdots & & \vdots \\ a_{m1} & \cdots & a_{mn} \end{pmatrix} \begin{pmatrix} x_1 \\ \vdots \\ x_n \end{pmatrix} = \begin{pmatrix} y_1 \\ \vdots \\ y_m \end{pmatrix}$$

with the agreement that $y_i = \sum_{j=1}^{n} a_{ij} x_j$ for each $i = 1, \ldots, m$.

An important corollary of Theorem 12.1 is given as the following remark.

Remark 12.1 Let $t: V \to V$ be a linear mapping and $\{v^1, \ldots, v^n\}$ a fixed basis of V. Then t is completely and uniquely described, relative to the chosen basis, by a set of n^2 scalars a_{ij}.

Following our previous convention, it would be convenient to consider the n^2 scalars a_{ij} as forming a square array of n rows and n columns.

12.2 Matrix over a Field

Given a linear mapping from a vector space to a vector space and a pair of fixed bases (one basis is enough if $t: V \to V$), we can describe it by a rectangular array of mn scalars (a square array of n^2 scalars when $t: V \to V$). Knowing now the meaning of the mentioned scalars *vis à vis* the linear mapping, that is, the linear-operator character of the array, we can give it a name.

Definition 12.1 A rectangular array of elements of a field F consisting of m rows and n columns is called an $m \times n$ matrix over F, and denoted

$$A = [a_{ij}] = \begin{pmatrix} a_{11} & \cdots & a_{1n} \\ \vdots & & \vdots \\ a_{m1} & \cdots & a_{mn} \end{pmatrix}.$$

We shall call a_{ij} the typical element of the matrix A. The indices i and j denote the row entry and the column entry, respectively.

Since a matrix represents a linear transformation (it is precisely a detailed description of a linear transformation) it is clear that we can learn much about the mathematical properties of the latter by examining the mathematical properties of the former. This will be one of our tasks in our study.

12.3 Basic Operations with Matrices

Suppose we have two linear mappings t_1 and t_2. Furthermore, suppose t_1 and t_2 are represented by the matrices A and B, respectively. A first important consequence of Theorem 12.1 is that it provides us with a simple way of performing the operations $t_1 + t_2$, kt_1, for $k \in F$, $t_2 t_1$ and $t_1 t_2$, by executing the corresponding matrix operations $A + B$, kA, BA, and AB, respectively.

Henceforth, the dimension $m \times n$ of a matrix will be called the order of the matrix.

Definition 12.2 Let both $A = [a_{ij}]$ and $B = [b_{ij}]$ be $m \times n$ matrices. Then $A = B$ if and only if $a_{ij} = b_{ij}$, for every $i = 1, \ldots, m$ and every $j = 1, \ldots, n$.

Sum of Matrices

The sum of the mappings t_1 and t_2 exists only if they are both defined for the same spaces. Accordingly, the sum of the matrices A and B is defined only if they are of the same order.

Definition 12.3 Let $A = [a_{ij}]$, $B = [b_{ij}]$ and $S = [s_{ij}]$ be three $m \times n$ matrices. Then $A + B = S$ if and only if $a_{ij} + b_{ij} = s_{ij}$ for every pair of indices (i, j).

Example

$$\begin{pmatrix} 2 & 0 & -1 \\ 5 & 3 & 0 \\ 0 & 4 & 0 \end{pmatrix} + \begin{pmatrix} 1 & 3 & 1 \\ -5 & 4 & 0 \\ 1 & -2 & 3 \end{pmatrix} = \begin{pmatrix} 3 & 3 & 0 \\ 0 & 7 & 0 \\ 1 & 2 & 3 \end{pmatrix}.$$

Scalar Multiple of a Matrix

Definition 12.4 Let $A = [a_{ij}]$ be an $m \times n$ matrix and k an element of F. Then the k-multiple of A is the $m \times n$ matrix $S = [s_{ij}]$ with $s_{ij} = ka_{ij}$ for every pair (i, j).

Example

$$3\begin{pmatrix} 3 & 0 & 1 \\ \frac{1}{3} & 2 & -5 \\ 1 & 0 & 0 \end{pmatrix} = \begin{pmatrix} 9 & 0 & 3 \\ 1 & 6 & -15 \\ 3 & 0 & 0 \end{pmatrix}.$$

Product of Matrices

Given t_1 and t_2, the existence of a composite mapping $t_2 t_1$ requires t_1 to map into the domain of t_2. Analogously, the existence of $t_1 t_2$ requires t_2 to map into the domain of t_1. And in general, we know $t_2 t_1 \neq t_1 t_2$.

Consider for instance the mappings $t_1 : V \to W$ and $t_2 : U \to V$ with $d(U) = m, d(V) = p$ and $d(W) = n$, respectively. The composite transformation $t_1 t_2 : U \to W$, thus, exists. If t_1 and t_2 are represented by the $p \times n$ matrix B and the $m \times p$ matrix A, then the composite mapping $t_1 t_2$ is represented by the $m \times n$ matrix AB. A condition for AB to be defined is that A and B be "conformable for multiplication," that is, that A have a number of columns equal to the number of rows of B.

In general, given two matrices A and B of order $m \times n$ and $p \times q$, respectively, AB is defined if $n = p$ and BA is defined if $q = m$. This corresponds to what was said regarding the linear mappings. Moreover, even if AB and BA are both defined, in general $AB \neq BA$.

Definition 12.5 Let $A = [a_{ij}]$ and $B = [b_{jk}]$ be two matrices of order $m \times p$ and $p \times n$, respectively. Then $AB = C$, where $C = [c_{ik}]$ is an $m \times n$ matrix with element $c_{ik} = \sum_{j=1}^{p} a_{ij} b_{jk}$, for every $i = 1, \ldots, m$ and $k = 1, \ldots, n$.

Example 1

$$AB = \begin{pmatrix} a_{11} & a_{12} & a_{13} \\ a_{21} & a_{22} & a_{23} \end{pmatrix} \begin{pmatrix} b_{11} & b_{12} \\ b_{21} & b_{22} \\ b_{31} & b_{32} \end{pmatrix}$$

$$= \begin{pmatrix} a_{11}b_{11} + a_{12}b_{21} + a_{13}b_{31} & a_{11}b_{12} + a_{12}b_{22} + a_{13}b_{32} \\ a_{21}b_{11} + a_{22}b_{21} + a_{23}b_{31} & a_{21}b_{12} + a_{22}b_{22} + a_{23}b_{32} \end{pmatrix}$$

$$= \begin{pmatrix} \sum_{j=1}^{3} a_{1j}b_{j1} & \sum_{j=1}^{3} a_{1j}b_{j2} \\ \sum_{j=1}^{3} a_{2j}b_{j1} & \sum_{j=1}^{3} a_{2j}b_{j2} \end{pmatrix} = \begin{pmatrix} c_{11} & c_{12} \\ c_{21} & c_{22} \end{pmatrix} = C.$$

Note that the product BA is also defined.

Example 2

$$AB = \begin{pmatrix} 1 & 2 \\ 3 & 4 \\ 5 & 6 \end{pmatrix} \begin{pmatrix} 0 & 1 \\ 2 & 0 \end{pmatrix} = \begin{pmatrix} 1 \cdot 0 + 2 \cdot 2 & 1 \cdot 1 + 2 \cdot 0 \\ 3 \cdot 0 + 4 \cdot 2 & 3 \cdot 1 + 4 \cdot 0 \\ 5 \cdot 0 + 6 \cdot 2 & 5 \cdot 1 + 6 \cdot 0 \end{pmatrix}$$

$$= \begin{pmatrix} 0 + 4 & 1 + 0 \\ 0 + 8 & 3 + 0 \\ 0 + 12 & 5 + 0 \end{pmatrix} = \begin{pmatrix} 4 & 1 \\ 8 & 3 \\ 12 & 5 \end{pmatrix}.$$

Note that the product BA is not defined.

Example 3

$$AB = \begin{pmatrix} 1 & 2 \\ 3 & 4 \end{pmatrix} \begin{pmatrix} a & b \\ c & d \end{pmatrix} = \begin{pmatrix} a + 2c & b + 2d \\ 3a + 4c & 3b + cd \end{pmatrix},$$

$$BA = \begin{pmatrix} a & b \\ c & d \end{pmatrix} \begin{pmatrix} 1 & 2 \\ 3 & 4 \end{pmatrix} = \begin{pmatrix} a + 3b & 2a + 4b \\ c + 3d & 2c + 4d \end{pmatrix}.$$

The multiplication rule is pictured in the following equation:

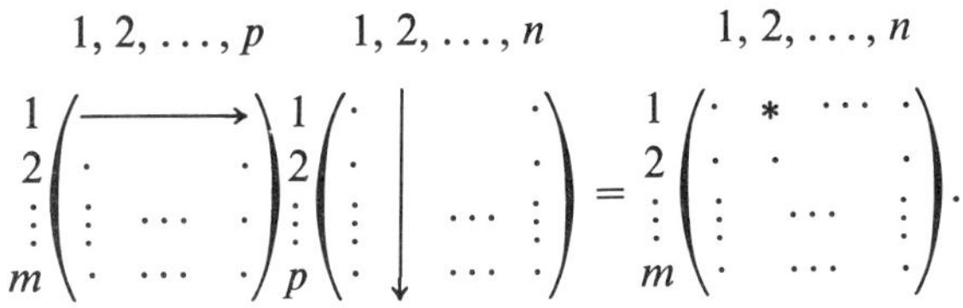

From the corresponding properties of linear transformations we can now derive a series of properties of matrix operations. These are the subject of the three following remarks.

Remark 12.2 Let A, B, and C be $m \times n$ matrices and let Z represent the $m \times n$ zero matrix. Then

(1) $A + B = B + A$;
(2) $A + (B + C) = (A + B) + C$;
(3) for any $k \in F$, $k(A + B) = kA + kB$;
(4) for any $k_1, k_2 \in F$, $(k_1 + k_2)A = k_1 A + k_2 A$;
(5) $A + Z = A$;
(6) $A + (-1)A = Z$;
(7) $0.A = Z$.

Remark 12.3 Let the matrices A, B, and C be conformable for the corresponding operations. Then

(1) $A(B + C) = AB + AC$;
(2) $(A + B)C = AC + BC$;
(3) $(AB)C = A(BC)$.

Remark 12.4 Let the matrices A, B, and C be conformable for multiplication and let Z be the zero matrix. Then

(1) $AB = Z$ does not imply $A = Z$ or $B = Z$;
(2) $AB = AC$ does not imply $B = C$;
(3) in general, as already seen, $AB \neq BA$.

Partitioning of Matrices and Operations with Partitioned Matrices

Sometimes it is convenient to distinguish certain sets of elements of a matrix as forming submatrices within the given matrix. This process is called partitioning of a matrix.

To illustrate the idea let A be the 5×4 matrix,

$$A = \begin{pmatrix} a_{11} & a_{12} & a_{13} & a_{14} \\ a_{21} & a_{22} & a_{23} & a_{24} \\ a_{31} & a_{32} & a_{33} & a_{34} \\ a_{41} & a_{42} & a_{43} & a_{44} \\ a_{51} & a_{52} & a_{53} & a_{54} \end{pmatrix} = \left(\begin{array}{ccc|c} a_{11} & a_{12} & a_{13} & a_{14} \\ a_{21} & a_{22} & a_{23} & a_{24} \\ \hline a_{31} & a_{32} & a_{33} & a_{34} \\ a_{41} & a_{42} & a_{43} & a_{44} \\ a_{51} & a_{52} & a_{53} & a_{54} \end{array}\right).$$

The matrix A has been partitioned into the 2×3 matrix A_{11}, the 2×1 matrix A_{12}, the 3×3 matrix A_{21}, and the 3×1 matrix A_{22}:

$$A_{11} = \begin{pmatrix} a_{11} & a_{12} & a_{13} \\ a_{21} & a_{22} & a_{23} \end{pmatrix}, \qquad A_{12} = \begin{pmatrix} a_{14} \\ a_{24} \end{pmatrix},$$

$$A_{21} = \begin{pmatrix} a_{31} & a_{32} & a_{33} \\ a_{41} & a_{42} & a_{43} \\ a_{51} & a_{52} & a_{53} \end{pmatrix}, \qquad A_{22} = \begin{pmatrix} a_{34} \\ a_{44} \\ a_{54} \end{pmatrix}.$$

Hence, we can write A as a matrix the elements of which are the submatrices A_{11}, A_{12}, A_{21}, and A_{22}, that is,

$$A = \begin{pmatrix} A_{11} & A_{12} \\ A_{21} & A_{22} \end{pmatrix}.$$

If two or more matrices are partitioned in such a manner that the resulting submatrices are conformable for the corresponding operations, then our definitions of addition and multiplication can be easily extended. We illustrate this point next.

Block Addition

The case of sum of matrices is very simple. Definition 12.3 is applied to the elements (submatrices) A_{ij} and B_{ij} of the partitions. The essential requirement is, of course, that A_{ij} and B_{ij} be conformable for the corresponding (i, j).

If two matrices A and B are partitioned as follows,

$$A = \begin{pmatrix} A_{11} & A_{12} \\ A_{21} & A_{22} \\ A_{31} & A_{32} \end{pmatrix}, \qquad B = \begin{pmatrix} B_{11} & B_{12} \\ B_{21} & B_{22} \\ B_{31} & B_{32} \end{pmatrix},$$

with A_{ij} and B_{ij} conformable for each pair (i, j), then the sum can be written

$$A + B = \begin{pmatrix} A_{11} + B_{11} & A_{12} + B_{12} \\ A_{21} + B_{21} & A_{22} + B_{22} \\ A_{31} + B_{31} & A_{32} + B_{32} \end{pmatrix}.$$

Block Multiplication

Definition 12.5 can be straightforwardly extended to cover the case of block multiplication, provided the involved matrices have been partitioned in a way which guarantees conformability for multiplication. Let A and B be partitioned as indicated below and suppose that A_{ij} and B_{ij} are conformable for every pair (i, j).

$$A = \begin{pmatrix} A_{11} & A_{12} & A_{13} \\ A_{21} & A_{22} & A_{23} \end{pmatrix}, \qquad B = \begin{pmatrix} B_{11} & B_{12} \\ B_{21} & B_{22} \\ B_{31} & B_{32} \end{pmatrix}.$$

Then,

$$AB = \begin{pmatrix} A_{11}B_{11} + A_{12}B_{21} + A_{13}B_{31} & A_{11}B_{12} + A_{12}B_{22} + A_{13}B_{32} \\ A_{21}B_{11} + A_{22}B_{21} + A_{23}B_{31} & A_{21}B_{12} + A_{22}B_{22} + A_{23}B_{32} \end{pmatrix}.$$

Examples Let A, B, and C be partitioned as shown below. Find (1) $A + B$; (2) AC:

$$A = \left(\begin{array}{cc|c} 2 & 0 & 3 \\ 1 & 3 & 5 \\ \hline 1 & 5 & 4 \end{array}\right), \qquad B = \left(\begin{array}{cc|c} 5 & 1 & 2 \\ 7 & -2 & 1 \\ \hline 0 & 0 & -4 \end{array}\right),$$

$$C = \left(\begin{array}{ccc|c} 1 & 2 & 3 & 0 \\ 4 & 5 & 6 & 1 \\ \hline 1 & 1 & 1 & 1 \end{array}\right).$$

$$(1)\quad A + B = \begin{pmatrix} \begin{pmatrix} 2+5 & 0+1 \\ 1+7 & 3-2 \end{pmatrix} & \begin{pmatrix} 3+2 \\ 5+1 \end{pmatrix} \\ (1+0 \quad 5+0) & (4-4) \end{pmatrix} = \left(\begin{array}{cc|c} 7 & 1 & 5 \\ 8 & 1 & 6 \\ \hline 1 & 5 & 0 \end{array}\right).$$

$$
(2)\quad AC = \left(\begin{array}{c|c}
\begin{pmatrix}2 & 0\\1 & 3\end{pmatrix}\begin{pmatrix}1 & 2 & 3\\4 & 5 & 6\end{pmatrix} + \begin{pmatrix}3\\5\end{pmatrix}(1\ \ 1\ \ 1) & \begin{pmatrix}2 & 0\\1 & 3\end{pmatrix}\begin{pmatrix}0\\1\end{pmatrix} + \begin{pmatrix}3\\5\end{pmatrix}(1) \\ \hline
(1\ \ 5)\begin{pmatrix}1 & 2 & 3\\4 & 5 & 6\end{pmatrix} + (4)\,(1\ \ 1\ \ 1) & (1\ \ 5)\begin{pmatrix}0\\1\end{pmatrix} + (4)(1)
\end{array}\right)
$$

$$
= \left(\begin{array}{c|c}
\begin{pmatrix}2 & 4 & 6\\13 & 17 & 21\end{pmatrix} + \begin{pmatrix}3 & 3 & 3\\5 & 5 & 5\end{pmatrix} & \begin{pmatrix}0\\3\end{pmatrix} + \begin{pmatrix}3\\5\end{pmatrix} \\ \hline
(21\ \ 27\ \ 33) + (4\ \ 4\ \ 4) & (5) + (4)
\end{array}\right)
$$

$$
= \left(\begin{array}{c|c}
\begin{pmatrix}5 & 7 & 9\\18 & 22 & 26\end{pmatrix} & \begin{pmatrix}3\\8\end{pmatrix} \\ \hline
(25\ \ 31\ \ 37) & (9)
\end{array}\right) = \begin{pmatrix}5 & 7 & 9 & 3\\18 & 22 & 26 & 8\\25 & 31 & 37 & 9\end{pmatrix}.
$$

Of course, we could have achieved identical result by multiplying A and C without partition.

In Chapter 11 we discussed some special mappings. Since a linear mapping can be represented by the operator matrix, it is necessary now to dedicate some time to investigating the properties of the special matrices associated with those special transformations.

12.4 The Transpose of a Matrix

Suppose we are given a linear mapping $t : U \to V$. The transpose of t, as we know, is a mapping $t^{\mathrm{T}} : V^* \to U^*$, where V^* and U^* are the dual spaces of V and U, respectively. This follows from Definition 11.17. Now, if t has been represented relative to a pair of bases by a matrix $A = [a_{ij}]$, we wonder about the form of the matrix which would represent t^{T}, and which will be called the transpose of A, denoted A^{T}. We answer this question in the following theorem:

Theorem 12.2 Let $\{u^1, \ldots, u^n\}$ and $\{v^1, \ldots, v^m\}$ be bases of the spaces U and V, respectively, and $\{q_1, \ldots, q_n\}$ and $\{r_1, \ldots, r_m\}$ the dual bases of the dual spaces U^* and V^* respectively. Then, if $t : U \to V$ is represented, relative to $\{u^1, \ldots, u^n\}$ and $\{v^1, \ldots, v^m\}$, by the matrix $A = [a_{ij}]$, its transpose $t^{\mathrm{T}} : V^* \to U^*$ is represented, relative to $\{r_1, \ldots, r_m\}$ and $\{q_1, \ldots, q_n\}$ by the matrix $A^{\mathrm{T}} = [b_{ij}]$, where $b_{ij} = a_{ji}$.

Proof Let us define, following the pattern of Theorem 11.6, $q_k(u^j) = \delta_{kj}$ and $r_i(v^k) = \delta_{ik}$. Now, if relative to $\{u^1, \ldots, u^n\}$ and $\{v^1, \ldots, v^m\}$ the $m \times n$ matrix $A = [a_{ij}]$ represents $t : U \to V$, then t^{T}, $V^* \to U^*$ is represented, relative to $\{r_1, \ldots, r_m\}$ and $\{q_1, \ldots, q_n\}$, by an $n \times m$ matrix which will be called $A^{\mathrm{T}} = [b_{jk}]$. We have, hence,

$$t^{\mathrm{T}}(r_1) = \sum_{k=1}^{n} b_{k1} q_k, \ldots, t^{\mathrm{T}}(r_m) = \sum_{k=1}^{n} b_{km} q_k,$$

that is,

$$t^{\mathrm{T}}(r_i) = \sum_{k=1}^{n} b_{ki} q_k \qquad \text{for} \qquad i = 1, \ldots, m.$$

Consider now the vector u^j of the basis of U. According to our previous result and the definition of t^{T}, we have

$$t^{\mathrm{T}}(r_i(u^j)) = r_i(t(u^j)) = r_i \sum_{k=1}^{m} a_{kj} v^k = \sum_{k=1}^{m} a_{kj} r_i(v^k) = \sum_{k=1}^{m} a_{kj} \delta_{ik} = a_{ij}.$$

Now, on the other hand,

$$t^{\mathrm{T}}(r_i(u^j)) = \left(\sum_{k=1}^{n} b_{ki} q_k\right)(u^j) = \sum_{k=1}^{n} b_{ki} \delta_{kj} = b_{ji}.$$

Thus $b_{ji} = a_{ij}$. Therefore, the element of the place (i, j) in the $n \times m$ matrix A^{T} is equal to the element of the place (j, i) of the $m \times n$ matrix A.

The importance of Theorem 12.2 lies in the fact that, knowing the matrix A which represents the mapping $t : U \to V$ relative to a pair of bases, it specifies in general terms the form of the matrix which represents the transpose mapping $t^{\mathrm{T}} : V^* \to U^*$ relative to the dual bases.

Our results may now be condensed in the following definition:

Definition 12.6 Let $A = [a_{ij}]$ be an $m \times n$ matrix. Then, the transpose of A, denoted $A^{\mathrm{T}} = [b_{ij}]$, is an $n \times m$ matrix with $b_{ij} = a_{ji}$.

Example

$$A = \begin{pmatrix} 1 & 2 & 3 & 4 \\ 5 & 6 & 7 & 8 \\ 9 & 10 & 11 & 12 \end{pmatrix}, \qquad A^{\mathrm{T}} = \begin{pmatrix} 1 & 5 & 9 \\ 2 & 6 & 10 \\ 3 & 7 & 11 \\ 4 & 8 & 12 \end{pmatrix}.$$

From the corresponding properties of the transpose of a linear transformation and the operator character of a transpose matrix we derive the following results.

Remark 12.5 Let A, B, and C be matrices over F. Then

(1) $(A^{\mathrm{T}})^{\mathrm{T}} = A$;
(2) $(kA)^{\mathrm{T}} = kA^{\mathrm{T}}$, for any $k \in F$;
(3) $(A + B)^{\mathrm{T}} = A^{\mathrm{T}} + B^{\mathrm{T}}$;
(4) $(AB)^{\mathrm{T}} = B^{\mathrm{T}}A^{\mathrm{T}}$;
(5) $(ABC)^{\mathrm{T}} = C^{\mathrm{T}}B^{\mathrm{T}}A^{\mathrm{T}}$.

From what was said so far it is clear that a vector $a = (a_1, \ldots, a_n)$ may be regarded as a $1 \times n$ matrix. If we take this standpoint, then we also have to recognize the existence of a transpose a^{T} of a. Since a^{T} is an $n \times 1$ matrix we indicate this fact by writing it as a column vector.

Thus, for the sake of consistency, we have to distinguish in the future between "row vectors" and "column vectors." The distinction is not irrelevant and has some bearing in relation to vector multiplication. We shall agree on the following convention. If both a and b are row vectors we shall express the dot product (see Chapter 10) as

$$ab^{\mathrm{T}} = (a_1 \quad \cdots \quad a_n)\begin{pmatrix} b_1 \\ \vdots \\ b_n \end{pmatrix} = \sum_{i=1}^{n} a_i b_i .$$

The dot product is of course a scalar or, if we like to think in terms of matrices, a 1×1 matrix.

On the other hand, $b^{\mathrm{T}}a$ is an $n \times n$ matrix

$$\begin{pmatrix} b_1 \\ \vdots \\ b_n \end{pmatrix}(a_1 \quad \cdots \quad a_n) = \begin{pmatrix} b_1 a_1 & \cdots & b_1 a_n \\ \vdots & & \vdots \\ b_n a_1 & \cdots & b_n a_n \end{pmatrix}$$

so that clearly $ab^{\mathrm{T}} \neq b^{\mathrm{T}}a$.

12.5 Square Matrices

Among the different types of matrices, the class of square matrices will play a very relevant role in the following chapters.

Definition 12.7 A matrix $A = [a_{ij}]$ is said to be an nth-order square matrix if it consists of n rows and n columns.

In Theorem 11.10 it was shown that the set T of all linear transformations $t: V \to V$ forms a linear algebra L_{A}. Consider $d(V) = n$ and suppose M represents the set of all nth-order square matrices. We already know (see Theorem 12.1 and especially Remark 12.1) that it is possible to construct a one-to-one mapping from T to M. Since this mapping is an isomorphism, as will be proved in the next theorem, the set M also forms a linear algebra M_A.

Theorem 12.3 The set M of all nth-order square matrices over F forms a linear algebra M_A isomorphic to the linear algebra L_A of all linear mappings on the n-dimensional space V over F.

Proof Assume $\{v^1, \dots, v^n\}$ is a basis of V. Then, relative to this basis, we have for $t_1, t_2 \in L_A$ the representative matrices $A, B \in M$, respectively. This association forms a one-to-one mapping since the matrix representation is unique relative to that basis.

Suppose

$$t_1(v^i) = \sum_{j=1}^{n} a_{ij} v^j, \qquad i = 1, \dots, n \qquad \text{and} \qquad t_2(v^i) = \sum_{j=1}^{n} b_{ij} v^j, \qquad i = 1, \dots, n.$$

First, for every $i = 1, \dots, n$,

$$(t_1 + t_2)(v^i) = t_1(v^i) + t_2(v^i) = \sum_{j=1}^{n} a_{ij} v^j + \sum_{j=1}^{n} b_{ij} v^j = \sum_{j=1}^{n} (a_{ij} + b_{ij}) v^j.$$

Thus, for $t_1 + t_2 \in L_A$ we have $A + B \in M$. Second, for every $i = 1, \dots, n$ and $k \in F$,

$$kt_1(v^i) = k \sum_{j=1}^{n} a_{ij} v^j = \sum_{j=1}^{n} ka_{ij} v^j.$$

Hence, for $kt_1 \in L_A$ we have $kA \in M$. Finally, for $i = 1, \dots, n$,

$$(t_2 t_1)(v^i) = t_2(t_1(v^i)) = \sum_{j=1}^{n} b_{kj} \left(\sum_{k=1}^{n} a_{ik} v^j \right) = \sum_{j=1}^{n} \sum_{k=1}^{n} a_{ik} b_{kj} v^j.$$

So, for $t_2 t_1 \in L_A$ we have $AB \in M$.

Therefore the mapping preserves all the operations and hence is an isomorphism; that is, M is isomorphic to the linear algebra L_A and constitutes itself, therefore, a linear algebra M_A.

Definition 12.8 Let $A = [a_{ij}]$ be an nth-order square matrix. Then the elements a_{ii} are called diagonal elements and the sum $\sum_{i=1}^{n} a_{ii} = \mathrm{tr}(A)$ is said to be the trace of A.

The proofs of the following properties are suggested as exercises.

Remark 12.6 Let A and B be nth-order square matrices. Then

(1) $\mathrm{tr}(A) = \mathrm{tr}(A^T)$;
(2) $\mathrm{tr}(kA) = k\,\mathrm{tr}(A)$;
(3) $\mathrm{tr}(A + B) = \mathrm{tr}(A) + \mathrm{tr}(B)$;
(4) $\mathrm{tr}(AB) = \mathrm{tr}(BA)$.

The Identity Matrix

Definition 12.9 An nth-order square matrix $I = [\delta_{ij}]$, where δ_{ij} is the Kronecker delta, is called an identity matrix.

An nth-order identity matrix represents, of course, the identity transformation in an n-dimensional space. The properties enumerated below follow immediately. The subscripts used indicate the order of the corresponding identity matrix. For instance,

$$I_2 = \begin{pmatrix} 1 & 0 \\ 0 & 1 \end{pmatrix}.$$

Remark 12.7 Let $A = [a_{ij}]$ be an $m \times n$ matrix. Then

(1) $I_m A = AI_n = I_m AI_n = A$;
(2) for k matrices I_n, $I_n + \cdots + I_n = kI_n$ and $I_n \cdot \cdots \cdot I_n = I_n^k = I_n$;
(3) $\operatorname{tr}(I_n) = n$.

Nonsingular and Inverse Matrices

Nonsingular and inverse transformations have been studied in the previous chapters. Since a nonsingular matrix is square—it describes a transformation $t: V \to V$—we can extend to it the properties we already know for nonsingular transformations. This is a consequence of the isomorphic character of L_A and M_A. We shall prove, however, some theorems concerning these types of matrices. Redundant though it might appear, the procedure serves to show us an alternative way to deal with these problems.

Definition 12.10 Let $A = [a_{ij}]$ be an nth-order square matrix. Then A is said to be nonsingular if and only if a matrix A^{-1} exists, called the inverse of A, such that $AA^{-1} = A^{-1}A = I$.

Theorem 12.4 The inverse of any square matrix A, if it exists, is unique.

Proof Suppose B and C are two inverses of A. This would mean $AB = CA = I$. Now $CAB = C(AB) = (CA)B = CI = IB$. Therefore $C = B$.

Theorem 12.5 Let A and B be two nonsingular square matrices such that the product AB is defined. Then $(AB)^{-1} = B^{-1}A^{-1}$.

Proof By definition we have $(AB)^{-1}(AB) = (AB)(AB)^{-1} = I$. Also, on the one hand, $(B^{-1}A^{-1})(AB) = B^{-1}(A^{-1}A)B = B^{-1}IB = B^{-1}B = I$ and on the other hand, $(AB)(B^{-1}A^{-1}) = A(B^{-1}B)A^{-1} = AIA^{-1} = AA^{-1} = I$. Therefore $(AB)^{-1} = B^{-1}A^{-1}$.

12.6 Miscellaneous Square Matrices

In the following chapters and in relation to somewhat different problems we shall find it helpful to work with some special types of square matrices. Here we list a few of these matrices and mention, in general without providing the corresponding proofs, some of their relevant properties.

Scalar Matrices

Definition 12.11 An nth-order square matrix $A = [k\delta_{ij}]$, where $k \in F$ and δ_{ij} is the Kronecker delta, is called a scalar matrix.

Example

$$A = \begin{pmatrix} 5 & 0 & 0 \\ 0 & 5 & 0 \\ 0 & 0 & 5 \end{pmatrix}, \qquad B = \begin{pmatrix} k & 0 \\ 0 & k \end{pmatrix}.$$

Remark 12.8 Let a square matrix A and a scalar matrix $B = [k\delta_{ij}]$ bc comformable for multiplication. Then $AB = BA = kA$.

That the identity matrix I is a special case of scalar matrix is obvious. Any scalar matrix $A = [k\delta_{ij}]$ can be written $A = kI$.

Diagonal Matrices

Definition 12.12 An nth-order square matrix A is said to be diagonal if and only if $a_{ij} = 0$ whenever $i \neq j$.

Example

$$A = \begin{pmatrix} 2 & 0 \\ 0 & 1 \end{pmatrix}, \qquad B = \begin{pmatrix} a & 0 \\ 0 & b \end{pmatrix}, \qquad C = \begin{pmatrix} 1 & 0 & 0 \\ 0 & 2 & 0 \\ 0 & 0 & 3 \end{pmatrix}.$$

To refer to a diagonal matrix we shall also use the notation $D = \text{diag}\,(a_{11}, \ldots, a_{nn})$.

Remark 12.9 Let A and B be diagonal matrices of order n. Then $A + B$ and $AB = BA$ are also diagonal matrices.

Triangular Matrices

Definition 12.13 A square matrix $A = [a_{ij}]$ is said to be upper triangular (lower triangular) if $a_{ij} = 0$ whenever $i > j$ $(i < j)$. Triangular matrices with $a_{ii} = 0$ for all i are called strictly triangular.

Example Note that A and B are upper triangular and strictly upper triangular, respectively,

$$A = \begin{pmatrix} a_{11} & a_{12} & a_{13} & a_{14} \\ 0 & a_{22} & a_{23} & a_{24} \\ 0 & 0 & a_{33} & a_{34} \\ 0 & 0 & 0 & a_{44} \end{pmatrix}, \qquad B = \begin{pmatrix} 0 & a_{12} & a_{13} & a_{14} \\ 0 & 0 & a_{23} & a_{24} \\ 0 & 0 & 0 & a_{34} \\ 0 & 0 & 0 & 0 \end{pmatrix}.$$

Commutative and Anticommutative Matrices

Definition 12.14 The square matrices A and B are said to be commutative (anticommutative) if $AB = BA$ (if $AB = -BA$).

Remark 12.10 A diagonal matrix of order n is commutative to any square matrix of the same order.

As we have seen, the above result also applies to scalar matrices.

Example Note that the matrices A and B are commutative while the matrices C and D are anticommutative:

$$A = \begin{pmatrix} 1 & 2 \\ 2 & 1 \end{pmatrix}, \quad B = \begin{pmatrix} 3 & 4 \\ 4 & 3 \end{pmatrix}, \quad C = \begin{pmatrix} 0 & -i \\ i & 0 \end{pmatrix}, \quad D = \begin{pmatrix} i & 0 \\ 0 & -i \end{pmatrix}.$$

Idempotent and Nilpotent Matrices

Definition 12.15 An nth-order square matrix A is said to be idempotent if $A^2 = A \cdot A = A$. It is said to be nilpotent of index r if $A^r = Z$ and $A^{r-1} \neq Z$, where Z is the $n \times n$ zero matrix.

Example Observe that A is an idempotent matrix while B is nilpotent of index 2.

$$A = \begin{pmatrix} 1 & -1 \\ 0 & 0 \end{pmatrix}, \qquad B = \begin{pmatrix} 1 & -3 & -4 \\ -1 & 3 & 4 \\ 1 & -3 & -4 \end{pmatrix}.$$

Symmetric and Skew-Symmetric Matrices

Definition 12.16 A square matrix A is said to be symmetric (skew-symmetric) if and only if $A = A^{\mathrm{T}}$ (if and only if $-A = A^{\mathrm{T}}$).

Example The matrices A and B below are symmetric and skew-symmetric, respectively:

$$A = \begin{pmatrix} 1 & 2 & 3 \\ 2 & 4 & 6 \\ 3 & 6 & 5 \end{pmatrix}, \qquad B = \begin{pmatrix} 0 & -2 & 3 \\ 2 & 0 & 6 \\ -3 & -6 & 0 \end{pmatrix}.$$

Remark 12.11 Let $A = [a_{ij}]$ be a square matrix. Then the matrices $A + A^{\mathrm{T}} = [a_{ij} + a_{ji}]$ and $A - A^{\mathrm{T}} = [a_{ij} - a_{ji}]$ are symmetric and skew-symmetric, respectively.

It follows immediately that a skew-symmetric matrix can only have zeros in its main diagonal.

A property which directly results from Remark 12.11 is the content of the next remark.

Remark 12.12 Let A be any square matrix. Then $A = B + C$, where B is the symmetric matrix defined by $B = \frac{1}{2}(A + A^{\mathrm{T}})$ and C is the skew-symmetric matrix defined by $C = \frac{1}{2}(A - A^{\mathrm{T}})$.

Example Consider

$$A = \begin{pmatrix} 1 & 2 \\ 3 & 4 \end{pmatrix}.$$

(1) $A + A^{\mathrm{T}} = \begin{pmatrix} 2 & 5 \\ 5 & 8 \end{pmatrix}$ and $A - A^{\mathrm{T}} = \begin{pmatrix} 0 & -1 \\ 1 & 0 \end{pmatrix}$.

(2) $B = \frac{1}{2}(A + A^{\mathrm{T}}) = \begin{pmatrix} 1 & \frac{5}{2} \\ \frac{5}{2} & 4 \end{pmatrix}$,

$C = \frac{1}{2}(A - A^{\mathrm{T}}) = \begin{pmatrix} 0 & -\frac{1}{2} \\ \frac{1}{2} & 0 \end{pmatrix}$,

so that

$$B + C = A = \begin{pmatrix} 1 & 2 \\ 3 & 4 \end{pmatrix}.$$

Remark 12.13 Let A and B be nth-order symmetric matrices. Then the matrix $C = AB$ is symmetric if and only if $AB = BA$.

The following property will be rather important in the context of future discussions and it deserves a formal proof.

Theorem 12.6 Let A be an mth-order symmetric matrix and P any $m \times n$ matrix. Then the matrix $B = P^{\mathrm{T}}AP$ is symmetric.

Proof Consider $B = P^{\mathrm{T}}AP$. The transpose of B is $B^{\mathrm{T}} = (P^{\mathrm{T}}AP)^{\mathrm{T}} = P^{\mathrm{T}}A^{\mathrm{T}}(P^{\mathrm{T}})^{\mathrm{T}} = P^{\mathrm{T}}A^{\mathrm{T}}P = P^{\mathrm{T}}AP$. Therefore $B^{\mathrm{T}} = B$; that is, B is symmetric.

By simply replacing the term "symmetric" by "skew-symmetric" in Theorem 12.6 and following a similar method, one can prove the following property:

Remark 12.14 Let A be an mth-order skew-symmetric matrix and P any $m \times n$ matrix. Then the matrix $B = P^{\mathrm{T}}AP$ is skew-symmetric.

Hermitian Matrices

If a transformation is defined between complex spaces then it is clear that the representative matrix will contain elements belonging to the complex field.

Definition 12.17 A matrix $A = [z_{ij}]$ where z_{ij} is a complex number is called a complex matrix and the matrix $\bar{A} = [\bar{z}_{ij}]$, where $\bar{z}_{ij}$ is the conjugate of the element z_{ij} is called the conjugate matrix of A.

Example

$$A = \begin{pmatrix} 3+i2 & 1 & 0 \\ 7 & 5 & 4+i3 \\ 4 & 6 & 5+i2 \end{pmatrix}, \qquad \bar{A} = \begin{pmatrix} 3-i2 & 1 & 0 \\ 7 & 5 & 4-i3 \\ 4 & 6 & 5-i2 \end{pmatrix}.$$

For the definition which follows we have to consider not only the matrices A and $\bar{A}$ but also the transpose of the conjugate, $\bar{A}^{\mathrm{T}}$. The following properties follow immediately.

Remark 12.15 Let A and B be complex matrices. Then

(1) $\overline{(A+B)} = \bar{A} + \bar{B}$;
(2) $\overline{(AB)} = \bar{A}\,\bar{B}$;
(3) $(\bar{A})^{\mathrm{T}} = \overline{(A^{\mathrm{T}})}$.

Definition 12.18 Let $A = [z_{ij}]$ be a square matrix over the complex field. Then A is called Hermitian (skew Hermitian) if $A = \bar{A}^{\mathrm{T}}$ (if $A = -\bar{A}^{\mathrm{T}}$).

Example Note that the matrices A and B below are Hermitian and skew-Hermitian, respectively.

$$A = \begin{pmatrix} 3 & 1-i & 5 \\ 1+i & 1 & i \\ 5 & -i & 0 \end{pmatrix}, \qquad B = \begin{pmatrix} i & 2-i & 3 \\ -2-i & i4 & i \\ -3 & i & 0 \end{pmatrix}.$$

A skew-Hermitian matrix can only have zeros or pure imaginary numbers as diagonal elements.

The proofs of the following remarks are quite simple and left as exercises.

Remark 12.16 Let $A = [z_{ij}]$ be a complex square matrix. Then the matrices $A + \bar{A}^{\mathrm{T}} = [z_{ij} + \bar{z}_{ji}]$ and $A - \bar{A}^{\mathrm{T}} = [z_{ij} - \bar{z}_{ji}]$ are Hermitian and skew-Hermitian, respectively.

Remark 12.17 Let A be any complex square matrix. Then $A = B + C$, where B is the Hermitian matrix defined by $B = \frac{1}{2}(A + \bar{A}^{\mathrm{T}})$ and C is the skew-Hermitian matrix defined by $C = \frac{1}{2}(A - A^{\mathrm{T}})$.

EXERCISES

E.12.1 Prove Remark 12.1.

E.12.2 Prove Remark 12.2.

E.12.3 For the matrices

$$A = \begin{pmatrix} 1 & 1 & 1 \\ 0 & 1 & 1 \\ 0 & 0 & 1 \end{pmatrix}, \qquad B = \begin{pmatrix} -1 & 1 & -1 \\ 0 & -1 & 1 \\ 0 & 0 & -1 \end{pmatrix},$$

$$M = \begin{pmatrix} a & b & c \\ d & e & f \\ g & h & j \end{pmatrix}, \qquad N = \begin{pmatrix} 1 & 0 & 2 \\ 0 & 3 & 0 \\ 4 & 0 & 5 \end{pmatrix},$$

compute the indicated operations: (a) AB; (b) BA; (c) A^2; (d) $A^2 + AB - 2B$; (e) $(A - I)(B - I)$; (f) $(M + N)A - (M + N)B$.

E.12.4 For the matrices

$$A = \begin{pmatrix} 1 & 1 & 1 & 1 \\ 2 & 1 & 3 & 4 \\ 0 & 1 & 2 & 3 \end{pmatrix}, \qquad B = \begin{pmatrix} 3 & 1 & 2 & 5 \\ 0 & 2 & 1 & 0 \\ 4 & 3 & 1 & 2 \end{pmatrix},$$

$$M = \begin{pmatrix} 1 & 2 & 3 \\ 4 & 5 & 6 \\ 7 & 8 & 9 \end{pmatrix}, \qquad N = \begin{pmatrix} 1 & 2 & 3 & 4 \\ 5 & 6 & 7 & 8 \\ 9 & 10 & 11 & 12 \end{pmatrix},$$

compute the indicated operations: (a) AB^{T}; (b) $(AB^{\mathrm{T}})^{\mathrm{T}}$; (c) $B^{\mathrm{T}}A$; (d) $(B^{\mathrm{T}}A)^{\mathrm{T}}$; (e) $AB^{\mathrm{T}} - M$; (f) $B^{\mathrm{T}}A - N$; (g) $(AB^{\mathrm{T}} - M)^{\mathrm{T}}$.

E.12.5 Prove Statement (4) in Remark 12.5.

E.12.6 Show in detail that the set of all 3×3 matrices over R forms a linear algebra isomorphic to the linear algebra of all linear mappings on R^3.

Hint: Use Theorem 12.3.

E.12.7 Prove Statements (1) and (3) in Remark 12.6.

E.12.8 Give examples of a commutative and an anticommutative matrix.

E.12.9 Give two examples of idempotent matrices.

E.12.10 Prove Remark 12.12.

PROBLEMS

P.12.1 The input–output systems of Problems P.1.4 and P.1.5 (Chapter 1) can be represented by means of the $n \times n$ input–output matrix $A = [a_{ij}]$, and the $m \times n$ input–output matrix $B = [b_{ij}]$, respectively. Using the concept of "independent subeconomies" explain the meaning of the input–output systems described by each of the matrices:

(a) $$A = \begin{pmatrix} A_{11} & A_{12} \\ A_{21} & A_{22} \end{pmatrix}, \qquad A = \begin{pmatrix} A_{11} & A_{12} \\ Z & A_{22} \end{pmatrix}, \qquad A = \begin{pmatrix} A_{11} & Z \\ Z & A_{22} \end{pmatrix},$$

where every submatrix indicated A_{ij} $(i, j = 1 \text{ or } 2)$ is different from Z, A_{11} is $n_1 \times n_1$, A_{22} is $n_2 \times n_2$, and $n_1 + n_2 = n$; and by each of the matrices:

(b) $$B = \begin{pmatrix} B_{11} & B_{12} \\ B_{21} & B_{22} \end{pmatrix}, \qquad B = \begin{pmatrix} B_{11} & B_{12} \\ Z & B_{22} \end{pmatrix}, \qquad B = \begin{pmatrix} B_{11} & Z \\ Z & B_{22} \end{pmatrix},$$

where every submatrix indicated B_{ij} $(i, j = 1 \text{ or } 2)$ is different from Z, B_{11} is $m_1 \times n_1$, B_{22} is $m_2 \times n_2$, $m_1 + m_2 = m$, and $n_1 + n_2 = n$.

P.12.2 A vector $v \in R^n$ is positive if all its components are positive; it is semipositive if all its components are nonnegative and at least one is positive; it is nonnegative if all its components are nonnegative. Point out some properties of the associated matrix A, such that the linear mapping $t: R^n \to R^n$ maps positive vectors into (a) only positive vectors; (b) only semipositive vectors; (c) nonnegative vectors.

P.12.3 Let A and B be $m \times n$ matrices. Give examples to show that if the two matrices have identical number of vectors in their maximal linearly independent set of row vectors, then the corresponding linear transformations have the same rank.

P.12.4 Suppose we want to test the hypothesis that some given T joint observations on $1 + r$ variables $y, x_1, \ldots, x_r, (y_1, \ldots, y_T, x_{11}, \ldots, x_{1r}, \ldots, x_{T1}, \ldots, x_{Tr})$ are related by a process responding to the linear equation $y = \alpha + \beta_1 x_1 + \beta_2 x_2 + \cdots + \beta_r x_r + \varepsilon$, where ε is a random disturbance.

Note that for each observation period $t = 1, \ldots, T$ we can write $y_t = \beta_0 x_{t0} + \beta_1 x_{t1} + \beta_2 x_{t2} + \cdots + \beta_r x_{tr} + e_t$ where $x_{t0} \equiv 1$, and thus $\beta_0 = \alpha$. We obtain the system of T equations

$$\begin{pmatrix} y_1 \\ \vdots \\ y_T \end{pmatrix} = \begin{pmatrix} x_{10} & \cdots & x_{1r} \\ \vdots & & \vdots \\ x_{T0} & \cdots & x_{Tr} \end{pmatrix} \begin{pmatrix} \beta_0 \\ \vdots \\ \beta_r \end{pmatrix} + \begin{pmatrix} e_1 \\ \vdots \\ e_T \end{pmatrix},$$

that is,

$$y = XB + e.$$

The ordinary least squares estimator of B, b, is computed by minimizing the square of the disturbance e, that is, by solving for b after calculating $dS/dB = \theta$ where $S = e^{\mathrm{T}}e = (y - Xb)^{\mathrm{T}}(y - Xb)$.

Prove that $b = (X^{\mathrm{T}}X)^{-1}X^{\mathrm{T}}y$.

Hint: Goldberger [1].

P.12.5 In the preceding problem we may write $b = R^{-1}X^{\mathrm{T}}y$, and thus $Rb = X^{\mathrm{T}}y$. Write explicitly the matrix R and the expression for each element r_{ij}. Interpret the case when R is singular.

NOTES

The classical work on matrices is by Gantmacher [1]. A quite different though very interesting approach to the study of matrices can be found in the work of Bellman [1]. These two works are obligatory reference books for economists. So is Karlin's book [1].

An important reference book for computation methods is that of Faddeeva [1]. For a comprehensive treatment of matrices which covers many topics of real interest to economists the reader is referred to Pease [1]. See also the Notes to Chapter 10.

CHAPTER 13

RANK, EQUIVALENCE, AND SIMILARITY

Three main problems will concern us in this chapter. In the first place we shall study the notion of the rank of a matrix. As one might imagine, this idea is closely related to the concept of the rank of a linear transformation already analyzed in Chapter 11.

The second and the third questions on which we shall focus our attention are related to the first problem, as we shall have an opportunity to see. As we know, a real or complex linear transformation and a pair of fixed bases determine a unique representative matrix. This means, however, that an infinite number of matrices will be associated with one and the same linear mapping, each of them corresponding to a different pair of bases chosen from a set of infinite possibilities. The question to be answered may be formulated as follows: What is the common feature of all these matrices related to the same mapping? We shall provide a general answer first. Afterward, we shall particularize it to the case of a square matrix.

13.1 The Rank of a Matrix

The rank of a linear transformation has been defined as the dimension of its range space. Now we begin our study with the definition of the rank of a matrix. The relation between the two concepts will be the subject of a theorem soon.

For the sake of notational consistency, we shall distinguish first the row rank of A, denoted $r'(A)$, the column rank of A, denoted $r''(A)$, and the rank of A, denoted $r(A)$. Very soon, however, we shall be able to simplify our notation.

Definition 13.1 Let $A = [a_{ij}]$ be an $m \times n$ matrix. Then $r'(A)$, the row rank of A, is the number of vectors in any maximal independent subset of row vectors of A and $r''(A)$, the column rank of A, is the number of vectors in any maximal independent subset of column vectors of A.

Example Consider the matrix,

$$A = \begin{pmatrix} 1 & 2 & 3 \\ 0 & 2 & 2 \\ 1 & 4 & 5 \end{pmatrix}.$$

Observe the row vectors A^i, $A^1 = (1, 2, 3)$, $A^2 = (0, 2, 2)$, $A^3 = (1, 4, 5)$. Since $A^3 = A^1 + A^2$ we have $r'(A) = 2$. For the column vectors A_j we have $A_1 = (1, 0, 1)^{\mathrm{T}}$, $A_2 = (2, 2, 4)^{\mathrm{T}}$, and $A_3 = (3, 2, 5)^{\mathrm{T}}$. Here too, since $A_3 = A_1 + A_2$, we have $r''(A) = 2$.

In the previous example we observed that for the given A, the row and column ranks were equal. In the next theorem we show that this did not happen by chance.

Theorem 13.1 Let $A = [a_{ij}]$ be any $m \times n$ matrix. Then $r'(A) = r''(A)$.

Proof Call $A^1, \ldots, A^m$ the m row vectors of A and $A_1, \ldots, A_n$ its n column vectors. The set $\{A^1, \ldots, A^m\}$ generates a subspace of F_n, the space of n-tuples of elements of F. We call it the row space of A. The set $\{A_1, \ldots, A_n\}$ on the other hand, generates a subspace of F_m, the space of m-tuples of elements of F. We call it the column space of A.

Assume now $r'(A) = r$ and select a basis $\{v^1, \ldots, v^r\} = \{(b_{11}, \ldots, b_{1n}), \ldots, (b_{r1}, \ldots, b_{rn})\}$. For suitable scalars we can then write

$$A^1 = \sum_{j=1}^{r} k_{1j} v^j; \ldots ; A^m = \sum_{j=1}^{r} k_{mj} v^j,$$

that is,

$$\begin{aligned} A^1 &= (k_{11}b_{11}, \ldots, k_{11}b_{1n}) + \cdots + (k_{1r}b_{r1}, \ldots, k_{1r}b_{rn}) \\ &\vdots \\ A^m &= (k_{m1}b_{11}, \ldots, k_{m1}b_{1n}) + \cdots + (k_{mr}b_{r1}, \ldots, k_{mr}b_{rn}). \end{aligned}$$

As a consequence we also have

$$\begin{aligned} a_{1j} &= k_{11}b_{1j} + \cdots + k_{1r}b_{rj} \\ &\vdots \\ a_{mj} &= k_{m1}b_{1j} + \cdots + k_{mr}b_{rj}, \end{aligned}$$

that is, for each column vector A_j, $j = 1, \ldots, n$,

$$A_j = b_{1j}(k_{11}, \ldots, k_{m1})^{\mathrm{T}} + \cdots + b_{rj}(k_{1r}, \ldots, k_{mr})^{\mathrm{T}}.$$

Every column vector of A can thus be expressed as a linear combination of the r vectors $(k_{11}, \ldots, k_{m1})^{\mathrm{T}}, \ldots, (k_{1r}, \ldots, k_{mr})^{\mathrm{T}}$. If S^{c} represents the column space, then clearly $d(S^{\mathrm{c}}) \leq r$. Equivalently, $r''(A) \leq r'(A)$.

With an analogous method, but beginning with the column vectors $A_1, \ldots, A_n$, we can show that the row space S^{r} has $d(S^{\mathrm{r}}) \leq r$, where r is now the column rank $r''(A)$. From here it follows that $r'(A) \leq r''(A)$. The two results together lead to the conclusion that $r'(A) = r''(A)$.

We can now summarize the results of our investigation in a new definition:

Definition 13.2 Let $A = [a_{ij}]$ be an $m \times n$ matrix. Then the rank of A, denoted $r(A)$, is the value common to its row and column rank.

We can pass now to studying the relationship between the range $r(t) = d(R_t)$ of a linear transformation t and the range $r(A)$ of its associated matrix A.

Theorem 13.2 Let t: $U \to V$ be a linear transformation and A its $m \times n$ representative matrix for a chosen pair of bases. Then $r(t) = r(A)$.

Proof Assume $\{u^1, \ldots, u^n\}$ is a basis of U. Any vector $v \in R_t$ is a linear combination of the column vectors of A. Therefore, any maximal independent subset of column vectors of A is a basis of R_t, that is, $r(A) = r(t)$.

Theorem 13.3 Let $A = [a_{ij}]$ be an $m \times n$ matrix. Then $r(A) = r(A^{\mathrm{T}})$.

Proof From Chapter 11 we know that $r(t) = r(t^{\mathrm{T}})$. From Theorem 13.2 we clearly have $r(t) = r(A)$ and $r(t^{\mathrm{T}}) = r(A^{\mathrm{T}})$. Therefore $r(A) = r(A^{\mathrm{T}})$.

Theorem 12.3 together with Theorem 13.2 lead immediately to the following result:

Remark 13.1 Let A be an nth-order nonsingular matrix. Then $r(A) = n$.

We close this section with two properties of the product of matrices.

Theorem 13.4 Let A and B be conformable for multiplication. Then $r(AB) \leq r(A)$ and $r(AB) \leq r(B)$.

Proof Let S^c represent the column space of A and $S^{c'}$ the column space of AB. Since each column of AB is represented by a linear combination AB_j, it lies in S^c. Thus $S^{c'}$ is contained in S^c and therefore, $r(AB) \leq r(A)$.

Let S^r be the row space of B and $S^{r'}$ the row space of AB. Each row of AB is represented by a linear combination A^iB. Hence, it lies on the row space of B and so $S^{r'}$ is contained in S^r. Therefore, $r(AB) \leq r(B)$.

The next theorem, the proof of which is left as an exercise, can be approached with a reasoning close to that used in Theorem 13.4.

Theorem 13.5 Let B be an nth-order matrix with $r(B) = n$. Let A and C be two arbitrary matrices of order $m \times n$ and $n \times p$, respectively. Then, $r(AB) = r(A)$ and $r(BC) = r(C)$.

13.2 Matrix Equivalence

We have pointed out more than once that a matrix provides a unique description of a linear mapping, but that this uniqueness should be understood in relative terms, that is, in relation to a fixed basis or pair of bases. This simply means that the same linear transformation will in general be represented by a different matrix when a different basis or pair of bases is chosen.

In spite of this fact, we may very well expect some "essential feature" common to all matrices associated with a particular mapping. And it is from this point of view that all these matrices may be regarded as equivalent since each of them serves to represent, for different bases, the same transformation.

Furthermore, since all equivalent matrices will share the "essential feature" mentioned above, we may become interested in the possibility of discovering a canonical matrix for the equivalence class under consideration, that is, a very simple, and if possible unique, matrix representing that "essential feature" as its dominant property. To these questions we shall dedicate the present and the next section.

Elementary Operations and Elementary Matrices

The rank of a matrix, as we know, is given by the number of row or column vectors in a maximal linearly independent subset of its row or column vectors. Operations which when performed on the rows and columns of a matrix do not modify the rank of it are usually called elementary operations. They alter, so to speak, the surface of the matrix but leave intact this "essential feature" which is the rank.

There are three basic operations which preserve the rank when performed on the rows of a matrix. They are called elementary row operations and are (1) interchange of two rows, denoted H_{ik}; (2) multiplication of a row by a scalar, denoted $H_i(c)$; (3) addition to a row of another row multiplied by a scalar, denoted $H_{ik}(c)$.

Example Consider the matrix A below and notice that the matrices B, C, and D have been obtained by means of performing on A the elementary row operations H_{23}, $H_2(3)$, and $H_{12}(-2)$, respectively,

$$A = \begin{pmatrix} 1 & 4 & 7 \\ 2 & 5 & 8 \\ 3 & 6 & 9 \end{pmatrix}, \qquad B = \begin{pmatrix} 1 & 4 & 7 \\ 3 & 6 & 9 \\ 2 & 5 & 8 \end{pmatrix},$$

$$C = \begin{pmatrix} 1 & 4 & 7 \\ 6 & 15 & 24 \\ 3 & 6 & 9 \end{pmatrix}, \qquad D = \begin{pmatrix} -3 & -6 & -9 \\ 2 & 5 & 8 \\ 3 & 6 & 9 \end{pmatrix}.$$

Analogous operations can of course be performed on the columns of a matrix. The three elementary column operations are (1) interchange of two columns, denoted K_{hj}; (2) multiplication of a column by a scalar, denoted $K_j(c)$; (3) addition to a column of another column multiplied by a scalar, denoted $K_{hj}(c)$.

The inverse of an elementary operation is simply an operation which undoes the effects of the former. For the inverses of the three elementary row operations and the three elementary column operations we write H_{ik}^{-1}, $H_i^{-1}(c)$, $H_{ik}^{-1}(c)$ and K_{hj}^{-1}, $K_j^{-1}(c)$, $K_{hj}^{-1}(c)$, respectively.

Definition 13.3 An elementary row (column) matrix is a matrix which results from the application of an elementary row (column) operation to the identity matrix.

Since there are three types of elementary row (column) operations it is clear that there are three pure types of elementary row (column) matrices. The elementary matrices will be represented by the same letters as the operations.

Example Observe the identity matrix I_3 below and the elementary row matrices obtained by means of the elementary row operations H_{12}, $H_3(c)$, $H_{23}(c)$:

$$I_3 = \begin{pmatrix} 1 & 0 & 0 \\ 0 & 1 & 0 \\ 0 & 0 & 1 \end{pmatrix}; \quad \begin{pmatrix} 0 & 1 & 0 \\ 1 & 0 & 0 \\ 0 & 0 & 1 \end{pmatrix}; \quad \begin{pmatrix} 1 & 0 & 0 \\ 0 & 1 & 0 \\ 0 & 0 & c \end{pmatrix}; \quad \begin{pmatrix} 1 & 0 & 0 \\ 0 & 1 & c \\ 0 & 0 & 1 \end{pmatrix}.$$

An nth-order identity matrix has a linearly independent set of n row (column) vectors. As we can see, elementary row (column) operations are unable to create linear dependence. The following result is thus straightforward.

Theorem 13.6 Every elementary matrix is nonsingular.

The usefulness of elementary matrices is that they provide a means of performing elementary operations on a given matrix. When a matrix is subject not to one but to a sequence of elementary operations this is especially the case.

Remark 13.2 An elementary row (column) operation can be performed on an $m \times n$ matrix A by premultiplying (postmultiplying) A by the corresponding $m \times m$ $(n \times n)$ elementary matrix.

Example Observe the matrix A below. By performing the row operation H_{13} on I_3 and premultiplying then A by the resulting elementary matrix we are able to perform H_{13} on A.

$$A = \begin{pmatrix} 1 & 2 & 3 \\ 4 & 5 & 6 \\ 7 & 8 & 9 \end{pmatrix}, \qquad I_3 = \begin{pmatrix} 1 & 0 & 0 \\ 0 & 1 & 0 \\ 0 & 0 & 1 \end{pmatrix}, \qquad H_{13} = \begin{pmatrix} 0 & 0 & 1 \\ 0 & 1 & 0 \\ 1 & 0 & 0 \end{pmatrix};$$

$$H_{13}A = \begin{pmatrix} 0 & 0 & 1 \\ 0 & 1 & 0 \\ 1 & 0 & 0 \end{pmatrix} \begin{pmatrix} 1 & 2 & 3 \\ 4 & 5 & 6 \\ 7 & 8 & 9 \end{pmatrix} = \begin{pmatrix} 7 & 8 & 9 \\ 4 & 5 & 6 \\ 1 & 2 & 3 \end{pmatrix}.$$

In general, to perform s elementary row operations on a matrix A we have

$$H^s \cdots H^3 H^2 H^1 A = PA = B,$$

where H^i is an elementary matrix.

Analogously, to perform r elementary column operations on A we have

$$AK^1 K^2 K^3 \cdots K^r = AQ = B,$$

where K^i is an elementary matrix.

Of course, a matrix can be subject to a sequence of elementary row and column operations at the same time. In such a case we write,

$$H^s \cdots H^2 H^1 A K^1 K^2 \cdots K^r = PAQ = B.$$

Row Equivalence and Canonical Form

Definition 13.4 Let A and B be two $m \times n$ matrices. Then B is said to be row equivalent to A if and only if B can be obtained by performing a finite number of elementary row operations on A.

It is easy to see that row equivalence is an equivalence relation in the sense of Chapter 2 for it satisfies the properties of reflexivity, transitivity, and symmetry.

Example Observe that B is row equivalent to A for it has been obtained by performing the operations $H_2(-1)$ and $H_{13}(-3)$ on A:

$$A = \begin{pmatrix} 3 & -5 & 18 & 9 \\ -1 & 2 & -8 & 4 \\ 2 & -4 & 16 & -8 \end{pmatrix}, \qquad B = \begin{pmatrix} 0 & 1 & -6 & 3 \\ 1 & -2 & 8 & -4 \\ 2 & -4 & 16 & -8 \end{pmatrix}.$$

A very important property of row equivalent matrices is given in the following theorem:

Theorem 13.7 Let the $m \times n$ matrices A and B be row equivalent. Then $r(A) = r(B)$.

Proof From Remark 13.2 we know we can write $H^s \cdots H^3H^2H^1A = PA = B$. From Theorem 13.6 we know every matrix H^k, $k = 1, \ldots, s$ is nonsingular. We have, on one hand, $r(H^s \cdots H^3H^2H^1A) = r(B)$. On the other hand, from Theorem 13.5 $r(H^s \cdots H^3H^2H^1A) = r(PA) = r(A)$. Therefore $r(A) = r(B)$.

An "essential feature" of equivalent matrices is thus the equality of their ranks. Given an $m \times n$ matrix A we can obtain, applying elementary row operations, a set of equivalent matrices. All have the same rank and each of them represents the same linear transformation but relative to a different pair of fixed bases. Our interest is directed now to finding a canonical form for row equivalence, that is, the simplest matrix to which we might arrive after a finite number of elementary row operations. Such a matrix is unique and, as we said in another place, contains the "essential feature" of the linear mapping. But, before discovering the canonical form, let us introduce an intermediate step.

Definition 13.5 An $m \times n$ matrix E row equivalent to a given matrix A is called an echelon matrix if it possesses the following properties: (1) If the rank of A is r, then only the first r rows of E are nonzero. (2) The first element different from zero in each nonzero row is 1; its place is always in a column to the right of the first element different from zero of any preceding row.

A row-reduced echelon form is not unique. Nevertheless, it has the important property that its number of nonzero rows is $r = r(E) = r(A)$. An echelon form provides thus an easy way to discover the rank of a given matrix.

Example Consider the matrix A of the previous example. After applying the sequence of elementary row operations $H_2(-1)$, $H_{12}(-3)$, $H_{32}(-2)$, and H_{12} one obtains the matrix E which, according to Definition 13.5, is a row-reduced echelon form of A. We can immediately see that $r(A) = r(E) = 2$. Notice that, using different operations, we would have obtained a different echelon form,

$$A = \begin{pmatrix} 3 & -5 & 18 & -9 \\ -1 & 2 & -8 & 4 \\ 2 & -4 & 16 & -8 \end{pmatrix}, \qquad E = \begin{pmatrix} 1 & -2 & 8 & -4 \\ 0 & 1 & -6 & 3 \\ 0 & 0 & 0 & 0 \end{pmatrix}.$$

By enlarging our sequence of elementary row operations we finally arrive at the unique row-equivalent matrix described in the next definition.

Definition 13.6 An $m \times n$ matrix C, row equivalent to a given matrix A, is said to be the row-reduced canonical form of A if it is an echelon matrix [properties (1) and (2) of Definition 13.5]. In addition (3) the first element different from zero in each nonzero row is the only nonzero element in its column.

Example After applying the sequence of elementary row operations $H_2(-1)$, $H_{12}(-3)$, $H_{32}(-2)$, H_{12}, and $H_{12}(2)$ to our known matrix A, we obtain its row-reduced canonical form C,

$$A = \begin{pmatrix} 3 & -5 & 18 & -9 \\ -1 & 2 & -8 & 4 \\ 2 & -4 & 16 & -8 \end{pmatrix}, \qquad C = \begin{pmatrix} 1 & 0 & -4 & 2 \\ 0 & 1 & -6 & 3 \\ 0 & 0 & 0 & 0 \end{pmatrix}.$$

Note that the requirement concerning property (3) of Definition 13.6 guarantees the uniqueness of the canonical matrix C.

An important result which was already apparent in the preceding discussion is the following:

Theorem 13.8 Let A be an $n \times n$ nonsingular matrix. Then A is row equivalent to I.

Proof Since A is nonsingular, it is equivalent to a row-reduced echelon matrix with full rank. But this matrix is by definition the identity matrix.

Column Equivalence and Canonical Form

All that has been said regarding row equivalence can be applied to column equivalence by only changing the word row for column in the preceding statements.

Definition 13.7 Let A and B be two $m \times n$ matrices. Then B is said to be column equivalent to A if and only if B can be obtained by performing a finite number of elementary column operations on A.

The next theorem is analogous to Theorem 13.7. The proof is similar and thus omitted.

Theorem 13.9 Let the $m \times n$ matrices A and B be column equivalent. Then $r(A) = r(B)$.

The parallel between row equivalence and column equivalence is obvious and all that has been said in relation to the former is applicable to the latter. We can, of course, also refer to column-reduced echelon matrices and to the column-reduced canonical form. Now we can generalize our ideas.

General Equivalence and Canonical Form

Definition 13.8 Let A and B be two $m \times n$ matrices. Then B is said to be equivalent to A, denoted $B \sim A$, if and only if B can be obtained from A by a finite number of elementary row and column operations.

The content of the following remark is equivalent to Definition 13.8.

Remark 13.3 A matrix B is equivalent to a matrix A if and only if two nonsingular matrices P and Q exist such that $B = PAQ$, where $P = H^s \cdots H^3 H^2 H^1$ and $Q = K^1 K^2 K^3 \cdots K^r$.

In studying row equivalence, we have seen that any matrix A can be reduced to a canonical form C. Since the canonical matrix C is unique, little would be obtained by continuing to manipulate its rows. But very often we can go further in the transformation of the matrix by applying also a sequence of elementary column operations. We obtain in this way the unique matrix defined below.

Definition 13.9 An $m \times n$ matrix N equivalent to a given matrix A with $r(A) = r$ is said to be the general canonical form under equivalence of A or also the normal form of A if it possesses the following properties: (1) The first r elements a_{ij}, with $i = j$, are equal to 1; (2) every other element of the matrix is zero.

From the definition it is clear that a general canonical form under equivalence contains an r-order identity matrix in the top left corner if $r < m$ and $r < n$, in the top r rows if $r < m$ and $r = n$, and in the left r columns if $r = m$ and $r < n$. It is equal to I_r if $m = n = r$.

Using Z to represent a zero matrix and by means of the notation introduced when discussing partitioned matrices, we can see that a normal form can take one of the following appearances:

$$I_r, \qquad (I_r \quad Z), \qquad \begin{pmatrix} I_r \\ Z \end{pmatrix}, \qquad \begin{pmatrix} I_r & Z \\ Z & Z \end{pmatrix}.$$

In some cases we can arrive at the normal form by using, exclusively, elementary row or column operations. However, this is not possible in general and, as a rule, the two types of elementary operations will have to be used.

Example We refer, for the last time, to the matrix A of previous examples. Its normal form N has been obtained after a sequence of elementary row operations $H_2(-1)$, $H_{12}(-3)$, $H_{32}(-2)$, H_{12}, and $H_{12}(2)$ and column operations $K_{34}(2)$, $K_{42}(-3)$, and $K_{41}(-2)$.

$$A = \begin{pmatrix} 3 & -5 & 18 & -9 \\ -1 & 2 & 8 & 4 \\ 2 & -4 & 16 & -8 \end{pmatrix}, \qquad N = \begin{pmatrix} 1 & 0 & 0 & 0 \\ 0 & 1 & 0 & 0 \\ 0 & 0 & 0 & 0 \end{pmatrix}.$$

As we already know, $r(A) = r(N) = 2$.

A special situation is that of a nonsingular matrix A. From Theorem 13.8 we know that a nonsingular matrix is row equivalent to the identity matrix. It is easy to prove, following an analogous method that a nonsingular matrix is also column equivalent to the identity matrix.

A more general and useful result (the usefulness of it will become clear in the last section of Chapter 14) is the content of the next theorem. The proof follows the same pattern as those of the previous theorems of this chapter and is left as an exercise. We shall come back to it in Chapter 14, Section 14.4.

Theorem 13.10 A matrix A is nonsingular if and only if it is equivalent to the identity matrix.

Linear Mappings and Equivalent Matrices

A matrix represents a linear transformation relative to a pair of bases. On the other hand, a matrix may be transformed, via elementary operations, into an equivalent matrix. The new matrix represents the same transformation though relative to a different pair of bases. This suggests that the effect of elementary operations on a matrix is to adjust it, to adapt it to different pairs of bases. We summarize this concept in following remark the proof of which is left as an exercise.

Remark 13.4 Two different $m \times n$ matrices A and B represent the same linear mapping $t\colon U \to V$ relative to two different pairs of bases if and only if $A \sim B$.

13.3 Matrix Similarity

The notion of equivalence, as discussed in the preceding section, is quite general for it is defined on the basis of $m \times n$ matrices. As we know, such a matrix represents a linear mapping from an n-dimensional space U into an m-dimensional space V.

A very important special case appears when we consider a linear transformation from an n-dimensional space U into itself. A mapping of this class is of course represented by an nth-order square matrix and this matrix is unique, relative to a fixed basis of U. Subjecting the matrix to one or more elementary operations, we obtain equivalent square matrices. These matrices, in turn, represent the same mapping relative to different fixed bases. We have here a special type of equivalence, special because of the class of matrices involved. It is called similarity, explained in the following definition.

Definition 13.10 Let A and B be nth-order square matrices. Then A is said to be similar to B if and only if, for some nonsingular matrix P, $B = PAP^{-1}$.

The matrix P may of course be regarded as having the same meaning as the matrix P of the previous section. But here, different than the case of equivalence in general (see, for instance, Remark 13.3) the place of Q in $B = PAQ$ is simply occupied by the inverse of P.

Remember that the trace of a square matrix is the sum of its diagonal elements, that is, for the $n \times n$ matrix A $\text{tr}(A) = \sum_{i=1}^{n} a_{ii}$. Now the following property is specific for similar matrices.

Theorem 13.11 Let A and B be similar matrices. Then $\text{tr}(A) = \text{tr}(B)$.

Proof Assume A is similar to B. Then there is a nonsingular matrix P such that $B = PAP^{-1}$. We clearly have $\text{tr}(B) = \text{tr}(PAP^{-1})$. However, by Remark 12.6. $\text{tr}(PAP^{-1}) = \text{tr}(APP^{-1}) = \text{tr}(A)$. Therefore $\text{tr}(A) = \text{tr}(B)$.

The problem of matrix similarity will reappear later. For the moment, we conclude this brief introduction to the subject by stating the analog for square matrices of Remark 13.4.

Remark 13.5 Two different nth-order square matrices A and B represent the same linear mapping $t\colon U \to U$ relative to two different bases of U if and only if A is similar to B.

EXERCISES

E.13.1 Determine, by inspection, the rank of the following matrices:

$$A = \begin{pmatrix} 3 & 9 & 12 & 15 \\ 6 & 9 & 15 & 3 \\ 3 & 6 & 9 & 6 \end{pmatrix}, \qquad B = \begin{pmatrix} \frac{1}{2} & 1 & \frac{3}{2} \\ \frac{3}{2} & 1 & \frac{1}{2} \\ 1 & \frac{1}{2} & \frac{3}{2} \end{pmatrix},$$

$$C = \begin{pmatrix} -1 & -1-2i & -1-i \\ -1 & -1-i & i \\ 0 & -i & -1-2i \end{pmatrix}.$$

E.13.2 Prove Theorem 13.5.

E.13.3 Show that elementary matrices are nonsingular.

E.13.4 Reduce each of the following matrices to the row-equivalent canonical form:

$$A = \begin{pmatrix} -2 & 0 & 3 & 1 & -1 \\ 2 & 2 & 3 & 1 & 3 \\ 0 & 2 & 6 & 2 & 2 \\ -2 & 0 & 3 & 1 & -1 \end{pmatrix}, \qquad B = \begin{pmatrix} -1 & 6 & 4 & 1 & 1 \\ 3 & -4 & -6 & -4 & -2 \\ -6 & 0 & 9 & 9 & 3 \\ 4 & -2 & -6 & -4 & -2 \end{pmatrix}.$$

E.13.5 Reduce each of the following matrices to the normal form:

$$A = \begin{pmatrix} 2 & -2 & 3 & 1 \\ -2 & 4 & 5 & 5 \\ -2 & 3 & 1 & 2 \end{pmatrix}, \qquad B = \begin{pmatrix} 3 & 1 & 1 & 5 \\ 2 & 0 & 1 & 3 \\ 8 & 2 & 3 & 13 \end{pmatrix}.$$

E.13.6 For the following matrices find (a) a row-equivalent echelon matrix; (b) the row-equivalent canonical form; (c) the column-equivalent canonical form; (d) the normal form:

$$A = \begin{pmatrix} -10 & 6 & 20 & 6 \\ 6 & -2 & 0 & 2 \\ -4 & 2 & 5 & 1 \end{pmatrix}, \qquad B = \begin{pmatrix} 2 & 2 & 7 & 6 \\ 3 & 1 & 2 & 1 \\ 5 & 3 & 9 & 7 \end{pmatrix}.$$

E.13.7 Prove that an $n \times n$ nonsingular matrix has the identity matrix as its normal form.

E.13.8 Determine whether the following statement is right or wrong: Any nonsingular matrix A is similar to its inverse A^{-1}.

E.13.9 Prove that the transpose of any elementary matrix is an elementary matrix.

E.13.10 Prove that the transpose of any nonsingular $n \times n$ matrix is nonsingular.

Hint: Use the result of E.13.9 in the proof.

PROBLEMS

P.13.1 Suppose we are told that two economies are represented by two input–output matrices of the same order, equivalent but not equal. What economic information, if any, could be drawn from the mentioned properties of the matrices?

P.13.2 Prove: If a sequence of elementary column operations reduces A to I, then it also reduces $(A, I)^{T}$ to $(I, A^{-1})^{T}$.

P.13.3 Let X be the set of all matrices. Show that the relation "matrix equivalence" on X has the following property: The collection of all equivalence matrices is a partition of X.

Hint: See P.2.3.

P.13.4 Show that Theorem 13.1 can be reformulated as follows: "for every matrix A with row space S^{r} and column space S^{c}, $d(S^{r}) = d(S^{c})$."

P.13.5 In analogy to the notion of the null space of a linear transformation define the corresponding null space of a matrix. Then prove that for any $m \times n$ matrix A, the dimension of the null space of A plus the dimension of the column space of A is equal to n.

NOTES

See Notes to Chapter 10, especially Gantmacher [1].

CHAPTER 14

DETERMINANTS AND MATRIX INVERSION

This chapter is divided into four sections, the first two of which are dedicated to the study of determinants. In Section 14.1 a self-contained theory of determinants is presented, while in Section 14.2 the concern is with evaluation of determinants.

The subjects of the remaining two sections are of a different nature. In Section 14.3 we introduce the concept and discuss a few properties of the adjoint of a square matrix. In the last section, we make use of some knowledge already gained in previous chapters to discuss and justify some methods of matrix inversion.

14.1 The Determinant Function

Determinants are defined only for square matrices and thus it is this type of matrix which we shall have in mind during the current discussion.

Definition 14.1† Let M be the set of all nth-order square matrices over the field F. A determinant is a function $\det: M \to F$, which assigns to every $A \in M$ a scalar $\det(A) \in F$, satisfying following axioms:

(1) If $A^1, \ldots, A^k, \ldots, A^n$ are the n rows of A such that for any index k and $b, c \in F$, $A^k = bB^k + cC^k$, then $\det(A^1, \ldots, A^k, \ldots, A^n) = b \cdot \det(A^1, \ldots, B^k, \ldots, A^n) + c \cdot \det(A^1, \ldots, C^k, \ldots, A^n)$.
(2) If for any index k, $A^k = A^{k+1}$, then $\det(A) = 0$.
(3) For the identity matrix I, $\det(I) = 1$.

The first of the three axioms defines a determinant as a linear function of each of the row vectors of a matrix. The second axiom establishes that a matrix having two equal adjacent rows has image zero. The meaning of axiom (3) is obvious.

Basic Properties of Determinants‡

The axioms of the preceding definition will help us now in proving several theorems concerning properties of determinants. These properties, in turn, will be useful in regard to the properties of the different operations in which determinants may be involved.

In all the theorems to be studied here, we assume the existence of a determinant function $\det: M \to F$ and focus our attention on the particular value $\det(A)$ for a given matrix $A \in M$. As in Definition 14.1, we regard the determinant value for A as a linear function of each of the rows $A^1, \ldots, A^n$ of A. Notice, however, that this is by no means the only possibility. We could have defined $\det(A)$ as linear on each of the columns $A_1, \ldots, A_n$ of A as well. Then, to be consistent, our subsequent theorems would have to consider operations involving column vectors rather than row vectors. The results, of course, would be equivalent to those discussed here.

Theorem 14.1 If a row A^k of A is multiplied by a scalar c, then $\det(A)$ is multiplied by c.

Proof From axiom (1) in Definition 14.1, if $A^k = bB^k + cC^k$ then

$$\begin{aligned}&\det(A^1, \ldots, A^k, \ldots, A^n)\\&\quad = \det(A^1, \ldots, bB^k + cC^k, \ldots, A^n)\\&\quad = b\det(A^1, \ldots, B^k, \ldots, A^n) + c\det(A^1, \ldots, C^k, \ldots, A^n).\end{aligned}$$

† Adapted from Finkbeiner [1].

‡ Theorems 14.1–14.8 in this subsection have been adapted from Finkbeiner [1].

Now setting $b = 0$ we have $A^k = cC^k$ and

$$c \det(A^1, \ldots, C^k, \ldots, A^n) = \det(A^1, \ldots, cC^k, \ldots, A^n) = \det(A^1, \ldots, A^k, \ldots, A^n).$$

Theorem 14.2 If for any row A^k of A we have $A^k = B^k + C^k$, then

$$\det(A) = \det(A^1, \ldots, B^k, \ldots, A^n) + \det(A^1, \ldots, C^k, \ldots, A^n).$$

Proof The theorem is immediately proved by writing $b = c = 1$ in axiom (1) of Definition 14.1.

Theorem 14.3 If for any row A^k of A we have $A^k = \theta$, then $\det(A) = 0$.

Proof The theorem is immediately proved by writing $b = c = 0$ in axiom (1) of Definition 14.1.

The next two theorems, though considering only special cases, are essential pieces in the proofs of Theorems 14.7 and 14.8.

Theorem 14.4 If to any row A^k of A we add a multiple of an adjacent row, then the value of $\det(A)$ does not change.

Proof Consider the rows A^k and A^{k+1}. By axiom (2) we have

$$\begin{aligned}
&\det(A^1, \ldots, A^k + cA^{k+1}, \ldots, A^n) \\
&\quad = \det(A^1, \ldots, A^k, \ldots, A^n) + \det(A^1, \ldots, cA^{k+1}, \ldots, A^n) \\
&\quad = \det(A^1, \ldots, A^k, \ldots, A^n) + c \det(A^1, \ldots, A^{k+1}, A^{k+1}, \ldots, A^n) \\
&\quad = \det(A) + 0 = \det(A).
\end{aligned}$$

Theorem 14.5 If two adjacent rows of A are interchanged, only the sign but not the value of $\det(A)$ changes.

Proof Consider the rows A^k and A^{k+1}. We have

$$\det(A) = \det(A^1, \ldots, A^k, \ldots, A^n) = \det(A^1, \ldots, A^k + A^{k+1}, \ldots, A^n),$$

by Theorem 14.4. Consider the new row $A^k + A^{k+1}$ and substract it from A^{k+1}. By the same theorem we can write

$$\begin{aligned}
\det(A) &= \det(A^1, \ldots, A^k + A^{k+1}, A^{k+1} - A^k - A^{k+1}, \ldots, A^n) \\
&= \det(A^1, \ldots, A^k + A^{k+1}, -A^k, \ldots, A^n).
\end{aligned}$$

Adding the row $-A^k$ to $A^k + A^{k+1}$, and still by Theorem 14.4, we get

$$\begin{aligned}
\det(A) &= \det(A^1, \ldots, A^k + A^{k+1} - A^k, -A^k, \ldots, A^n) \\
&= \det(A^1, \ldots, A^{k+1}, -A^k, \ldots, A^n) \\
&= -\det(A^1, \ldots, A^{k+1}, A^k, \ldots, A^n).
\end{aligned}$$

Theorem 14.6 If any two rows of A are equal then $\det(A) = 0$.

Proof Suppose we have $A^k = A^{k+2}$ in A. Then

$$\begin{aligned}\det(A^1, \ldots, A^k, A^{k+1}, A^{k+2}, \ldots, A^n)\\ = -\det(A^1, \ldots, A^k, A^{k+2}, A^{k+1}, \ldots, A^n) = 0,\end{aligned}$$

by axiom (2) of Definition 14.1. This result can be extended to rows $A^k = A^{k+i}$.

The previous three results will allow us now to generalize some properties of $\det(A)$.

Theorem 14.7 If to any row A^k of A a multiple of any other row is added, then $\det(A)$ does not change.

Proof Consider adding cA^{k+2} to A^k. Then

$$\begin{aligned}&\det(A^1, \ldots, A^k + cA^{k+2}, A^{k+1}, A^{k+2}, \ldots, A^n)\\ &\quad = -\det(A^1, \ldots, A^k + cA^{k+2}, A^{k+2}, A^{k+1}, \ldots, A^n)\\ &\quad = -\det(A^1, \ldots, A^k, A^{k+2}, A^{k+1}, \ldots, A^n)\\ &\qquad - c\det(A^1, \ldots, A^{k+2}, A^{k+2}, A^{k+1}, \ldots, A^n)\\ &\quad = \det(A) + 0 = \det(A),\end{aligned}$$

by Theorems 14.5 and 14.6. As in the previous theorem, we can generalize to adding cA^{k+i} to A^k.

Theorem 14.8 If any two rows of A are interchanged, the sign but not the value of $\det(A)$ changes.

Proof Suppose we interchange A^k and A^{k+2}. Then

$$\begin{aligned}\det(A) &= \det(A^1, \ldots, A^k, A^{k+1}, A^{k+2}, \ldots, A^n)\\ &= -\det(A^1, \ldots, A^{k+1}, A^k, A^{k+2}, \ldots, A^n)\\ &= \det(A^1, \ldots, A^{k+1}, A^{k+2}, A^k, \ldots, A^n)\\ &= -\det(A^1, \ldots, A^{k+2}, A^{k+1}, A^k, \ldots, A^n).\end{aligned}$$

The more general result of the theorem easily follows.

Clearly, Theorems 14.4 and 14.5 are only special cases of Theorems 14.7 and 14.8, respectively.

The Form of a Determinant

Up to now we have been concerned with the concept and the properties of a determinant function. Nothing has been said about the questions concerning the existence and uniqueness of such a function for each set of nth-order square matrices.

An existence theorem can be formulated and proved, in a relatively easy way, by exploiting our previous results on the properties of a determinant function. The question of the uniqueness of such a function is intimately associated with the question of the form of a determinant. It amounts to proving that an explicit form exists such that for any matrix A in the set of all $n \times n$ matrices, $\det(A)$ is uniquely determined by the elements of A.

We analyze these two questions in the following theorem.

Theorem 14.9† Let M be the set of all nth-order square matrices over F. Then a unique determinant function $\det: M \to F$ exists such that for every $A \in M$,

$$\det(A) = \sum_p (\operatorname{sgn} p) a_{1p(1)} a_{2p(2)} \cdots a_{np(n)},$$

where (1) the sum runs over all permutations $p = p(1)\, p(2) \cdots p(n) = j_1 j_2 \cdots j_n$; (2) $\operatorname{sgn} p = +1$ or $\operatorname{sgn} p = -1$, according to whether the number of inversions in a permutation is even or odd.

Proof We first assume that a determinant function exists for every n and show that it is unique. Consider the $n \times n$ identity matrix I. From axiom (3) of Definition 14.1 we can write $\det(I) = \det(e^1, \ldots, e^n) = 1$, where $e^1, \ldots, e^n$ are the unit row vectors $(1, 0, \ldots, 0), \ldots, (0, \ldots, 0, 1)$, respectively. Since $AI = A$, clearly $\det(AI) = \det(A)$. Now any row k of A can be written

$$A^k = \sum_{j=1}^{n} A^k(e^j) = \sum_{j=1}^{n} a_{kj} e^j.$$

Hence, $\det(A) = \det(A^1, \ldots, A^n) = \det(\sum_{j=1}^n a_{1j} e^j, \ldots, \sum_{j=1}^n a_{nj} e^j)$ or, using axiom (1) of Definition 14.1,

$$\begin{aligned} \det(A) &= \sum \det(a_{1j_1} e^{j_1}, a_{2j_2} e^{j_2}, \ldots, a_{nj_n} e^{j_n}) \\ &= \sum (a_{1j_1} a_{2j_2} \cdots a_{nj_n}) \det(e^{j_1}, e^{j_2}, \ldots, e^{j_n}). \end{aligned}$$

The sum is now over all sequences $j_1 j_2 \cdots j_n$ for the integers $1, 2, \ldots, n$. From Theorem 14.6 we know that the only nonzero determinants of the preceding sum are those for which the indices $j_1, j_2, \ldots, j_n$ are all different. So, we only have to sum over all permutations $p = j_1 j_2 \cdots j_n = p(1)p(2) \cdots p(n)$. Furthermore, invoking Theorem 14.8 it is easy to show that for a permutation p with an even (odd) number of inversions $\det(e^{j_1}, e^{j_2}, \ldots, e^{j_n}) = 1\ (-1)$. Thus we can write $\det(A) = \sum_p (\operatorname{sgn} p) a_{1p(1)} a_{2p(2)} \cdots a_{np(n)}$. So any determinant function is of this form.

Next we show that a function with the described form exists for every n, and that such a function satisfies the three axioms of Definition 14.1. We proceed by induction on n. First, if $n = 1$, $A = (a)$ and $\det(A) = a$. The theorem

† Adapted from Finkbeiner [1].

is trivially true. Assume it is also valid for $(n-1) \times (n-1)$ matrices, and call $|A_{ij}|$ the "signed" determinant obtained by multiplying by $(-1)^{i+j}$ the determinant of the $(n-1)$th-order matrix which remains from deleting the ith row and jth column of A. It is easy to prove by induction (the proof is left as exercise) that for the $n \times n$ matrix A,

$$\det(A) = \sum_{p} (\operatorname{sgn} p) a_{1p(1)} a_{2p(2)} \cdots a_{np(n)} = \sum_{j=1}^{n} a_{ij} |A_{ij}|.$$

This expression, as we shall show next, satisfies Definition 14.1.

Let us begin with axiom (1). Suppose we have three $n \times n$ matrices A, B, and C such that $A^i = B^i = C^i$ for $i = 1, \ldots, k-1, k+1, \ldots, n$ and $A^k = bB^k + cC^k$. For fixed $i \neq k$, $a_{ij}|A_{ij}| = bb_{ij}|B_{ij}| + cc_{ij}|C_{ij}|$. For fixed $i = k$, $a_{kj}|A_{kj}| = bb_{kj}|B_{kj}| + cc_{kj}|C_{kj}|$. Note that in the latter case we have $|A_{kj}| = |B_{kj}| = |C_{kj}|$. Thus we can write $\det(A) = \sum_{j=1}^{n} a_{ij}|A_{ij}| = \sum_{j=1}^{n} (bb_{ij}|B_{ij}| + cc_{ij}|C_{ij}|) = b\det(B) + c\det(C)$. We now consider axiom (2). Suppose $A^k = A^{k+1}$ and, for the sake of simplicity of the proof, let us choose $i \neq k$, $i \neq k+1$. Then $\det(A) = \sum_{j=1}^{n} a_{ij}|A_{ij}|$. But each determinant $|A_{ij}| = 0$ because it contains two adjacent equal rows. Thus $\det(A) = 0$. Finally, let us look at axiom (3). Consider the nth-order identity matrix I_n. We have $\det(I_n) = \sum_{j=1}^{n} a_{ij}|A_{ij}| = \det(I_{n-1}) = 1$. Therefore, there exists a unique determinant function with the prescribed form.

Special Determinants

We have studied up to now three general subjects; the basic idea of what a determinant function is, its form, and some of the essential properties of a determinant. It will be convenient, therefore, to focus our attention on some special cases, that is, on the properties of the determinants of particular types of matrices.

In the following pages we use both the notations $\det(A)$ and $|A|$ to express the determinant of a matrix A.

Theorem 14.10 Let $H^s, \ldots, H^2, H^1$ be elementary nth-order matrices and A any $n \times n$ matrix. Then $|H^s \cdots H^2 H^1 A| = |H^s| \cdots |H^2| \cdot |H^1| \cdot |A|$.

Proof Take first the matrix H^1. By definition H^1 is obtained by performing on I one of the elementary row operations H_{ik}, $H_i(c)$, or $H_{ik}(c)$. Since $|I| = 1$, we can invoke Theorems 14.8, 14.1, and 14.7 to claim that the determinant of H^1 will be $|H^1| = -1$, $|H^1| = c$, and $|H^1| = 1$, respectively. Moreover, as we know, premultiplying A by H^1 merely means performing the corresponding row operation on A. On the one hand, $|IA| = |I| \cdot |A| = |A|$. On the other hand, using for each case the corresponding theorem we get $|H^1 A| = -|A|$, $|H^1 A| = c|A|$, and $|H^1 A| = |A|$. Thus, in general,

$|H^1A| = |H^1| \cdot |A|$. Now, by induction, $|H^s \cdots H^2H^1A| = |H^s| \cdot |H^{s-1} \cdots H^2H^1A| = |H^s| \cdots |H^2| \cdot |H^1| \cdot |A|$.

Theorem 14.11 An nth-order square matrix A is nonsingular if and only if $\det(A) \neq 0$.

Proof If A is nonsingular, then it is equivalent to the identity matrix; that is, there is a sequence of elementary matrices $H^s, \ldots, H^2, H^1$ such that $H^s \cdots H^2H^1A = I$. Now we have, from the previous theorem, $|H^s \cdots H^2H^1A| = |H^s| \cdots |H^2| \cdot |H^1| \cdot |A| = |I| = 1$. Since for each $i = 1, 2, \ldots, s$ $|H^i| \neq 0$, we must have $|A| \neq 0$.

If A is singular, then it is equivalent to a row-reduced echelon matrix E with at least one zero row. $|E| = 0$ by Theorem 14.3. Thus we have $|H^s \cdots H^2H^1A| = |H^s| \cdots |H^2| \cdot |H^1| \cdot |A| = |E| = 0$. Since for each $i = 1, 2, \ldots, s$ $|H^i| \neq 0$, we must have $|A| = 0$.

Theorem 14.12 Let A and B be nth-order square matrices. Then $\det(AB) = \det(A) \cdot \det(B)$.

Proof Suppose A and B are nonsingular. Then A can be written as a product of elementary matrices $H^s \cdots H^1 \cdot I = A$ and it follows immediately from Theorem 14.10 that $|AB| = |H^s \cdots H^2H^1IB| = |H^s \cdots H^2H^1| \cdot |B| = |A| \cdot |B|$. If A is singular, so also is AB. We have then $|AB| = |A| \cdot |B| = 0$, $|B| = 0$.

The proof of the following result, left as an exercise, can be directly derived from Theorem 14.9.

Theorem 14.13 Let A be an $n \times n$ matrix. Then $|A| = |A^{\mathrm{T}}|$.

Theorem 14.14 Let A be a nonsingular square matrix. Then $|A^{-1}| = |A|^{-1}$.

Proof As we know $AA^{-1} = I$ and so $|AA^{-1}| = |I| = 1$. But $|AA^{-1}| = |A| \cdot |A^{-1}|$ by Theorem 14.12. Therefore $|A^{-1}| = |A|^{-1}$.

Theorem 14.15 Let A be similar to B. Then $|A| = |B|$.

Proof By definition of similarity there exists a nonsingular matrix P such that $B = PAP^{-1}$. From Theorem 14.12 we have $|B| = |PA| \cdot |P^{-1}| = |P| \cdot |A| \cdot |P^{-1}| = |A| \cdot |P| \cdot |P^{-1}|$. Now, from Theorem 14.14, $|P^{-1}| = |P|^{-1}$. Therefore $|B| = |A| \cdot |P| \cdot |P^{-1}| = |A| \cdot |P| \cdot |P|^{-1} = |A|$.

The following results are direct consequences of the properties of determinants and, in particular, of the theorem on the form of a determinant. The proofs are simple and suggested as exercises.

Remark 14.1 Let A be an nth-order square matrix and k any scalar. Then $|kA| = k^n|A|$.

Remark 14.2 Let A, B, C, and D be nth-order triangular, strictly triangular, Hermitian, and diagonal matrices, respectively. Then (1) $\det(A) = \prod_{i=1}^{n} a_{ii}$; (2) $\det(B) = 0$; (3) $\det(C) \in R$; $\det(D) = \prod_{i=1}^{n} a_{ii}$.

Minors and Complements

To complete this section we shall introduce some explicit definitions and generalizations of concepts which have already been presented in our discussions.

Definition 14.2 Let M_{ij} be the $(n-1) \times (n-1)$ submatrix which results from removing the ith row and jth column from the nth-order square matrix $A = [a_{ij}]$. Then (1) the determinant of M_{ij} is called the first minor or simply the minor of the element a_{ij} and denoted $\det(M_{ij}) = |M_{ij}|$; (2) the signed minor $(-1)^{1+j} \cdot |M_{ij}|$ is called the cofactor of the element a_{ij} and denoted $|A_{ij}|$.

Notice that $|M_{ij}|$ is merely the determinant of a submatrix of A but considered as if it were an independent matrix. Also note that the sign attached to $|M_{ij}|$ makes it a "piece" of $\det(A)$ in the terms of Theorem 14.9. As a matter of fact, we met the cofactor $|A_{ij}|$ in the proof of Theorem 14.9.

Example Observe the structure of the minors $|M_{11}|$, $|M_{12}|$, $|M_{22}|$, $|M_{32}|$ and the corresponding cofactors $|A_{11}|$, $|A_{12}|$, $|A_{22}|$, $|A_{32}|$ for the given matrix A.

$$A = \begin{pmatrix} a_{11} & a_{12} & a_{13} \\ a_{21} & a_{22} & a_{23} \\ a_{31} & a_{32} & a_{33} \end{pmatrix};$$

$$|M_{11}| = \begin{vmatrix} a_{22} & a_{23} \\ a_{32} & a_{33} \end{vmatrix}, \qquad |A_{11}| = |M_{11}|;$$

$$|M_{12}| = \begin{vmatrix} a_{21} & a_{23} \\ a_{31} & a_{33} \end{vmatrix}, \qquad |A_{12}| = -|M_{12}|;$$

$$|M_{22}| = \begin{vmatrix} a_{11} & a_{13} \\ a_{31} & a_{33} \end{vmatrix}, \qquad |A_{22}| = |M_{22}|;$$

$$|M_{32}| = \begin{vmatrix} a_{11} & a_{13} \\ a_{21} & a_{23} \end{vmatrix}, \qquad |A_{32}| = -|M_{32}|.$$

In the foregoing definition it is clear that the concepts of minor and cofactor have been specified relative to a single element of the matrix. A generalization of these concepts can be achieved if, instead, we define them relative to a set of elements forming a well-defined submatrix of the original matrix.

Definition 14.3 Let an $n \times n$ matrix A be partitioned into two submatrices

$$A_{j_1, j_2, \ldots, j_m}^{i_1, i_2, \ldots, i_m} \quad \text{and} \quad A_{j_{m+1}, j_{m+2}, \ldots, j_n}^{i_{m+1}, i_{m+2}, \ldots, i_n},$$

where (1) i denotes rows and j denotes columns; (2) m $(1 \leq m \leq n)$ of the row indices are arranged in the increasing order $i_1, i_2, \ldots, i_m$ and the same number of column indices are arranged in the increasing order $j_1, j_2, \ldots, j_m$ to form the first submatrix; (3) the remaining $n - m$ row indices are arranged in the increasing order $i_{m+1}, i_{m+2}, \ldots, i_n$ and the remaining column indices are arranged in the increasing order $j_{m+1}, j_{m+2}, \ldots, j_n$ to form the second matrix. Then the determinants

$$|A_{j_1, j_2, \ldots, j_m}^{i_1, i_2, \ldots, i_m}| \quad \text{and} \quad |A_{j_{m+1}, j_{m+2}, \ldots, j_n}^{i_{m+1}, i_{m+2}, \ldots, i_n}|$$

are said to be complementary minors of A and the signed minors

$$(-1)^p |A_{j_1, j_2, \ldots, j_m}^{i_1, i_2, \ldots, i_m}| \quad \text{and} \quad (-1)^q |A_{j_{m+1}, j_{m+2}, \ldots, j_n}^{i_{m+1}, i_{m+2}, \ldots, i_n}|$$

where

$$p = i_1 + i_2 + \cdots + i_m + j_1 + j_2 + \cdots + j_m$$

and

$$q = i_{m+1} + i_{m+2} + \cdots + i_n + j_{m+1} + j_{m+2} + \cdots + j_n$$

are said to be algebraic complements.

Example Let us consider the following 5×5 matrix:

$$A = \begin{pmatrix} a_{11} & a_{12} & a_{13} & a_{14} & a_{15} \\ a_{21} & a_{22} & a_{23} & a_{24} & a_{25} \\ a_{31} & a_{32} & a_{33} & a_{34} & a_{35} \\ a_{41} & a_{42} & a_{43} & a_{44} & a_{45} \\ a_{51} & a_{52} & a_{53} & a_{54} & a_{55} \end{pmatrix}.$$

Taking rows 2 and 5 and columns 1 and 3 for the first submatrix and rows 1, 3, and 4 and columns 2, 4, and 5 for the second submatrix we determine the complementary minors.

$$\left| A_{1,3}^{2,5} \right| = \begin{vmatrix} a_{21} & a_{23} \\ a_{51} & a_{53} \end{vmatrix} \quad \text{and} \quad \left| A_{2,4,5}^{1,3,4} \right| = \begin{vmatrix} a_{12} & a_{14} & a_{15} \\ a_{32} & a_{34} & a_{35} \\ a_{42} & a_{44} & a_{45} \end{vmatrix}.$$

Now $(-1)^{2+5+1+3} \cdot |A_{1,3}^{2,5}| = -A_{1,3}^{2,5}$ is the algebraic complement of $|A_{2,4,5}^{1,3,4}|$ whereas $(-1)^{1+3+4+2+4+5} \cdot |A_{2,4,5}^{1,3,4}| = -|A_{2,4,5}^{1,3,4}|$ is the algebraic complement of $|A_{1,3}^{2,5}|$.

It is easy to see that Definition 14.2 constitutes, as suggested earlier, a special case of Definition 14.3. If $m = 1$ and taking $i_1, j_1 = 1$, the submatrix $A_{j_1, j_2, \ldots, j_m}^{i_1, i_2, \ldots, i_m}$ becomes $A_1{}^1 = a_{11}$ with $|A_1{}^1| = a_{11}$. The complementary minor is the principal minor $|A_{j_2, \ldots, j_n}^{i_2, \ldots, i_n}| = |M_{11}|$ and the algebraic complement $(-1)^p|M_{11}|$, the cofactor $|A_{11}|$.

The concepts of principal minor, cofactor, complementary minor and algebraic complement will be useful in the next section, when discussing the problem of the evaluation of determinants.

14.2 Evaluation of Determinants

This section will be dedicated to a brief survey of the usual methods of evaluating determinants. Actually we shall not be concerned with strict computational matters but rather with giving a more practical interpretation to the abstract results of Theorem 14.9.

A determinant can be evaluated by direct application of our main result of Theorem 14.9. Although, in general, such a procedure will not seem to be recommendable from a practical point of view, a careful examination of it may serve to give us some insight into the concrete form of a determinant. It will serve also as an introduction to other more efficient methods.

Direct Method

By Theorem 14.9, the explicit form of det(A) is given by the expression $\sum_p (\text{sgn } p)\, a_{1p(1)}a_{2p(2)} \cdots a_{np(n)}$. The process of finding the value det(A) can be decomposed thereafter in the following steps:

(1) Form first the product of n elements, $a_{1p(1)} \cdot a_{2p(2)} \cdots a_{np(n)}$.

(2) Calculate the number of such products: $p = n! = n(n-1)(n-2) \cdots 2 \cdot 1$.

(3) Calculate the number of times that each element a_{ij} of A appears as the first, second, ..., nth element of the products, $n!/n$.

(4) Write down the $n!$ products leaving the column index open. Form the permutations of the integers $1, 2, \ldots, n$ and fill the column indices of each of the products preserving the order in which the integers appear in the corresponding permutation.

(5) Finally, count the number of inversions in each permutation and attach signs to the products.

(6) Sum the products.

Example Find the determinant of the matrix A below (see Table 14.1):

$$A = \begin{pmatrix} a_{11} & a_{12} & a_{13} \\ a_{21} & a_{22} & a_{23} \\ a_{31} & a_{32} & a_{33} \end{pmatrix}.$$

TABLE 14.1

Permutation	Inversions	Sign	Product
123	0	+	$a_{11} \cdot a_{22} \cdot a_{33}$
132	1	−	$a_{11} \cdot a_{23} \cdot a_{32}$
213	1	−	$a_{12} \cdot a_{21} \cdot a_{33}$
231	2	+	$a_{12} \cdot a_{23} \cdot a_{31}$
312	2	+	$a_{13} \cdot a_{21} \cdot a_{32}$
321	3	−	$a_{13} \cdot a_{22} \cdot a_{31}$

Finally, $\det(A) = a_{11}a_{22}a_{33} - a_{11}a_{23}a_{32} - a_{12}a_{21}a_{33} + a_{12}a_{23}a_{31} + a_{13}a_{21}a_{32} - a_{13}a_{22}a_{31}$.

Expansion Using Cofactors

From Theorem 14.9 we know that a determinant can be computed by means of a row (or column) expansion, that is, by summing the products of each element of a row (or column) by its corresponding cofactor. In the following example we illustrate how this procedure immediately follows from the direct method.

Example In the previous example we obtained $\det(A) = a_{11}a_{22}a_{33} - a_{11}a_{23}a_{32} - a_{12}a_{21}a_{33} + a_{12}a_{23}a_{31} + a_{13}a_{21}a_{32} - a_{13}a_{22}a_{31}$. Rearranging terms we can write $\det(A) = a_{11}(a_{22}a_{33} - a_{23}a_{32}) - a_{12}(a_{21}a_{33} - a_{23}a_{31}) + a_{13}(a_{21}a_{32} - a_{22}a_{31})$.

Notice this can also be written

$$\det(A) = a_{11}(-1)^2 \begin{vmatrix} a_{22} & a_{23} \\ a_{32} & a_{33} \end{vmatrix} + a_{12}(-1)^3 \begin{vmatrix} a_{21} & a_{23} \\ a_{31} & a_{33} \end{vmatrix} + a_{13}(-1)^4 \begin{vmatrix} a_{21} & a_{22} \\ a_{31} & a_{32} \end{vmatrix},$$

that is, $\det(A) = a_{11}|A_{11}| + a_{12}|A_{12}| + a_{13}|A_{13}| = \sum_{j=1}^{3} a_{1j}|A_{1j}|$.

The method can, of course, be used to evaluate determinants of higher order matrices by performing the necessary number of iterations.

Row or Column Replacement

This convenient method makes use of the property studied in Theorem 14.7 to replace a row (or column) by a new row (or column) having only one nonzero element. This facilitates the subsequent expansion of the determinant by means of the new row or column. The procedure can be reiterated as many times as necessary.

Example

$$\det(A) = \begin{vmatrix} 1 & 0 & -1 & 2 \\ 2 & 1 & 3 & -3 \\ 2 & 3 & -2 & 5 \\ 3 & 4 & 0 & 1 \end{vmatrix} = \begin{vmatrix} 1 & 0 & -1 & 2 \\ 0 & 1 & 5 & -7 \\ 0 & 3 & 0 & 1 \\ 0 & 4 & 3 & -5 \end{vmatrix}$$

$$= \begin{vmatrix} 1 & 5 & -7 \\ 3 & 0 & 1 \\ 4 & 3 & -5 \end{vmatrix} = \begin{vmatrix} 1 & 5 & -7 \\ 0 & -15 & 22 \\ 0 & 17 & 23 \end{vmatrix}$$

$$= \begin{vmatrix} -15 & 22 \\ -17 & 23 \end{vmatrix} = 29.$$

General Procedure: Laplace Expansion

It has been said in the previous section that a cofactor may be regarded as a special case of an algebraic complement. In the general case, an algebraic complement can be interpreted as the "cofactor" of a minor rather than of a single element of a matrix.

Following the idea of the previous paragraph, an alternative way to evaluate the determinant of A seems obvious. Instead of expanding $\det(A)$ by the cofactors of the elements of a row (or column), we may use in the expansion the algebraic complements of the minors of several rows (or columns).

The suggested expansion, known as the Laplace expansion, is clearly a generalization of the method which uses the cofactors of a row (or column).

The Laplace expansion of the determinant of an $n \times n$ matrix A can now be summarized in the following formula:

$$\det(A) = \sum_r (-1)^p \left| A^{i_1, \ldots, i_m}_{j_1, \ldots, j_m} \right| \cdot \left| A^{i_{m+1}, \ldots, i_n}_{j_{m+1}, \ldots, j_n} \right|,$$

where

(1) $p = i_1 + \cdots + i_m + j_1 + \cdots + j_m$;

(2) the sum $\sum_r$ extends over all combinations of n elements taken m at a time, that is,

$$r = C_n^{\ m} = \frac{n(n-1)\cdots(n-m+1)}{1 \cdot 2 \cdots m}.$$

Example We expand the same matrix of the previous example using minors of the first two rows. Note that we have $m = 2$ and $r = 4 \cdot \frac{3}{2} = 6$.

$$A = \begin{pmatrix} 1 & 0 & -1 & 2 \\ 2 & 1 & 3 & -3 \\ 2 & 3 & -2 & 5 \\ 3 & 4 & 0 & 1 \end{pmatrix};$$

$$(-1)^{1+2+1+2}|A_{1,2}^{1,2}| \cdot |A_{3,4}^{3,4}| = +\begin{vmatrix} 1 & 0 \\ 2 & 1 \end{vmatrix}\begin{vmatrix} -2 & 5 \\ 0 & 1 \end{vmatrix} = -2;$$

$$(-1)^{1+2+1+3}|A_{1,3}^{1,2}| \cdot |A_{2,4}^{3,4}| = -\begin{vmatrix} 1 & -1 \\ 2 & 3 \end{vmatrix}\begin{vmatrix} 3 & 5 \\ 4 & 1 \end{vmatrix} = +85;$$

$$(-1)^{1+2+1+4}|A_{1,4}^{1,2}| \cdot |A_{2,3}^{3,4}| = +\begin{vmatrix} 1 & 2 \\ 2 & -3 \end{vmatrix}\begin{vmatrix} 3 & -2 \\ 4 & 0 \end{vmatrix} = -56;$$

$$(-1)^{1+2+2+3}|A_{2,3}^{1,2}| \cdot |A_{1,4}^{3,4}| = +\begin{vmatrix} 0 & -1 \\ 1 & 3 \end{vmatrix}\begin{vmatrix} 2 & 5 \\ 3 & 1 \end{vmatrix} = -13;$$

$$(-1)^{1+2+2+4}|A_{2,4}^{1,2}| \cdot |A_{1,3}^{3,4}| = -\begin{vmatrix} 0 & 2 \\ 1 & -3 \end{vmatrix}\begin{vmatrix} 2 & -2 \\ 3 & 0 \end{vmatrix} = +12;$$

$$(-1)^{1+2+3+4}|A_{3,4}^{1,2}| \cdot |A_{1,2}^{3,4}| = +\begin{vmatrix} -1 & 2 \\ 3 & -3 \end{vmatrix}\begin{vmatrix} 2 & 3 \\ 3 & 4 \end{vmatrix} = +3.$$

Thus, $\det(A) = -2 + 85 - 56 - 13 + 12 + 3 = 29$.

14.3 The Adjoint of a Square Matrix

Definition 14.4 Let A be an nth-order square matrix. Then the adjoint of A, denoted adj A, is the transpose of the square matrix the elements of which are the cofactors of the elements of A.

For an $n \times n$ matrix

$$A = \begin{pmatrix} a_{11} & \cdots & a_{1n} \\ \vdots & & \vdots \\ a_{n1} & \cdots & a_{nn} \end{pmatrix};$$

the adjoint is thus the $n \times n$ matrix

$$\begin{pmatrix} |A_{11}| & \cdots & |A_{n1}| \\ \vdots & & \vdots \\ |A_{1n}| & \cdots & |A_{nn}| \end{pmatrix}.$$

Example

$$A = \begin{pmatrix} 1 & 2 & 3 \\ 0 & 1 & 2 \\ 3 & 0 & 1 \end{pmatrix}; \qquad \text{adj } A = \begin{pmatrix} 1 & -2 & 1 \\ 6 & -8 & -2 \\ -3 & 6 & 1 \end{pmatrix}.$$

An important property of the adjoint matrix is shown in the following theorem. The proof is left as an exercise and can be obtained as a generalization of the example following the theorem:

Theorem 14.16 Let A be an $n \times n$ matrix. Then $A \cdot \operatorname{adj} A = \operatorname{adj} A \cdot A = [|A|\delta_{ij}]$, where δ_{ij} is the Kronecker delta.

The matrix $[|A|\delta_{ij}]$ is, according to the notation used in Definition 12.11, a scalar matrix. The only nonzero elements are those of the main diagonal and these are all equal to the determinant of A. The following expressions are, of course, equivalent: $[|A|\delta_{ij}]$, $|A| \cdot I$, $\operatorname{diag}(|A|, \ldots, |A|)$.

Example Show that the theorem is true for a 3×3 matrix:

$$A = \begin{pmatrix} a_{11} & a_{12} & a_{13} \\ a_{21} & a_{22} & a_{23} \\ a_{31} & a_{32} & a_{33} \end{pmatrix}.$$

For the adjoint of the matrix A we have

$$\begin{pmatrix} a_{22}a_{33} - a_{23}a_{32} & a_{13}a_{32} - a_{12}a_{33} & a_{12}a_{23} - a_{13}a_{22} \\ a_{23}a_{31} - a_{21}a_{33} & a_{11}a_{33} - a_{13}a_{31} & a_{13}a_{21} - a_{11}a_{33} \\ a_{21}a_{32} - a_{22}a_{31} & a_{12}a_{31} - a_{11}a_{32} & a_{11}a_{22} - a_{12}a_{21} \end{pmatrix}.$$

It is easy to see now that every nondiagonal element of the matrix $\operatorname{adj} A \cdot A = A \cdot \operatorname{adj} A$ vanishes, whereas every diagonal element is $|A|$. We can thus write

$$A \cdot \operatorname{adj} A = \begin{pmatrix} |A| & 0 & 0 \\ 0 & |A| & 0 \\ 0 & 0 & |A| \end{pmatrix} = \det(A) \cdot \begin{pmatrix} 1 & 0 & 0 \\ 0 & 1 & 0 \\ 0 & 0 & 1 \end{pmatrix} = \operatorname{adj} A \cdot A.$$

This property of the adjoint matrix will be most significant in the context of the following section.

14.4 Methods of Matrix Inversion

Beginning in Chapter 12, we have had opportunities to study some of the properties of inverse matrices. Nothing has been said so far, however, about the methods by means of which the inverse of a given matrix can be computed.

We complete the present chapter with the introduction of three methods of matrix inversion. Here too, we must point out, our concern is not so much with the technical, computational side of the problem but rather with its theoretical implications. In particular, we would like to show how concepts and tools developed earlier can be of help to solve this new problem.

Inversion by Means of Elementary Operations

The first method we shall present here makes use of Theorem 13.10. As we know, a nonsingular matrix A is equivalent to the identity matrix I. This is the same as saying that, given A, there exist two matrices P and Q such that $PAQ = I$. Premultiplying by P^{-1} and postmultiplying by Q^{-1} we get $P^{-1}PAQQ^{-1} = P^{-1}IQ^{-1}$, that is, $A = P^{-1}Q^{-1}$. So finally, $A^{-1} = (P^{-1}Q^{-1})^{-1} = QP$.

We can take advantage of this property by remembering that P and Q represent products of elementary row matrices and elementary column matrices, respectively. In other words, $P = H^s \cdots H^2H^1$ and $Q = K^1K^2 \cdots K^r$.

The process of finding the inverse of A can be described in following scheme:

$$\begin{array}{c|c} I & A \\ \hline & I \end{array} \to \begin{array}{c} I \mid A \\ \hline \end{array} \to \begin{array}{|c} A \\ \hline I \end{array} \to \begin{array}{c|c} P & I \\ \hline & Q \end{array}.$$

(1) (2) (3) (4)

We form the array (1). Elementary row operations must be performed on I and A simultaneously, that is, on the "super row" (2). Analogously, elementary column operations must be performed on I and A simultaneously, that is, on the "super column" (3). At the end of the process, after the necessary number of elementary operations, A becomes the identity matrix and we also obtain P and Q as shown in array (4). We then have $QP = A^{-1}$.

Example

$$A = \begin{pmatrix} 1 & 2 & 3 \\ 0 & 1 & 1 \\ 3 & 6 & 10 \end{pmatrix} \qquad \begin{array}{ccc|ccc} 1 & 0 & 0 & 1 & 2 & 3 \\ 0 & 1 & 0 & 0 & 1 & 1 \\ 0 & 0 & 1 & 3 & 6 & 10 \\ \hline & & & 1 & 0 & 0 \\ & & & 0 & 1 & 0 \\ & & & 0 & 0 & 1 \end{array}$$

$$\begin{array}{ccc|ccc} 1 & 0 & 0 & 1 & 2 & 3 \\ 0 & 1 & 0 & 0 & 1 & 1 \\ -3 & 0 & 1 & 0 & 0 & 1 \\ \hline & & & 1 & 0 & 0 \\ \multicolumn{3}{c|}{H_{31}(-3)} & 0 & 1 & 0 \\ & & & 0 & 0 & 1 \end{array} \qquad \begin{array}{ccc|ccc} 1 & 0 & 0 & 1 & 2 & 3 \\ 3 & 1 & -1 & 0 & 1 & 0 \\ -3 & 0 & 1 & 0 & 0 & 1 \\ \hline & & & 1 & 0 & 0 \\ \multicolumn{3}{c|}{H_{23}(-1)} & 0 & 1 & 0 \\ & & & 0 & 0 & 1 \end{array}$$

$$\begin{array}{ccc|ccc} 1 & 0 & 0 & 1 & 2 & 0 \\ 3 & 1 & -1 & 0 & 1 & 0 \\ -3 & 0 & 1 & 0 & 0 & 1 \\ \hline & & & 1 & 0 & -3 \\ \multicolumn{3}{c|}{K_{31}(-3)} & 0 & 1 & 0 \\ & & & 0 & 0 & 1 \end{array} \qquad \begin{array}{ccc|ccc} 1 & 0 & 0 & 1 & 0 & 0 \\ 3 & 1 & 1 & 0 & 1 & 0 \\ -3 & 0 & 1 & 0 & 0 & 1 \\ \hline & & & 1 & -2 & -3 \\ \multicolumn{3}{c|}{K_{21}(-2)} & 0 & 1 & 0 \\ & & & 0 & 0 & 1 \end{array}$$

The inverse of A is now the matrix

$$A^{-1} = QP = \begin{pmatrix} 1 & -2 & -3 \\ 0 & 1 & 0 \\ 0 & 0 & 1 \end{pmatrix} \begin{pmatrix} 1 & 0 & 0 \\ 3 & 1 & -1 \\ -3 & 0 & 1 \end{pmatrix} = \begin{pmatrix} 4 & -2 & -1 \\ 3 & 1 & -1 \\ -3 & 0 & 1 \end{pmatrix}.$$

Inversion by Means of the Adjoint Method

As we know from Theorem 14.16, $A \cdot \text{adj}\, A = \text{adj}\, A \cdot A = \det(A) \cdot I$. Hence, $A \cdot \text{adj}\, A/\det(A) = I$, and finally, premultiplying by A^{-1}, $\text{adj}\, A/\det(A) = A^{-1}$. We have thus a new way to compute the inverse of a given matrix A.

Example

$$A = \begin{pmatrix} 1 & 2 & 3 \\ 0 & 1 & 1 \\ 3 & 6 & 10 \end{pmatrix}, \qquad \det(A) = 1;$$

$$A^{-1} = \frac{\text{adj}\, A}{|A|} = \begin{pmatrix} 4 & -2 & -1 \\ 3 & 1 & -1 \\ -3 & 0 & 1 \end{pmatrix}.$$

Note that in this particular example the inverse of A is the adjoint of A because the value of the determinant of A is one.

Inversion by Means of Partitioning

We complete this chapter by showing a method which is particularly useful for higher order square matrices and which makes use of matrix partitioning as developed in Section 12.3.

Suppose an $n \times n$ matrix A is partitioned,

$$A = \begin{pmatrix} A_{11} & A_{12} \\ A_{21} & A_{22} \end{pmatrix},$$

where A_{11}, A_{12}, A_{21}, A_{22} are, respectively, $n_1 \times n_1$, $n_1 \times n_2$, $n_2 \times n_1$, $n_2 \times n_2$ submatrices of A. Define further a matrix $B = A_{22} - A_{21}A_{11}^{-1}A_{12}$. We claim then that the inverse of A is the matrix

$$A^{-1} = \begin{pmatrix} A_{11}^{-1}(I + A_{12}B^{-1}A_{21}A_{11}^{-1}) & -A_{11}^{-1}A_{12}B^{-1} \\ -B^{-1}A_{21}A_{11}^{-1} & B^{-1} \end{pmatrix}.$$

The proof is very simple if we start from the fact that, if the assertion is true, then

$$A^{-1}A = \begin{pmatrix} A_{11}^{-1}(I + A_{12}B^{-1}A_{21}A_{11}^{-1}) & -A_{11}^{-1}A_{12}B^{-1} \\ -B^{-1}A_{21}A_{11}^{-1} & B^{-1} \end{pmatrix} \begin{pmatrix} A_{11} & A_{12} \\ A_{21} & A_{22} \end{pmatrix} = \begin{pmatrix} I & Z \\ Z & I \end{pmatrix}.$$

Performing the corresponding block operations we get

(1) first row of A^{-1} times first column of A,

$$\begin{aligned}&A_{11}^{-1}(I + A_{12}B^{-1}A_{21}A_{11}^{-1})A_{11} - A_{11}^{-1}A_{12}B^{-1}A_{21}\\&\quad = A_{11}^{-1}IA_{11} + A_{11}^{-1}A_{12}B^{-1}A_{21}A_{11}^{-1}A_{11} - A_{11}^{-1}A_{12}B^{-1}A_{21}\\&\quad = I + A_{11}^{-1}A_{12}B^{-1}A_{21} - A_{11}^{-1}A_{12}B^{-1}A_{21} = I;\end{aligned}$$

(2) first row of A^{-1} times second column of A,

$$\begin{aligned}&A_{11}^{-1}(I + A_{12}B^{-1}A_{21}A_{11}^{-1})A_{12} - A_{11}^{-1}A_{12}B^{-1}A_{22}\\&\quad = A_{11}^{-1}IA_{12} + A_{11}^{-1}A_{12}B^{-1}A_{21}A_{11}^{-1}A_{12} - A_{11}^{-1}A_{12}B^{-1}A_{22}\\&\quad = A_{11}^{-1}A_{12}(I + B^{-1}A_{21}A_{11}^{-1}A_{12} - B^{-1}A_{22})\\&\quad = A_{11}^{-1}A_{12}(I + B^{-1}(A_{21}A_{11}^{-1}A_{12} - A_{22}))\\&\quad = A_{11}^{-1}A_{12}(I + B^{-1}(-B))\ (\text{since } -B = A_{21}A_{11}^{-1}A_{12} - A_{22}) = Z;\end{aligned}$$

(3) second row of A^{-1} times first column of A,

$$\begin{aligned}&-B^{-1}A_{21}A_{11}^{-1}A_{11} + B^{-1}A_{21}\\&\quad = -B^{-1}A_{21}I + B^{-1}A_{21}\\&\quad = -B^{-1}A_{21} + B^{-1}A_{21} = Z;\end{aligned}$$

(4) second row of A^{-1} times second column of A

$$\begin{aligned}-B^{-1}A_{21}A_{11}^{-1}A_{12} + B^{-1}A_{22} &= B^{-1}(A_{22} - A_{21}A_{11}^{-1}A_{12})\\&= B^{-1}B\ (\text{since } B = A_{22} - A_{21}A_{11}^{-1}A_{12}) = I.\end{aligned}$$

Thus, summarizing,

$$A^{-1}A = I = \begin{pmatrix} I & Z \\ Z & I \end{pmatrix}.$$

Example

$$A = \begin{pmatrix} 1 & 2 & 3 \\ 0 & 1 & 1 \\ 3 & 6 & 10 \end{pmatrix} = \begin{pmatrix} A_{11} & A_{12} \\ A_{21} & A_{22} \end{pmatrix}$$

where

$$A_{11} = \begin{pmatrix} 1 & 2 \\ 0 & 1 \end{pmatrix}, \qquad A_{12} = \begin{pmatrix} 3 \\ 1 \end{pmatrix}, \qquad A_{21} = (3 \quad 6), \qquad A_{22} = (10).$$

We calculate first

$$A_{11}^{-1} = \begin{pmatrix} 1 & -2 \\ 0 & 1 \end{pmatrix}, \qquad B = (10) - (36)\begin{pmatrix} 1 & -2 \\ 0 & 1 \end{pmatrix}\begin{pmatrix} 3 \\ 1 \end{pmatrix} = 1$$

and

$$B^{-1} = 1.$$

Therefore

$$A^{-1} = \left(\begin{array}{c|c} \dfrac{\begin{pmatrix}1 & -2\\0 & 1\end{pmatrix}\left[\begin{pmatrix}1 & 0\\0 & 1\end{pmatrix} + \begin{pmatrix}3\\1\end{pmatrix}(1)(3 \quad 6)\begin{pmatrix}1 & -2\\0 & 1\end{pmatrix}\right]}{-(1)(3 \quad 6)\begin{pmatrix}1 & -2\\0 & 1\end{pmatrix}} & \dfrac{-\begin{pmatrix}1 & -2\\0 & 1\end{pmatrix}\begin{pmatrix}3\\1\end{pmatrix}(1)}{(1)} \end{array} \right)$$

$$= \begin{pmatrix} 4 & -2 & -1\\ 3 & 1 & -1\\ -3 & 0 & 1 \end{pmatrix}.$$

EXERCISES

Determinants and Adjoints of Square Matrices

E.14.1 Prove Theorem 14.13.

E.14.2 Show that $|kA| = k^n|A|$.

E.14.3 Using the different evaluation methods discussed, compute the determinants of the following matrices:

$$A = \begin{pmatrix} 4 & 6 & -4 & 8\\ -2 & 4 & 0 & 5\\ 3 & 2 & 3 & 4\\ 3 & -2 & 1 & 2 \end{pmatrix}, \qquad B = \begin{pmatrix} 5 & 5 & 7 & -2\\ 4 & 6 & 4 & -4\\ 2 & 4 & 2 & 1\\ 1 & 0 & -1 & 2 \end{pmatrix}.$$

Hint: $|A| = 572$, $|B| = 144$.

E.14.4 Use the Laplace expansion method to evaluate the determinants of the following matrices.

$$\begin{pmatrix} 4 & -5 & 7 & -1 & -2\\ 3 & -3 & 4 & 1 & 0\\ 2 & -3 & -1 & -1 & -4\\ 1 & -4 & -3 & -2 & -5\\ 4 & -6 & -1 & 0 & -7 \end{pmatrix}, \qquad \begin{pmatrix} 1 & 2 & 2 & 3 & 2\\ 1 & 2 & 1 & 2 & 1\\ 1 & 1 & 1 & 1 & 2\\ 1 & 1 & 2 & 2 & 4\\ 1 & 2 & 2 & 1 & 1 \end{pmatrix}.$$

Hint: $|A| = 118$, $|B| = -2$.

E.14.5 Give examples to show that if a matrix A is triangular (strictly triangular) then $\det(A) = \prod_{i=1}^{n} a_{ii}$ [$\det(A) = 0$].

E.14.6 Let S be the matrix defined below, where A and C are square matrices. Show that $\det(S) = \det(A) \cdot \det(C)$.

$$S = \begin{pmatrix} A & B\\ Z & C \end{pmatrix}.$$

E.14.7 Let (a_1, a_2, a_3), (b_1, b_2, b_3), and (c_1, c_2, c_3) be three points of R^3 not lying on the same line. Show that the equation of the plane passing through the three points is given by

$$\begin{vmatrix} x_1 & x_2 & x_3 & 1 \\ a_1 & a_2 & q_3 & 1 \\ b_1 & b_2 & b_3 & 1 \\ c_1 & c_2 & c_3 & 1 \end{vmatrix} = 0.$$

E.14.8 Prove Theorem 14.16.

Hint: Generalize the result obtained for a 3×3 matrix.

Matrix Inversion

E.14.9 Show that if the $n \times n$ matrix A, which is to be inverted by means of partitioning, is symmetric, then its inverse reduces to the following:

$$A^{-1} = \begin{pmatrix} A_{11}^{-1}(I + A_{11}^{-1}A_{12})B^{-1}(A_{11}^{-1}A_{12})^{\mathrm{T}} & -A_{11}^{-1}A_{12}B^{-1} \\ -(A_{11}^{-1}A_{12}B^{-1})^{\mathrm{T}} & B^{-1} \end{pmatrix}.$$

E.14.10 Using the methods described in the chapter find the inverses of the following matrices and compare the results:

$$A = \begin{pmatrix} 5 & 1 \\ 5 & 2 \end{pmatrix}, \qquad B = \begin{pmatrix} 1 & 1 & -2 \\ 1 & 2 & 3 \\ 4 & 2 & 0 \end{pmatrix}, \qquad C = \begin{pmatrix} 5 & 10 & 8 & 4 \\ 5 & 11 & 7 & -4 \\ 2 & 5 & 2 & -3 \\ 2 & 0 & 12 & 17 \end{pmatrix}.$$

PROBLEMS

P.14.1 Every 2×2 matrix A determines a unique parallelepiped in R^2 (see Fig. 14.1) with vertices θ, A^1, A^2, and $A^1 + A^2$, where

$$A = \begin{pmatrix} A^1 \\ A^2 \end{pmatrix} = \begin{pmatrix} a_{11} & a_{12} \\ a_{21} & a_{22} \end{pmatrix}.$$

(a) Show that, as special cases, every 2×2 diagonal matrix determines a rectangle, every 2×2 scalar matrix a square, and the identity matrix the unit square. (b) Show that the area of any of these figures in R^2 is given by the determinant of the representation matrix.

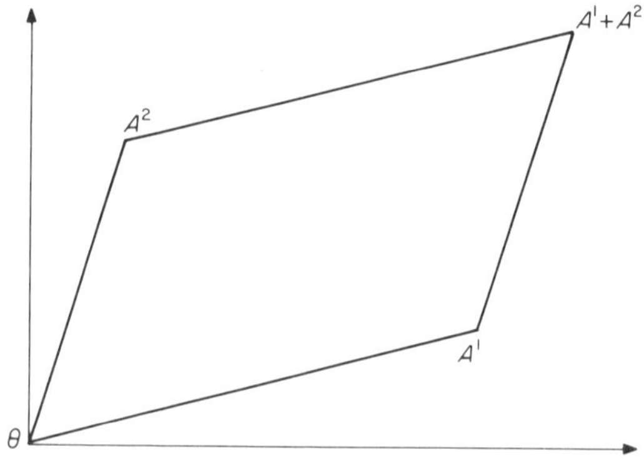

Figure 14.1

P.14.2 Observe that the three parallelepipeds in Fig. 14.2 have the same area. Interpret this in terms of matrix similarity, and the corresponding property of determinants of similar matrices.

$$A = \begin{pmatrix} 3 & 0 \\ 0 & 2 \end{pmatrix}$$

$$B = \begin{pmatrix} 3 & 2 \\ 0 & 2 \end{pmatrix}$$

$$C = \begin{pmatrix} 5 & 2 \\ 2 & 2 \end{pmatrix}$$

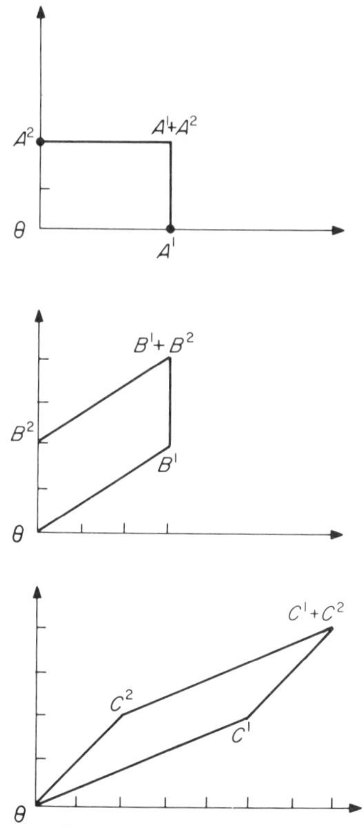

Figure 14.2

P.14.3 Generalize the results of P.14.1 and P.14.2 for (a) 3×3 matrices, and volumes in R^3; (b) $n \times n$ matrices, and volumes in R^n. To what extent is it justified to speak of the "volume" of a square matrix A?

P.14.4 Prove that the area of a triangle with vertices θ, (a_{11}, a_{12}, a_{13}), (a_{21}, a_{22}, a_{23}) is $\frac{1}{2}|AA^{\mathrm{T}}|^{1/2}$, with

$$A = \begin{pmatrix} a_{11} & a_{12} & a_{13} \\ a_{21} & a_{22} & a_{23} \end{pmatrix}.$$

P.14.5 Prove that the volume of the parallelepiped given by $A^1, A^2, \ldots, A^m$ is $|AA^{\mathrm{T}}|^{1/2}$, where

$$A = \begin{pmatrix} a_{11} & \cdots & a_{1n} \\ \vdots & & \vdots \\ a_{m1} & \cdots & a_{mn} \end{pmatrix}.$$

NOTES

This chapter, in particular the way of treating determinants, is based on the work of Finkbeiner [1]. For a different approach the reader should consult Hadley [1].

CHAPTER 15

LINEAR EQUATIONS SYSTEMS

15.1 Introduction

Suppose we have the following system of m linear equations in n unknowns:

$$
\begin{aligned}
a_{11}x_1 + \cdots + a_{1n}x_n &= b_1 \\
&\vdots \\
a_{m1}x_1 + \cdots + a_{mn}x_n &= b_m .
\end{aligned}
$$

Using our matrix notation, we could also write

$$
\begin{pmatrix} a_{11} & \cdots & a_{1n} \\ \vdots & & \vdots \\ a_{m1} & \cdots & a_{mn} \end{pmatrix} \begin{pmatrix} x_1 \\ \vdots \\ x_n \end{pmatrix} = \begin{pmatrix} b_1 \\ \vdots \\ b_m \end{pmatrix}
$$

or even

$$
Ax = b.
$$

Obviously A represents, in the latter expression, the $m \times n$ matrix of coefficients a_{ij}, x is the n-component column vector of unknowns, and b the m-component column vector of constants.

The foregoing system describes, relative to a fixed pair of bases $\{u^1, \ldots, u^n\}$ in the n-dimensional space U and $\{v^1, \ldots, v^m\}$ in the m-dimensional space V, a linear transformation $t: U \to V$. In very general terms, the question of solving the system $Ax = b$ is identical to the question of finding the unknown vector (or vectors) $x \in U$ which is (or are) carried into the vector $b \in V$ by the operator A. The link between the theories of linear mappings, matrices and linear equations systems is now complete.

We shall deal first with some general problems. This we do in Section 15.2 by studying the general case, that is, the nonhomogeneous linear system. Then, in Section 15.3, we focus our attention on the special case of the homogeneous linear system. Section 15.4 has again a general character. There we study a fundamental property of linear systems.

But before doing all that, let us introduce the notion of equivalence into our present context.

Equivalent Systems

Definition 15.1 Let $Ax = b$ and $Px = q$ be two linear systems of m equations in n unknowns. Then they are said to be equivalent if $[A\,|\,b] \sim [P\,|\,q]$, that is, if their augmented coefficient matrices are equivalent.

The definition uses the concept of matrix equivalence as the decisive criterion. Knowing how a given matrix can be reduced to an equivalent one, and taking account of the matrix representation of systems introduced at the beginning of the chapter, we immediately discover the three possible operations which yield equivalent systems: (1) interchange of two equations; (2) multiplication of an equation by a scalar; (3) sum to one equation of another equation multiplied by a scalar. These, of course, may be directly obtained by performing the corresponding elementary row operations on the augmented coefficient matrix.

The concept of system equivalence immediately suggests to us a method of finding the solution or solutions of a given system of linear equations. Suppose the original system is given by $Ax = b$. By an appropriate number of elementary row operations, we can transform the augmented matrix $[A\,|\,b]$ into the augmented matrices $[E\,|\,h]$ or $[C\,|\,k]$ of the equivalent linear systems $Ex = h$ or $Cx = k$, respectively. Here E and C are an echelon matrix and the canonical matrix row equivalent to A. The solution vector or vectors x can then be directly obtained from the forms $Ex = h$ or $Cx = k$. This fact also suggests the alternative definition of systems equivalence which is given next.

Definition 15.2 Two linear systems of m equations in n unknowns are said to be equivalent if they have the same solution.

The mathematical fact of the existence of equivalent systems poses interesting and sometimes difficult problems to different branches of science, including economics. This is especially true for those disciplines, for instance, econometrics, concerned with the statistical estimation of models which represent particular "real-world" systems.

A competitive market, for instance, may be conceived as a system of linear equations representing the behavior of consumers and suppliers of a well-specified commodity. Usually, the equations represent the quantity of the commodity a consumer demands or a supplier offers at each price level. In this way, quantity and price are simultaneously determined by the interplay of all equations involved.

The econometrician aims at estimating the different equations of the system, that is, at "discovering," within well-established probabilistic limits, the magnitude of the coefficients of the postulated explanatory equations. For doing this, whatever estimation method he may use, he has to base his work on statistical information about quantity and prices which is supposed to be generated by the same system he is studying. In other words, he uses the solution—or sequence of solutions—of the system to "discover" the probabilistic values of its coefficients. It is only understandable that the existence of equivalent systems makes this task, under circumstances, quite difficult. For instance, knowing the solution vector x, it is not necessarily a straightforward process to decide whether it has been produced by the market system $Ax = b$ or by the equivalent though different market system $Mx = n$. The discussion of this type of identification problems constitutes an important chapter of econometrics. A very simple, nonstatistical illustration is given next.

Example Assume we know that the "correct" aggregate demand and supply functions in a market are represented by the equations

$$x + a_1 p + b_1 = 0 \qquad \text{and} \qquad x + a_2 p + b_2 = 0,$$

respectively, where x represents the demanded or supplied quantity of a given commodity and p its price.

Note that neither the demand nor the supply functions are completely identified. For instance, we may multiply the demand and the supply functions by $\frac{1}{5}$ and $\frac{4}{5}$, respectively and present the added result as a "demand function." Analogously, we may multiply the demand and the supply functions by $\frac{2}{7}$ and $\frac{5}{7}$, respectively, and call their sum a "supply function."

The "false" demand and supply functions are now

$$x + \frac{a_1 + 4a_2}{5} p + \frac{b_1 + 4b_2}{5} = 0, \qquad x + \frac{2a_1 + 5a_2}{7} p + \frac{2b_1 + 5b_2}{7} = 0,$$

respectively. They are obviously different than those of the original system. However, the two systems give the same equilibrium price, that is, they have the same (price) solution.

15.2 General Case: The Nonhomogeneous System

Definition 15.3 A linear system $Ax = b$ is said to be nonhomogeneous if $b \neq \theta$, that is, if $b_i \neq 0$ for at least one $i = 1, \ldots, m$.

We know what it means, within the framework of the theory of linear mappings, to solve the linear system $Ax = b$. With this concept in mind we also know, still making use of our knowledge about linear mappings, that three alternatives are open to the system $Ax = b$: (1) It is possible that there is no vector x satisfying the system, that is, no vector x in U which is carried by the operator A into the vector b of V. (2) It is also possible that there is exactly one $x \in U$ for which b is the image under the mapping described by A. (3) Finally (and this less obvious case will be clarified soon) there is a third possibility that an infinite number of vectors x satisfy the system.

In the first case mentioned above, we shall speak of an inconsistent system, meaning thereby a system which admits no solution at all. An inconsistent system does not present any particular interest and the first question we have to ask ourselves—before attempting to find solutions—is whether the system under consideration would really justify our concern and effort, that is, whether it is consistent.

Theorem 15.1 Let $Ax = b$ be a nonhomogeneous linear system. Then the system is consistent, that is, has a solution, if and only if $r(A) = r([A \mid b])$, where $[A \mid b]$ is the "augmented matrix,"

$$[A \mid b] = \begin{pmatrix} a_{11} & \cdots & a_{1n} & b_1 \\ \vdots & & \vdots & \vdots \\ a_{m1} & \cdots & a_{mn} & b_m \end{pmatrix}.$$

Proof On the one hand, a solution will exist if and only if the vector b is in the range space of the transformation represented by A, that is, if $b \in R_t$. On the other hand, R_t is generated by the columns of A. Hence, if a solution exists, b must be a linear combination of the columns of A. In other words, if a solution exists, we must also have $r(A) = r([A \mid b])$, and vice versa.

Once the question of consistency has been positively answered, we confront the second and third alternatives mentioned above. We study these two cases of consistent systems separately.

Linear Systems with a Unique Solution

We begin the study with the first of the two consistent cases. Our question concerns the conditions under which it is guaranteed that a nonhomogeneous linear system of equations will have one and only one solution.

Theorem 15.2 Let $Ax = b$ be a consistent linear system with $r(A) = m = n$. Then the system has a unique solution.

Proof Since $m = n$, by hypothesis, it is clear that the operator A is a square matrix. Furthermore A is nonsingular by the same hypothesis and has a unique inverse A^{-1}. Therefore, the unique solution of $Ax = b$ is given by $x = A^{-1}b$.

A very useful result concerning a system $Ax = b$ with $r(A) = m = n$ is the so-called "Cramer's rule," surely well known to the reader. We give the formal proof next.

Theorem 15.3 Let $Ax = b$ be a linear system with $\det(A) \neq 0$. Then the system has a unique solution given by $x_j = \det(A_1, \ldots, A_{j-1}, b, A_{j+1}, \ldots, A_n)/\det(A)$, for each $j = 1, \ldots, n$.

Proof If $\det(A) \neq 0$, A is nonsingular and A^{-1} exists. The system has a unique solution according to Theorem 15.2. Consider $A^{-1} = \operatorname{adj} A/\det(A)$. We can write $x = A^{-1}b = (\operatorname{adj} A/\det(A))b$. For the jth row of A^{-1}, we have $(A_{1j}, \ldots, A_{nj})/\det(A)$. Taking $b = (b_1, \ldots, b_n)^{\mathrm{T}}$, we get $x_j = (b_1 A_{1j} + \cdots + b_n A_{nj})/\det(A)$. From Theorem 14.9 we can write $b_1 A_{1j} + \cdots + b_n A_{nj} = \det(A_1, \ldots, A_{j-1}, b, A_{j+1}, \ldots, A_n)$. Therefore, for every j, $x_j = \det(A_1, \ldots, A_{j-1}, b, A_{j+1}, \ldots, A_n)/\det(A)$.

Linear Systems with an Infinite Number of Solutions

We come now to our third and last case. The two previous theorems established the conditions under which a system is consistent and those under which the solution is unique. These suggest to us also the case in which a system will have an infinite number of solutions.

Theorem 15.4 Let $Ax = b$ be a consistent linear system with $r(A) < n$. Assigning arbitrary values to $n - r(A)$ chosen unknowns, the remaining $r(A)$ unknowns are uniquely determined.

Proof By choosing and assigning fixed values to a set of $n - r(A)$ unknowns, one reduces the original system to a system satisfying Theorem 15.2. Since the rank of the coefficient matrix of the reduced system is $r(A)$, the uniqueness of the solution is guaranteed. However, since there is an infinite number of possibilities of assigning values to $n - r(A)$ unknowns, the system admits an infinite number of solutions. For each of the infinite number of possible assignments, there is, therefore, a uniquely determined solution.

One point of the above theorem ought to be clearly understood. In choosing and assigning values to $n - r(A)$ of the unknowns we are completely free. However, once this first step has been made, the remaining $r(A)$ unknowns are uniquely determined. In the particular case in which we assign the value zero to each of the $n - r(A)$ chosen unknowns, we call the solution of $Ax = b$ a "basic solution."

Example 1

$$\begin{aligned} x_1 + x_2 + x_3 &= 4, \\ 2x_1 + 5x_2 - 2x_3 &= 3, \\ x_1 + 7x_2 - 7x_3 &= 5. \end{aligned}$$

It is easy to see that $r(A) \neq r[(A \mid b])$ and the system is inconsistent. In particular,

$$[A \mid b] = \begin{pmatrix} 1 & 1 & 1 & 4 \\ 2 & 5 & -2 & 3 \\ 1 & 7 & -7 & 5 \end{pmatrix} \sim \begin{pmatrix} 1 & 1 & 1 & 4 \\ 0 & 3 & -4 & -5 \\ 0 & -6 & 8 & -1 \end{pmatrix}$$

$$\sim \begin{pmatrix} 1 & 1 & 1 & 4 \\ 0 & 1 & -\frac{4}{3} & -\frac{5}{3} \\ 0 & -6 & 8 & -1 \end{pmatrix} \sim \begin{pmatrix} 1 & 0 & \frac{7}{3} & \frac{17}{3} \\ 0 & 1 & -\frac{4}{3} & -\frac{5}{3} \\ 0 & 0 & 0 & -11 \end{pmatrix}.$$

Note that the last row gives us

$$0x_1 + 0x_2 + 0x_3 = -11.$$

Example 2

$$\begin{aligned} x_1 + 2x_2 + 3x_3 &= 2, \\ x_2 + x_3 &= 0, \\ 3x_1 + 6x_2 + 10x_3 &= 5; \end{aligned}$$

$r(A) = 3$, from the examples of Section 14.4, and obviously $r(A) = r([A \mid b])$.

$$A = \begin{pmatrix} 1 & 2 & 3 \\ 0 & 1 & 1 \\ 3 & 6 & 0 \end{pmatrix}, \qquad A^{-1} = \begin{pmatrix} 4 & -2 & -1 \\ 3 & 1 & -1 \\ -3 & 0 & 1 \end{pmatrix}.$$

We have thus

$$\begin{pmatrix} x_1 \\ x_2 \\ x_3 \end{pmatrix} = \begin{pmatrix} 4 & -2 & 1 \\ 3 & 1 & -1 \\ -3 & 0 & 1 \end{pmatrix} \begin{pmatrix} 2 \\ 0 \\ 5 \end{pmatrix} = \begin{pmatrix} 13 \\ 1 \\ -1 \end{pmatrix}.$$

Example 3

$$
\begin{aligned}
5x_1 + 6x_2 + 8x_3 &= 12,\\
4x_1 + 4x_2 + 4x_3 &= 6,\\
x_1 + 2x_2 + x_3 &= 1,\\
2x_1 + 3x_2 + 4x_3 &= 6.
\end{aligned}
$$

Performing the sequence of operations $H_{14}(-2)$, $H_{24}(-1)$, $H_{21}(-2)$, $H_{32}(-2)$, $H_{31}(-1)$, $H_{43}(-4)$, $H_{42}(-3)$, and $H_{41}(-2)$ transforms the augmented coefficient matrix

$$[A\,|\,b] = \begin{pmatrix} 5 & 6 & 8 & 12 \\ 4 & 4 & 4 & 6 \\ 1 & 2 & 1 & 1 \\ 2 & 3 & 4 & 6 \end{pmatrix}$$

into the matrix

$$[C\,|\,k] = \begin{pmatrix} 1 & 0 & 0 & 1 \\ 0 & 1 & 0 & 0 \\ 0 & 0 & 1 & 1 \\ 0 & 0 & 0 & 0 \end{pmatrix},$$

corresponding to the equivalent system,

$$\begin{pmatrix} 1 & 0 & 0 \\ 0 & 1 & 0 \\ 0 & 0 & 1 \\ 0 & 0 & 0 \end{pmatrix}\begin{pmatrix} x_1 \\ x_2 \\ x_3 \end{pmatrix} = \begin{pmatrix} 1 \\ 0 \\ 1 \\ 0 \end{pmatrix},$$

or simply,

$$\begin{pmatrix} 1 & 0 & 0 \\ 0 & 1 & 0 \\ 0 & 0 & 1 \end{pmatrix}\begin{pmatrix} x_1 \\ x_2 \\ x_3 \end{pmatrix} = \begin{pmatrix} 1 \\ 0 \\ 1 \end{pmatrix}$$

with solution $x = (1 \quad 0 \quad 1)^{\mathrm{T}}$.

Example 4

$$
\begin{aligned}
x_1 \qquad\quad + 3x_3 + 2x_4 &= 7,\\
x_1 + x_2 + 2x_3 + 2x_4 &= 10,\\
3x_1 + 3x_2 + 6x_3 + 7x_4 &= 30.
\end{aligned}
$$

After applying the sequence of operations $H_{32}(-3)$, $H_{21}(-1)$, and $H_{13}(-2)$ we get

$$[A\,|\,b] = \begin{pmatrix} 1 & 0 & 3 & 2 & 7 \\ 1 & 1 & 2 & 2 & 10 \\ 3 & 3 & 6 & 7 & 30 \end{pmatrix} \sim \begin{pmatrix} 1 & 0 & 3 & 0 & 7 \\ 0 & 1 & -1 & 0 & 3 \\ 0 & 0 & 0 & 1 & 0 \end{pmatrix} = [E\,|\,h].$$

The equivalent system is

$$\begin{pmatrix} 1 & 0 & 3 & 0 \\ 0 & 1 & -1 & 0 \\ 0 & 0 & 0 & 1 \end{pmatrix} \begin{pmatrix} x_1 \\ x_2 \\ x_3 \\ x_4 \end{pmatrix} = \begin{pmatrix} 7 \\ 3 \\ 0 \end{pmatrix},$$

that is,

$$\begin{aligned} x_1 \quad + 3x_3 \quad &= 7, \\ x_2 - \ x_3 \quad &= 3, \\ x_4 &= 0. \end{aligned}$$

We can thus write,

$$x_1 = 7 - 3x_3, \qquad x_2 = 3 + x_3, \qquad x_4 = 0.$$

Assigning the value $x_3 = a$, we find a solution

$$x = (7 - 3a, 3 + a, a, 0)^{\mathrm{T}}.$$

Example 5

$$\begin{aligned} 6x_1 + 3x_2 &= 12, \\ x_1 + 3x_2 &= \ 5, \end{aligned}$$

that is,

$$\begin{pmatrix} 6 & 3 \\ 1 & 2 \end{pmatrix} \begin{pmatrix} x_1 \\ x_2 \end{pmatrix} = \begin{pmatrix} 12 \\ 5 \end{pmatrix}.$$

Using Cramer's rule yields

$$x_1 = \frac{\det(b, A_2)}{\det(A)} = \frac{\begin{vmatrix} 12 & 3 \\ 5 & 2 \end{vmatrix}}{\begin{vmatrix} 6 & 3 \\ 1 & 2 \end{vmatrix}} = 1,$$

$$x_2 = \frac{\det(A_1, b)}{\det(A)} = \frac{\begin{vmatrix} 6 & 12 \\ 1 & 5 \end{vmatrix}}{\begin{vmatrix} 6 & 3 \\ 1 & 2 \end{vmatrix}} = 2.$$

Example 6 Suppose a production economy consists of n industries (also called activities), each of them producing one different commodity. We call x_i the ith commodity and x_{ij} the amount of the ith commodity used in the production of x_j. If the technology is assumed to be constant and linear, we may write, for the ith commodity, the production function, $x_i - \sum_{j=1}^{n} a_{ij} x_j = y_i$, where $a_{ij} = x_{ij}/x_j$, a technical coefficient, represents the amount of the ith commodity used in the production of one unit of the jth commodity, and y_i a surplus or "final demand."

The system can thus be written

$$\begin{pmatrix} x_1 \\ \vdots \\ x_n \end{pmatrix} - \begin{pmatrix} a_{11} & \cdots & a_{1n} \\ \vdots & & \vdots \\ a_{n1} & \cdots & a_{nn} \end{pmatrix} \begin{pmatrix} x_1 \\ \vdots \\ x_n \end{pmatrix} = \begin{pmatrix} y_1 \\ \vdots \\ y_n \end{pmatrix},$$

that is $x - Ax = y$, or also

$$\begin{pmatrix} 1 - a_{11} & -a_{12} & \cdots & -a_{1n} \\ -a_{21} & 1 - a_{22} & \cdots & -a_{2n} \\ \vdots & \vdots & & \vdots \\ -a_{n1} & -a_{n2} & \cdots & 1 - a_{nn} \end{pmatrix} \begin{pmatrix} x_1 \\ x_2 \\ \vdots \\ x_n \end{pmatrix} = \begin{pmatrix} y_1 \\ y_2 \\ \vdots \\ y_n \end{pmatrix},$$

that is $(I - A)x = y$.

The solution of the system, for a given final demand vector, is given by $x = (I - A)^{-1}y$. It indicates to us, for a given "technology" A, the production vector which would satisfy a final demand vector y. This simple linear model of the structure of an economy is known in economics as the open Leontief model.

15.3 Special Case: The Homogeneous System

As a special case of the theory of Section 15.2, we now discuss the concept of a homogeneous linear system.

Definition 15.4 A linear system $Ax = b$ is said to be homogeneous if $b = \theta$, that is, if $b_1, \ldots, b_m = 0$.

Theorem 15.5 A homogeneous linear system is always consistent.

Proof Consider the system $Ax = \theta$. By definition, θ is a linear combination of the columns of the matrix A. We obviously have $r(A) = r([A \mid \theta])$.

Theorem 15.6 Let $Ax = \theta$ be a homogeneous linear system. Then, if $r(A) = m = n$, the unique solution of the system is the trivial solution $x = \theta$.

Proof That the solution is unique is simply a consequence of Theorem 15.2. Notice that for a homogeneous system all solutions must lie in the kernel or null space K_t. Since in our present case the solution is unique and, on the other hand, we always have $\theta \in K_t$, we can show that the system can only be satisfied by $x_1 = x_2 = \cdots = x_n = 0$.

Note that for the given $t: U \to V$, $d(U) = r(t) + k(t) = n$. The dimension of the kernel is $k(t) = n - r(t)$ which, according to Theorem 13.2, can also be written $k(t) = n - r(A)$. But, since by hypothesis $r(A) = n$, we have $k(t) = 0$. This means that there is only $\theta \in K_t$.

An extension of the preceding result is the following.

Theorem 15.7 Let $Ax = \theta$ be a homogeneous linear system. The system has a nontrivial solution if and only if $r(A) < n$.

Proof First assume $r(A) < n$. Then $n - r(A) > 0$ and the kernel contains vectors other than θ. Since the system is homogeneous, every such vector is a solution.

Assume next there are vectors in the kernel other than θ. Then we must have $k(t) = n - r(A) > 0$, that is, $r(A) < n$.

Two corollaries follow from the preceding theorem. The first is a consequence of the fact that always $\theta \in K_t$.

Remark 15.1 The vector θ is always a solution of $Ax = \theta$.

The second corollary is a consequence of the fact that any linear combination of vectors in the kernel is itself a vector in the kernel.

Remark 15.2 If $r(A) < n$, then $Ax = \theta$ has a linearly independent solution set containing exactly $n - r(A)$ solutions and every linear combination of these is also a solution of the system.

Example 1

$$\begin{aligned} x_1 + 2x_2 + 3x_3 &= 0, \\ x_2 + x_3 &= 0, \\ 3x_1 + 6x_2 + 10x_3 &= 0; \end{aligned}$$

$$A = \begin{pmatrix} 1 & 2 & 3 \\ 0 & 1 & 1 \\ 3 & 6 & 10 \end{pmatrix}, \qquad r(A) = 3; \qquad A^{-1} = \begin{pmatrix} 4 & -2 & -1 \\ 3 & 1 & -1 \\ -3 & 0 & 1 \end{pmatrix}.$$

Since $x = A^{-1}\theta$ we clearly have $x = \theta$.

Example 2

$$\begin{aligned} 2x_1 - x_2 + 3x_3 &= 0, \\ 3x_1 + 2x_2 + x_3 &= 0, \\ x_1 - 4x_2 + 5x_3 &= 0. \end{aligned}$$

Thus

$$[A \mid \theta] = \begin{pmatrix} 2 & -1 & 3 & 0 \\ 3 & 2 & 1 & 0 \\ 1 & -4 & 5 & 0 \end{pmatrix} \sim \begin{pmatrix} 1 & -4 & 5 & 0 \\ 3 & 2 & 1 & 0 \\ 2 & -1 & 3 & 0 \end{pmatrix}$$

$$\sim \begin{pmatrix} 1 & -4 & 5 & 0 \\ 0 & 14 & -14 & 0 \\ 0 & 7 & -7 & 0 \end{pmatrix} \sim \begin{pmatrix} 1 & -4 & 5 & 0 \\ 0 & 1 & -1 & 0 \\ 0 & 1 & -1 & 0 \end{pmatrix}$$

$$\sim \begin{pmatrix} 1 & 0 & 1 & 0 \\ 0 & 1 & -1 & 0 \\ 0 & 0 & 0 & 0 \end{pmatrix}.$$

We have therefore

$$\begin{aligned} x_1 \qquad + x_3 &= 0, \\ x_2 - x_3 &= 0. \end{aligned}$$

Now setting $x_3 = -a$, we get $x_1 = -x_2 = -x_3 = a$.

Example 3 A very simple kind of "general equilibrium model"—the closed Leontief model—can be obtained by setting $y = \theta$ in the system of the last example of Section 15.2. We can make then the following interpretation: (1) Production takes place by using inputs generated only by the system itself; (2) production is "consumed" within the same system, so that for all i, $y_i = 0$ and $x_i - \sum_{j=1}^{n} a_{ij} x_j = 0$; (3) labor and other personal services constitute "industries" of the system in the sense that they transform the inputs they receive from the other industries (food, clothing, and so forth) into outputs they provide to the other industries (labor services, and so forth).

The closed Leontief model can then be written

$$(I - A)x = \theta, \qquad \text{or also} \qquad Ax = x.$$

15.4 A Fundamental Property of Linear Systems of Equations

At this stage we know enough about the solutions of the linear systems $Ax = b$ and $Ax = \theta$ as to be able to study a very important property which concerns the link between the two.

Consider any consistent nonhomogeneous linear system $Ax = b$. It is clear that by replacing the m-component vector b by the m-component vector θ we obtain a homogeneous system which, with the exception of the replaced vector, is identical with the original one. We shall call this $Ax = \theta$ the homogeneous system "associated" with $Ax = b$.

$Ax = b$ was supposed to be consistent so that it must have one solution at least. Its associated homogeneous system always has a solution, at least the trivial one. The two systems being related as they are, we wonder how this relation will express itself in terms of the solutions. The following theorem provides us with the basis for the answer.

Theorem 15.8 Let $Ax = b$ be a nonhomogeneous linear system and let x^{p} be a solution of it. Then x^{c} is a solution of $Ax = b$ if and only if $x^{\mathrm{c}} = x^{\mathrm{p}} + x^0$ for some x^0 in the kernel K_t.

Proof First assume that both x^{c} and x^{p} are solutions of $Ax = b$. Then $Ax^{\mathrm{c}} = Ax^{\mathrm{p}} = b$, that is, $Ax^{\mathrm{c}} - Ax^{\mathrm{p}} = A(x^{\mathrm{c}} - x^{\mathrm{p}}) = b - b = \theta$, and so $x^{\mathrm{c}} - x^{\mathrm{p}} = x^0 \in K_t$. Conversely, assume $x^{\mathrm{c}} = x^{\mathrm{p}} + x^0$ for some $x^0 \in K_t$. Then $Ax^{\mathrm{c}} = A(x^{\mathrm{p}} + x^0) = Ax^{\mathrm{p}} + Ax^0 = b + \theta = b$.

The result of the theorem leads us to following definition:

Definition 15.5 Let x^{p} be any solution of the nonhomogeneous linear system $Ax - b$ and x^0 a solution of its associated homogeneous system $Ax = \theta$. Then the expressions $x^{\mathrm{c}} = x^0 + x^{\mathrm{p}}$, x^0 and x^{p} are also called the complete solution, the characteristic solution and a particular solution of $Ax = b$, respectively.

The vector x^0 may, of course, be called the "complete solution" of $Ax = \theta$. But the qualification of "complete" really does not add very much to the concept.

If $x^{\mathrm{p}1}$ and $x^{\mathrm{p}2}$ are two distinct particular solutions of $Ax = b$, we must have $Ax^{\mathrm{p}1} = b$ and $Ax^{\mathrm{p}2} = b$. Hence $Ax^{\mathrm{p}1} - Ax^{\mathrm{p}2} = A(x^{\mathrm{p}1} - x^{\mathrm{p}2}) = Ax^0 = \theta$. In this case, since $x^{\mathrm{p}1} \neq x^{\mathrm{p}2}$, $x^0 = x^{\mathrm{p}1} - x^{\mathrm{p}2}$ is a nontrivial solution of $Ax = \theta$. Conversely, if x^0 is a nontrivial solution of $Ax = \theta$ and x^{p} a particular solution $x^{\mathrm{c}} = x^0 + x^{\mathrm{p}}$ is a different solution of $Ax = b$, as the previous theorem has shown.

This property is essential for linear systems. More general versions of the theorem play important roles in other branches of mathematics, for instance in the theory of ordinary linear differential equations.

EXERCISES

E.15.1 Show that the following two systems are equivalent:

$$
\begin{aligned}
3x_1 + 5x_2 + 2x_3 + x_4 &= 23, \\
-x_1 + 3x_2 - 7x_3 + 6x_4 &= 8, \\
2x_1 - x_2 + 4x_3 - 5x_4 &= -8, \\
4x_1 + 2x_2 + x_3 + 3x_4 &= 23;
\end{aligned}
$$

$$\begin{aligned} 5x_1 + 4x_2 + 6x_3 - 4x_4 &= 15, \\ 2x_1 + 8x_2 - 5x_3 + 7x_4 &= 31, \\ 4x_1 + 7x_2 - x_3 + 2x_4 &= 23, \\ 5x_2 - 2x_3 - x_4 &= 0. \end{aligned}$$

E.15.2 Show whether the following system is consistent:

$$\begin{aligned} 4x_1 + x_2 - x_3 &= 6, \\ 8x_1 + 2x_2 - 2x_3 &= 9, \\ 9x_1 + 4x_2 - 5x_3 &= 8. \end{aligned}$$

E.15.3 Using matrix inversion find the unique solution of the system:

$$\begin{aligned} x_1 + 8x_3 + 5x_4 &= 6, \\ x_1 + x_2 - 3x_3 - 4x_4 &= -1, \\ 2x_1 + 6x_2 - 10x_3 - 4x_4 &= -2, \\ 4x_2 - 4x_3 + 4x_4 &= 6. \end{aligned}$$

Hint: $x = (2, \frac{1}{5}, 0, \frac{4}{5})^T$.

E.15.4 Using Cramer's rule find the solution of the system below. Is the solution unique?

$$\begin{aligned} 7x_1 + 5x_2 + x_3 &= 14, \\ 10x_1 + 7x_2 + 3x_3 &= 15, \\ 12x_1 + 9x_2 + 4x_3 &= 18. \end{aligned}$$

Hint: $x = (1, 2, -3)^T$.

E.15.5 Find the complete and the basic solution of the following system:

$$\begin{aligned} 3x_1 + 5x_2 + 5x_3 + 2x_4 &= 10, \\ 2x_1 + 4x_2 + 4x_3 + x_4 &= 6, \\ 2x_1 + 2x_2 + 3x_3 &= 10. \end{aligned}$$

Hint: Basic solution $x = (5, -3, 2, 0)^T$.

E.15.6 Show that the linear system $Ax = b$ has a solution if and only if the vector b lies in the subspace generated by the columns of the matrix A.

E.15.7 Prove that the homogeneous system $Ax = \theta$ has a nontrivial solution if and only if $\det(A) = 0$.

E.15.8 Suppose $Ax = \theta$ consists of a linearly independent set of m equations in n unknowns. Prove that the dimension of its "solution space" is $n - m$.

E.15.9 Give the number of nontrivial solutions for each of the following systems:

$$\begin{aligned} x_1 + 2x_2 + 3x_3 &= 0, \\ x_2 + x_3 &= 0, \\ 3x_1 + 6x_2 + 10x_3 &= 0; \end{aligned}$$

$$\begin{aligned} 3x_1 + 8x_2 + 5x_3 &= 0, \\ 2x_1 + 5x_2 + 2x_3 &= 0, \\ 3x_1 + 9x_2 + 9x_3 &= 0, \\ 2x_1 + 7x_2 + 10x_3 &= 0. \end{aligned}$$

E.15.10 Find the characteristic solution, a particular solution, and the complete solution for each of the following systems:

$$\begin{aligned} x_1 + 3x_2 + 3x_3 &= 1, \\ x_1 + 4x_2 + 3x_3 &= 2, \\ x_1 + 3x_2 + 4x_3 &= 3; \end{aligned}$$

$$\begin{aligned} 5x_1 + 3x_2 + 13x_3 &= 7, \\ 5x_1 + 5x_2 + 10x_3 &= 5, \\ -3x_1 - x_2 - 9x_3 &= -5. \end{aligned}$$

PROBLEMS

P.15.1 Prove: If $Ax = b$ has a solution, then the system $yA = \theta$, $yb = 1$ cannot have a solution.

Hint: See Chapter 18.

P.15.2 Consider the Leontief system $(I - A)x = y$. Prove that if θ is the only solution of the associated homogeneous system $(I - A)x = \theta$, then $(I - A)x = y$ has a unique and different solution for every different y.

P.15.3 Suppose $(I - A)x = y$ has no independent subeconomy (A should have the form discovered in problem P.12.1; see also Chapter 19, Section 19.1). According to a well-known theorem (Hawkins and Simon [1]), $(I - A)x = y$ has a solution $x \gg \theta$ for any $y > \theta$ if and only if all principal minors of $|I - A|$ are positive. Prove the necessary character of the condition.

Hint: Use induction on n, the number of equations. Prove for $n = 1$; assume the theorem holds for $2, 3, \ldots, n - 1$ equations. Finally, show it also holds for n equations. Use Cramer's rule to express the x_i.

P.15.4 It can be shown that associated with the Leontief "physical" system (1) $(I - A)x = y$ there is a "value" system (2) $p(I - A) = v$, called the dual of (1), where $p = (p_1, \ldots, p_n)$ is a price vector, and $v = (v_1, \ldots, v_n)$ a vector of the value added in each industry. Discover a necessary condition for a solution $p \gg \theta$ for any $v > \theta$.

P.15.5 Let the coefficients $a_{11} = 0.30$, $a_{12} = 0.20$, $a_{21} = 0.25$, and $a_{22} = 0.15$ define an open two-sector Leontief economy where labor is the only primary factor. (a) If final demand is given (in some physical units) by $y_1 = 60$, $y_2 = 42$, and the total value added in each sector (in dollars) by $v_1 = 60$, $v_2 = 175$, compute the solutions $x \gg \theta$ and $p \gg \theta$ under the assumption that the wage rate is 1 dollar/hr. (b) What is the total value produced by each sector (in dollars)? (c) What are the input requirements x_{ij} of each sector (in physical units) and its labor requirement coefficients a_{lj} (in man-hours per unit of output)?

Hint: Find $(I - A)^{-1}$ first.

NOTES

See Notes to Chapter 10, especially Hadley [1]. The literature on input–output economics is very extensive and we give here only a few important titles. The classic reference is by Leontief [1]. Very good treatments of the subject are given by Allen [1], Lange [1], and Dorfman *et al.* [1]. For more abstract and formal treatments see Gale [1, 3], Karlin [1], Koopmans [2], Lancaster [1], Morishima [1], and Nikaido [4]. For the relations between Walrasian general equilibrium theory and Leontief input–output economics a study by Morishima [1] is fundamental. The author includes in its extensive bibliography most of the relevant papers and articles on the subject. The reader should also consult Notes in Chapter 19 regarding nonnegative matrices.

CHAPTER 16

CHARACTERISTIC VECTORS, DIAGONALIZATION, AND TRIANGULARIZATION

We here take up the line of investigation initiated in Chapter 13, in particular in its last section. However, we shall be not so much interested in the question of similarity in general, but rather in the conditions under which a given square matrix is similar to a diagonal or a triangular matrix.

The topics we shall be confronting here have a twofold interest to us. On the one hand, we shall learn something new about what one might call the " intrinsic properties " of a square matrix. Indeed, the richness of information contained in a particular square matrix is a most remarkable fact.

On the other hand, there is a practical reason for our current concern. Diagonal and triangular matrices, as it is apparent, are very simple structures and thus most convenient to deal with. The task of studying the properties of a given square matrix may be greatly simplified if we can reduce it to the task of studying the properties of a similar diagonal or triangular matrix. And in this sense, the question of whether a matrix is reducible to a similar diagonal or triangular matrix is most relevant.

The following two sections are devoted to the analysis of some square-matrix properties and constitute the core of the chapter. Afterward, in Section 16.3, we use some of these properties to approach the issues of the diagonalization and triangularization of a matrix. More special questions are studied in Section 16.4.

16.1 Characteristic Polynomials, Roots, and Vectors

An n-square matrix is related to a linear mapping defined on an n-dimensional space V. We initiate our study, therefore, by focusing our attention on a given mapping $t\colon V \to V$ and a related square matrix A.

A transformation t on V carries one vector of the space into a vector of the same space. Selecting an image vector under the mapping t, we may wonder about the vector which is mapped by t into it. Let us discuss briefly some of the questions which might be interesting in this respect.

(1) We may be given a fixed vector $b \in V$ and our concern is to discover which $x \in V$ are mapped into it. The topic was discussed in the previous chapter as the problem of the solution of the equations system $t(x) = b$ or, using matrix notation, $Ax = b$.

(2) As a special situation of the preceding problem (investigated in Chapter 15 as well) we have the case $t(x) = \theta$, that is, $Ax = \theta$.

(3) An important case is that of the vector which is mapped into itself. The study of $t(x) = x$, or $Ax = x$, is nothing but the study of a fixed point. We have contemplated this question, and certainly in much more general terms, in Chapter 7. To some extent it will also concern us here, but only in so far as it is included, as a special case, in the problem which follows.

(4) Finally consider the case of a vector $x \in V$ which is mapped by t into a scalar multiple of itself. If λ is a scalar we may write $t(x) = \lambda x$ or, in matrix notation, $Ax = \lambda x$. Note that if $\lambda = 1$, we obtain the fixed-point problem.

The preceding classification gives us some insight into the links relating the four problems. It is apparent that only the fourth problem awaits to be analyzed at this stage. It is, indeed, the central subject of this chapter.

Definition 16.1 Let $t\colon V \to V$ be a linear mapping on the space V over F. Then a nonzero vector $x \in V$ and a scalar $\lambda \in F$ are called, respectively, a characteristic vector (or eigenvector) of t and a characteristic value (or eigenvalue) of t if $t(x) = \lambda x$. The set of all characteristic values of t is called the spectrum of t.

Theorem 16.1 Let λ be a characteristic value of t. Then the set of all characteristic vectors of t associated with λ forms a subspace V_λ of V.

Proof Consider $x, y \in V_\lambda$. By definition, we have $t(x) = \lambda x$, $t(y) = \lambda(y)$. Thus, for $a, b \in F$, we may write $t(ax + by) = at(x) + bt(y) = a\lambda x + b\lambda y = \lambda(ax + by)$. Thus $ax + by \in V_\lambda$. In particular, for $a = b = 0$, $ax + by = \theta \in V_\lambda$. Therefore V_λ is a subspace of V.

We can state now the matrix analog of Definition 16.1. We suppose, of course, that the referred to square matrix A is related to the mapping t in the usual sense.

Definition 16.2 A nonzero vector x and a scalar λ are called, respectively, a characteristic vector (or eigenvector) of A and a characteristic value (or eigenvalue) of A if $Ax = \lambda x$. The set of all characteristic values of A is called the spectrum of A.

Let us find a more explicit form for the expression $t(x) = \lambda x$. First, we have a very important property.

Theorem 16.2 A scalar $\lambda \in F$ is a characteristic value of a linear mapping $t: V \to V$ if and only if the linear mapping $(t - \lambda i): V \to V$, where i represents the identity mapping, is singular.

Proof If λ is a characteristic value of t, then there exists a vector $x \neq \theta$ such that $t(x) = \lambda x = \lambda i(x) = (\lambda i)(x)$. We can also write $t(x) - (\lambda i)(x) = (t - \lambda i)(x) = \theta$. Therefore $(t - \lambda i)$ is singular. If $(t - \lambda i)$ is singular, then there is an $x \neq \theta$ such that $(t - \lambda i)(x) = \theta$. Then $t(x) - \lambda i(x) = \theta$ and we have $t(x) = \lambda i(x) = \lambda x$.

A corollary of Theorem 16.2 follows.

Remark 16.1 Let x be a characteristic vector of $t: V \to V$. Then x is in the kernel of $(t - \lambda i): V \to V$.

The preceding remark together with Theorem 16.1 leads us to a further result.

Remark 16.2 Let $t: V \to V$ be a linear transformation. Then the space V_λ of all characteristic vectors of t belonging to λ is the kernel of $(t - \lambda i)$.

In the matrix formulation, it is clear that a characteristic vector x is a solution of the n homogeneous equations in n unknowns $Ax = \lambda x$, that is, $(A - \lambda I)x = \theta$. The system has a nontrivial solution only if $r(A - \lambda I) < n$. that is, if $\det(A - \lambda I) = 0$. We can thus write

$$\det(A - \lambda I) = \begin{vmatrix} a_{11} - \lambda & a_{12} & \cdots & a_{1n} \\ a_{21} & a_{22} - \lambda & \cdots & a_{2n} \\ \vdots & \vdots & \vdots & \vdots \\ a_{n1} & a_{n2} & \cdots & a_{nn} - \lambda \end{vmatrix} = 0.$$

Note that the expanded determinant gives a polynomial of degree n in λ. Representing the sum of products by b_i we have $\det(A - \lambda I) = (-1)^n\lambda^n + b_1\lambda^{n-1} + \cdots + b_{n-1}\lambda + b_n = 0$.

From the fundamental theorem of algebra, we know there are n numbers λ, not necessarily distinct, which satisfy the latter equation. These numbers may be real or complex, even if the b_i, that is, the elements a_{ij} of A, belong to the real field. In any case, the λ's are the roots—distinct or multiple, real or complex—of a polynomial of degree n. We can summarize these results in the following definition and remark.

Definition 16.3 Let A be an nth-order square matrix. Then the expressions $\det(A - \lambda I)$ and $\det(A - \lambda I) = 0$ are called the characteristic polynomial of A and characteristic equation of A, respectively.

Remark 16.3 Let $\det(A - \lambda I) = 0$ be the characteristic equation of A. Then the roots of $\det(A - \lambda I) = 0$ are the characteristic values of A.

Example Note that the matrix A below has the characteristic vectors $x^1 = (1, 1)^T$ and $x^2 = (\frac{1}{2}, 1)^T$ associated with the characteristic values $\lambda_1 = 3$ and $\lambda_2 = 2$, respectively.

$$A = \begin{pmatrix} 4 & -1 \\ 2 & 1 \end{pmatrix};$$

$$|A - \lambda I| = \begin{vmatrix} 4 - \lambda & -1 \\ 2 & 1 - \lambda \end{vmatrix} = (4 - \lambda)(1 - \lambda) + 2 = \lambda^2 - 5\lambda + 6.$$

Thus $\lambda_1 = 3$ and $\lambda_2 = 2$. For $\lambda_1 = 3$ we have

$$Ax^1 = \lambda_1 x^1 = \begin{pmatrix} 4 & -1 \\ 2 & 1 \end{pmatrix}\begin{pmatrix} x_1 \\ x_2 \end{pmatrix} = 3\begin{pmatrix} x_1 \\ x_2 \end{pmatrix},$$

that is,

$$\begin{aligned} 4x_1 - x_2 &= 3x_1, \\ 2x_1 + x_2 &= 3x_2. \end{aligned}$$

Arbitrarily putting $x_2 = 1$ we get $x_1 = 1$, that is, $x^1 = (1, 1)^T$. For $\lambda_2 = 2$ we get

$$Ax^2 = \lambda_2 x^2 = \begin{pmatrix} 4 & -1 \\ 2 & 1 \end{pmatrix}\begin{pmatrix} x_1 \\ x_2 \end{pmatrix} = 2\begin{pmatrix} x_1 \\ x_2 \end{pmatrix},$$

that is,

$$\begin{aligned} 4x_1 - x_2 &= 2x_1, \\ 2x_1 + x_2 &= 2x_2. \end{aligned}$$

Putting $x_2 = -2$ we get $x_1 = -1$, that is, $x^2 = (-1, -2)^T$.

The system $(A - \lambda I)x = \theta$ is obviously homogeneous and therefore the characteristic vectors x are only determined up to a scalar multiple. Any vector kx, for $k \in F$, will satisfy the system if the vector x does. Sometimes it is convenient to eliminate this indeterminacy by working with the "normalized characteristic vector" $\hat{x} = \|x\|^{-1} \cdot x$, where $\|x\|$ is the length of x. Clearly, $\|\hat{x}\| = 1$.

There is another point which is worth discussing here. From the way in which we write the equation $(A - \lambda I)x = \theta$, it is easy to see that the characteristic vectors x are column vectors. We may, however, consider instead the equation $y(A - \lambda I) = \theta$. Written in this form it follows immediately that the characteristic vectors y represent row vectors.

In both cases the characteristic values of A are identical, for they are the roots of the same equation $\det(A - \lambda I) = 0$. Nevertheless, we may wonder, for a given λ, what are the relationships between the row vectors y and the column vectors x.

Observe that for the characteristic values λ_k, $k = 1, \ldots, n$, the n expressions $Ax^k = \lambda_k x^k$ can be written $AQ = Q\Lambda$, where Q is an nth-order square matrix having x^k as its kth column while Λ is the diagonal matrix with λ_k in the kth diagonal place. Furthermore, if the set of n characteristic vectors $\{x^1, \ldots, x^n\}$ is linearly independent, then the matrix Q is nonsingular and we have $Q^{-1}AQ = \Lambda$. We shall analyze very soon (see Theorem 16.3) the conditions under which this will happen. For the time being let us simply assume that Q^{-1} exists.

Similarly, for the same characteristic values λ_k, $k = 1, \ldots, n$, we may condense the n expressions $y^k A = \lambda_k y^k$ by means of $PA = \Lambda P$. Here P is an nth-order square matrix with y^k as the kth row. Here too, if the set $\{y^1, \ldots, y^n\}$ is linearly independent, then P is nonsingular. Assuming this is the case, we may write $PAP^{-1} = \Lambda$.

Henceforth, from the two different expressions for Λ, $Q = P^{-1}$ and thus $PQ = I$. We have therefore, for adequately chosen scalar multiples, $x^k y^k = 1$ and $x^i y^j = 0$ for every $i \neq j$.

Let us summarize this result.

Remark 16.4 Let A be an $n \times n$ matrix. Suppose from the set of characteristic row and column vectors we can form the linearly independent sets $\{y^1, \ldots, y^n\}$ and $\{x^1, \ldots, x^n\}$, respectively. Then a characteristic row vector is normal to the characteristic column vector corresponding to the same characteristic value and orthogonal to the characteristic column vector corresponding to any other characteristic value.

Example Consider the same A of the previous example.

$$A = \begin{pmatrix} 4 & -1 \\ 2 & 1 \end{pmatrix}; \qquad (4 - \lambda)(1 - \lambda) + 2 = 0; \qquad \lambda_1 = 3, \qquad \lambda_2 = 2.$$

Let us first take λ_1 and then λ_2:

$$y^1 A = \lambda_1 y^1; \qquad \text{so} \qquad (y_1, y_2)\begin{pmatrix} 4 & -1 \\ 2 & 1 \end{pmatrix} = 3(y_1, y_2);$$

that is

$$\begin{aligned} 4y_1 + 2y_2 &= 3y_1, \\ -y_1 + \ \ y_2 &= 3y_2. \end{aligned}$$

Taking $y_2 = -1$ we get $y_1 = 2$. Thus $y^1 = (2, -1)$,

$$y^2 A = \lambda_2 y^2 = (y_1, y_2)\begin{pmatrix} 4 & -1 \\ 2 & 1 \end{pmatrix} = 2(y_1, y_2);$$

that is

$$\begin{aligned} 4y_1 + 2y_2 &= 2y_1, \\ -y_1 + \ \ y_2 &= 2y_2. \end{aligned}$$

Taking $y_2 = -1$, we get $y_1 = 1$. Thus $y^2 = (1, -1)$.

From the previous and present example we can form the matrices P and Q. We have

$$Q = \begin{pmatrix} 1 & -1 \\ 1 & -2 \end{pmatrix}, \qquad P = \begin{pmatrix} 2 & -1 \\ 1 & -1 \end{pmatrix}.$$

Notice that Q is the inverse of P so that $PQ = I$. Of course, this fact depends on the choice we made of the scalar multiples. Provided the matrix is invertible, however, we can always make the choice so that $Q = P^{-1}$ or $Q^{-1} = P$.

But thus far, we have only assumed the linear independence of a set $\{x^1, \ldots, x^n\}$ of characteristic vectors of A. This is the most relevant case in our study and the one we shall be almost exclusively concerned with. Let us, nevertheless, investigate the conditions under which we can guarantee this will happen.

Theorem 16.3 Let an $n \times n$ matrix A have n distinct characteristic values $\lambda_1, \ldots, \lambda_n$. Then the corresponding n characteristic column (row) vectors form a linearly independent set $\{x^1, \ldots, x^n\}$ $(\{y^1, \ldots, y^n\})$.

Proof We prove the theorem for the column vectors. An identical proof is valid for the row vectors. For the n equations $Ax^k = \lambda_k x^k$, $k = 1, \ldots, n$, we can write $AQ = Q\Lambda$. Assume the n characteristic values λ_k, $k = 1, \ldots, n$ are all distinct, but suppose the set of characteristic vectors $\{x^1, \ldots, x^n\}$, where x^i is associated with λ_i, is linearly dependent. Hence $r(Q) < n$ and the homogeneous system $Qx = \theta$ has a nontrivial solution, that is, there exists an $x \neq \theta$ such that $Qx = \theta$. We can also write $AQx = Q\Lambda x$. Since $Qx = \theta$, we have $AQx = Q\Lambda x = \theta$. This means that $x' = \Lambda x$ is a solution of $Qx' = Q\Lambda x = \theta$.

Observe that $x_i' = \lambda_i x_i$. Since $r(Q) < n$, x and x' can differ in not more that $n - r(Q)$ components. This implies that $r(Q)$ of the characteristic values of Q must be equal, thus violating our assumption about n distinct values λ_k.

The characteristic values of a real matrix are the roots of the characteristic polynomial of the matrix. A polynomial may have, as we know, pairs of complex conjugate roots. It is easy to deduce that the complex characteristic values of a real matrix will give rise to complex characteristic vectors. This is the subject of the next theorem.

Theorem 16.4 Let A be an $n \times n$ matrix over the real field. If A has a complex characteristic value λ, then its conjugate $\bar{\lambda}$ is also a characteristic value of A and if the complex characteristic vector x is associated with λ, its conjugate $\bar{x}$ is a characteristic vector associated with $\bar{\lambda}$.

Proof Suppose λ and x are the complex characteristic value and corresponding characteristic vector of the real matrix A. We then have $Ax = \lambda x$. Taking conjugates, we must have $\bar{A}\bar{x} = \bar{\lambda}\bar{x}$. But since A is real, we have $A = \bar{A}$. Therefore, $\bar{A}\bar{x} = \bar{\lambda}\bar{x}$.

From our study of determinants we remember that if A is a triangular or diagonal matrix, then $\det(A) = \prod_{i=1}^{n} a_{ii}$. This property leads us immediately to the following important result.

Remark 16.5 The characteristic values of triangular or diagonal matrices are their diagonal elements.

Triangular, and especially diagonal matrices, are very convenient forms. They are extremely simple and have their characteristic values on the main diagonal. Besides, it is obvious that the rank of a triangular or diagonal matrix is simply equal to the number of nonzero diagonal elements.

The following properties can be easily derived from previous theorems and our knowledge of determinants. Proofs are left as exercises.

Remark 16.6 Let A be an $n \times n$ matrix. Then (1) $\det(A) = \prod_{i=1}^{n} \lambda_i$; (2) A and A^{T} have the same characteristic values; (3) if A^{-1} exists, its characteristic values are the reciprocals λ_i^{-1}, $i = 1, \ldots, n$ of the characteristic values of A; (4) if x is a characteristic vector belonging to the characteristic value λ_k of A, it is also a characteristic vector belonging to the characteristic value $\lambda_k{}^n$ of A^n, for any positive or negative rational number n.

Remark 16.7 For any two $n \times n$ matrices A and B, the products AB and BA have the same characteristic values.

From our preceding discussion it seems clear that the characteristic values of a square matrix A are the characteristic values of the linear transformation described by the matrix A. But the description provided by A, as we know, is only in relation to a fixed basis of the space on which the mapping is defined. We may then expect that if the characteristic values are intrinsic properties of the given linear transformation t, they must be the same for all matrices by means of which t may be described. That this is precisely the case is shown in the next theorem and subsequent remark.

Theorem 16.5 Let A and B be similar matrices. Then A and B have the same characteristic polynomial.

Proof Assume A and B are similar matrices. Then, by definition, there exists a nonsingular matrix P such that $A = PBP^{-1}$. Hence, we have $A - \lambda I = PBP^{-1} - \lambda I = P(B - \lambda I)P^{-1}$. This implies $\det(A - \lambda I) = \det\big(P(B - \lambda I)P^{-1}\big) = \det(P) \cdot \det(B - \lambda I) \cdot \det(P^{-1}) = \det(B - \lambda I)$. Therefore A and B have the same characteristic polynomial.

As a corollary of Theorem 16.5 the following remark is obvious.

Remark 16.8 Two similar matrices A and B have the same characteristic values and vectors.

We shall follow up this line of argument later in Section 16.3.

16.2 The Cayley–Hamilton Theorem

The concept of characteristic polynomial studied in the preceding section can be looked upon from a more general viewpoint than the one followed so far. This we shall do here, and the proof of a very important property will be our reward.

A polynomial over a field F can be expressed by $p(\lambda) = a_n \lambda^n + \cdots + a_1 \lambda + a_0$. If $t: V \to V$ is a linear mapping and $i: V \to V$ represents the identity transformation, a polynomial on t can be written $p(t) = a_n t^n + \cdots + a_1 t + a_0 i$. We shall say that t is a root or zero of $p(\lambda)$ if it happens that $p(t) = 0$.

As we know, $t: V \to V$ can be represented by means of a square matrix A. Using the identity matrix I to represent $i: V \to V$, we can define $p(\lambda)$ as a polynomial on A by $p(A) = a_n A^n + \cdots + a_1 A + a_0 I$. Here too, A will be called a root or zero of $p(\lambda)$ if $p(A) = 0$.

Example 1 Let A be the matrix of the last two examples and $p(\lambda) = \lambda^2 - 2\lambda + 3$.

Since

$$A = \begin{pmatrix} 4 & -1 \\ 2 & 1 \end{pmatrix},$$

we can write the corresponding polynomial on A, $p(A) = A^2 - 2A + 3I$,

$$p(A) = \begin{pmatrix} 4 & -1 \\ 2 & 1 \end{pmatrix}^2 - 2\begin{pmatrix} 4 & -1 \\ 2 & 1 \end{pmatrix} + 3\begin{pmatrix} 1 & 0 \\ 0 & 1 \end{pmatrix} = \begin{pmatrix} 9 & -3 \\ 6 & 0 \end{pmatrix}.$$

Clearly A is not a zero of $p(\lambda)$.

Example 2 Let A still be the same matrix but $p(\lambda) = \lambda^2 - 5\lambda + 6$. The polynomial on A is

$$p(A) = \begin{pmatrix} 4 & -1 \\ 2 & 1 \end{pmatrix}^2 - 5\begin{pmatrix} 4 & -1 \\ 2 & 1 \end{pmatrix} + 6\begin{pmatrix} 1 & 0 \\ 0 & 1 \end{pmatrix} = \begin{pmatrix} 0 & 0 \\ 0 & 0 \end{pmatrix}.$$

Here A is obviously a zero of $p(\lambda)$.

It is convenient to pay some attention to the results obtained in the two examples above. In the first example, A is not a zero of the polynomial. It should be observed that the polynomial used in this case has been arbitrarily chosen. This is not the case in the second example. Note that there we used the polynomial $\lambda^2 - 5\lambda + 6$, which happens to be the characteristic polynomial of the matrix A. The result is that A is a zero of its characteristic polynomial. This did not happen by chance. On the contrary, Example 2 only illustrates a property of a square matrix A which is the subject of the Cayley–Hamilton theorem. We prove it next.

Theorem 16.6 Let A be an $n \times n$ matrix. Then A is a zero of its characteristic polynomial $\det(A - \lambda I)$.

Proof Consider the $n \times n$ matrix A and denote by $|A - \lambda I| = \lambda^n + b_{n-1}\lambda^{n-1} + \cdots + b_1\lambda + b_0$ its characteristic polynomial. Let $\operatorname{adj}(A - \lambda I)$ be the adjoint matrix of $(A - \lambda I)$. The element in the place (i, j) of $\operatorname{adj}(A - \lambda I)$ is the cofactor of the element in the place (j, i) of $(A - \lambda I)$, and as such is a polynomial of degree less than or equal to $n - 1$. Thus we can write $\operatorname{adj}(A - \lambda I) = \Gamma_{n-1}\lambda^{n-1} + \cdots + \Gamma_1\lambda + \Gamma_0$, where the matrices of the sum are

$$\begin{aligned}
&\Gamma_{n-1} = I; \\
&\Gamma_{n-2} = [\gamma_{ij}^{(n-2)}], && \text{where } \gamma_{ij}^{(n-2)} && \text{is a coefficient of } \lambda^{n-2}; \\
&\quad\vdots \\
&\Gamma_1 = [\gamma'_{ij}], && \text{where } \gamma'_{ij} && \text{is a coefficient of } \lambda; \\
&\Gamma_0 = [\gamma_{ij}], && \text{where } \gamma_{ij} && \text{is a coefficient of } 1.
\end{aligned}$$

From Chapter 14 we know that $\operatorname{adj}(A - \lambda I) \cdot (A - \lambda I) = |A - \lambda I| \cdot I$, that is,

$$(\Gamma_{n-1}\lambda^{n-1} + \cdots + \Gamma_1\lambda + \Gamma_0)(A - \lambda I) = (\lambda^n + b_{n-1}\lambda^{n-1} + \cdots + b_1\lambda + b_0) \cdot I.$$

Multiplying, rearranging terms and equating the coefficients of equal powers of λ we get

$$\begin{aligned} -\Gamma_{n-1} &= I, \\ \Gamma_{n-1}A - \Gamma_{n-2} &= b_{n-1}I, \\ \Gamma_{n-2}A - \Gamma_{n-3} &= b_{n-2}I, \\ &\vdots \\ \Gamma_1 A - \Gamma_0 &= b_1 I, \\ \Gamma_0 A &= b_0 I. \end{aligned}$$

Postmultiplying the preceding equations by the matrices A^n, A^{n-1}, $A^{n-2}, \ldots,$ A, I, respectively, yields

$$\begin{aligned} -\Gamma_{n-1}A^n &= A^n, \\ \Gamma_{n-1}A^n - \Gamma_{n-2}A^{n-1} &= b_{n-1}A^{n-1}, \\ \Gamma_{n-2}A^{n-1} - \Gamma_{n-3}A^{n-2} &= b_{n-2}A^{n-2}, \\ &\vdots \\ \Gamma_1 A^2 - \Gamma_0 A &= b_1 A, \\ \Gamma_0 A &= b_0 I. \end{aligned}$$

Adding all equations we finally get

$$Z = A^n + b_{n-1}A^{n-1} + b_{n-2}A^{n-2} + \cdots + b_1 A + b_0 I,$$

where Z is the zero matrix.

16.3 Diagonalization and Triangularization

In this section we study some practical problems related to the transformation of a square matrix A into a similar diagonal matrix Λ or a similar triangular matrix Δ.

The diagonalization problem has already been touched upon in Section 16.1 in reference to the expressions $Q^{-1}AQ = \Lambda$ and $PAP^{-1} = \Lambda$, where Q and P represent the matrices of column and row eigenvectors of A, respectively.

Two main questions ought to be answered. First it is necessary to determine under which circumstances a linear mapping t: $V \to V$ is representable by a diagonal matrix Λ or a triangular matrix Δ. Second, we have to study the conditions under which a matrix A is similar to a diagonal or a triangular matrix. The latter topic will then suggest to us practical ways of diagonalizing or triangularizing a matrix.

Similarity to a Diagonal Matrix and Diagonalization

The answer to the first question, so far as diagonalization is concerned, is almost obvious, in the light of our previous theory.

Theorem 16.7 Let $t\colon V \to V$ be a linear transformation and $d(V) = n$. Then t can be represented by a diagonal matrix $\Lambda = \text{diag}(\lambda_1, \ldots, \lambda_n)$ if and only if V has a basis of characteristic vectors of t.

Proof Suppose the set of characteristic vectors of t $\{x^1, \ldots, x^n\}$ is a basis of V. Then, as we know,

$$\begin{matrix} t(x^1) = \lambda_1 x^1 \\ \vdots \\ t(x^n) = \lambda_n x^n \end{matrix} = \begin{pmatrix} \lambda_1 & & 0 \\ & \ddots & \\ 0 & & \lambda_n \end{pmatrix} \begin{pmatrix} x^1 \\ \vdots \\ x^n \end{pmatrix},$$

so t is represented by $\Lambda = \text{diag}(\lambda_1, \ldots, \lambda_n)$. Conversely, suppose t is represented by $\Lambda = \text{diag}(\lambda_1, \ldots, \lambda_n)$. Then from the representation theorem, there exists a basis $\{x^1, \ldots, x^n\}$ such that Λ represents t relative to this basis. Therefore $t(x^1) = \lambda_1 x^1, \ldots, t(x^n) = \lambda_n x^n$ and $\{x^1, \ldots, x^n\}$ are characteristic vectors of t.

We can now answer the second question, namely that of the similarity of an $n \times n$ matrix A to a diagonal matrix Λ. This theorem is a direct result of the previous one.

Theorem 16.8 An $n \times n$ matrix A is similar to a diagonal matrix Λ if and only if it has a linearly independent set of n characteristic vectors.

Consider the characteristic equations $Ax^i = \lambda_i x^i$, $i = 1, \ldots, n$, with $\{x^1, \ldots, x^n\}$ linearly independent. Assembling the vectors x^i in a square matrix we have $Q = (x^1, \ldots, x^n)$. Thus $AQ = A(x^1, \ldots, x^n) = (Ax^1, \ldots, Ax^n) = (\lambda_1 x^1, \ldots, \lambda_n x^n) = (x^1, \ldots, x^n) \cdot \Lambda$, where $\Lambda = \text{diag}(\lambda_1, \ldots, \lambda_n)$. We can also write $AQ = Q\Lambda$. The columns of Q constitute a linearly independent set and therefore Q^{-1} exists. Premultiplying the latter equation by Q^{-1}, we get $Q^{-1}AQ = \Lambda = \text{diag}(\lambda_1, \ldots, \lambda_n)$. Conversely, if $Q^{-1}AQ = \Lambda$, then the columns of Q form a linearly independent set. By definition of Q these are the characteristic column vectors of A.

A similar conclusion would have been reached by using the matrix $P = (y^1, \ldots, y^n)^{\mathrm{T}}$ instead of Q in the preceding theorem. As we know $P^{-1} = Q$.

There are two practical implications of Theorem 16.8 which should be stressed. The matrix Λ has the characteristic values of A as its diagonal elements. The diagonalizing matrix $Q(P)$—called the modal matrix—has the characteristic column (row) vectors of A as its columns (rows).

Example Consider the matrices A, P, and Q from the previous examples. Since $P = Q^{-1}$,

$$Q^{-1}AQ = PAP^{-1} = \begin{pmatrix} 2 & -1 \\ 1 & -1 \end{pmatrix}\begin{pmatrix} 4 & -1 \\ 2 & 1 \end{pmatrix}\begin{pmatrix} 1 & -1 \\ 1 & -2 \end{pmatrix} = \begin{pmatrix} 3 & 0 \\ 0 & 2 \end{pmatrix} = \Lambda.$$

The diagonal elements of Λ are precisely the characteristic values of A.

Diagonalization of matrices is a very useful tool in the solution of many mathematical problems. An example with economic content is given at the end of the chapter.

Theorem 16.8 states a necessary and sufficient condition for diagonalization. Comparing this result with Theorem 16.3 it is clear that the latter establishes sufficient though not necessary conditions for diagonalization. A matrix with repeated characteristic values might still have a linearly independent set of n eigenvectors.

A matrix with repeated characteristic values may, of course, be nondiagonalizable. In the study of a specific problem, this may become a very unsatisfactory situation. Fortunately we possess, at least, an approximation solution for such cases. The discussion of this topic is beyond the scope of our current study and we confine ourselves to giving, without proof, the following approximation theorem due to Bellman [1].

Theorem 16.9 Let A be a matrix with repeated characteristic values. Then there exists a matrix B with distinct characteristic values such that the norm

$$\|A - B\| = \sum_{i=1}^{n} \sum_{j=1}^{n} |a_{ij} - b_{ij}| \leq \delta,$$

where δ is any positive number as small as is desired.

The usefulness of the theorem is apparent. If A does not have a linearly independent set of n characteristic vectors, which is a possible consequence of the presence of repeated eigenvalues, then A may be replaced by a matrix B having n different eigenvalues. B can be chosen as "close" to A as is desired.

Diagonalization of Real Symmetric Matrices

Real symmetric matrices play a considerable role in theoretical economics especially, though not only, in the context of neoclassical theory. We examine here the question of the diagonalization of this type of matrix. Let us study, first, some properties of their characteristic values and vectors.

Theorem 16.10 The characteristic values of an $n \times n$ real symmetric matrix are real.

Proof Let A be a real symmetric matrix. Contrary to what was stated, assume it has the complex conjugate characteristic values $\lambda = m + in$ and $\bar{\lambda} = m - in$. Define now the matrix $B = (A - \lambda I)(A - \bar{\lambda} I) = \big(A - (m + in)I\big) \times \big(A - (m - in)I\big) = A^2 - 2mA + (m^2 + n^2)I = (A - mI)^2 + n^2 I$. B is clearly real. On the other hand, it is singular since $\big(A - (m + in)I\big)$ is singular. Thus, there exists a real vector $x \neq \theta$ such that $Bx = \theta$. Multiplying $Bx = \theta$ by x^{T}, we get $x^{\mathrm{T}}Bx = x^{\mathrm{T}}(A - mI)^2 x + n^2 x^{\mathrm{T}} I x = x^{\mathrm{T}}(A - mI)^{\mathrm{T}}(A - mI)x + n^2 x^{\mathrm{T}} x = 0$. Since $(A - mI)x$ is a real vector we have $\big((A - mI)x\big)^{\mathrm{T}} \cdot (A - mI)x \geq 0$. Also $x \neq \theta$ implies $x^{\mathrm{T}}x > 0$. Therefore $n = 0$, and a complex characteristic value is excluded.

Theorem 16.11 Let A be a real symmetric matrix. Then the characteristic vectors belonging to different characteristic values of A are orthogonal.

Proof Assume λ_1 and λ_2 are two different characteristic values of A, and x^1 and x^2 are their respective eigenvectors. Then $Ax^1 = \lambda_1 x^1$ and $Ax^2 = \lambda_2 x^2$. Multiplying the two equations by $(x^2)^{\mathrm{T}}$ and $(x^1)^{\mathrm{T}}$, respectively, we get $(x^2)^{\mathrm{T}}Ax^1 = \lambda_1 (x^2)^{\mathrm{T}} x^1$ and $(x^1)^{\mathrm{T}}Ax^2 = \lambda_2 (x^1)^{\mathrm{T}} x^2$. Subtracting the second from the first expression we have $(x^2)^{\mathrm{T}}Ax^1 - (x^1)^{\mathrm{T}}Ax^2 = (\lambda_1 - \lambda_2)(x^1)^{\mathrm{T}}x^2$. Clearly $(x^1)^{\mathrm{T}}x^2 = (x^2)^{\mathrm{T}}x^1$. Also, since $A = A^{\mathrm{T}}$, $(x^2)^{\mathrm{T}}Ax^1 = (x^1)^{\mathrm{T}}Ax^2$. Hence $(\lambda_1 - \lambda_2)(x^1)^{\mathrm{T}}x^2 = 0$. Since by assumption $\lambda_1 \neq \lambda_2$, we have $(x^1)^{\mathrm{T}}x^2 = 0$ and x^1 is orthogonal to x^2.

An interesting corollary follows from the preceding theorem. Remember that a normalized vector $\hat{x}^i$ has length equal to unity, that is, $\|\hat{x}^i\| = 1$. Clearly then $(x^i)^{\mathrm{T}}x^i = 1$. We can now summarize this result.

Remark 16.9 Let x^i, $i = 1, \ldots, n$ represent normalized characteristic vectors corresponding to n different eigenvalues of a real symmetric matrix A. Then $(x^i)^{\mathrm{T}}x^j = \delta_{ij}$ where δ_{ij} is the Kronecker delta.

We shall conclude our discussion of this aspect of real symmetric matrices by focusing our attention on the diagonalization problem. Let us point out, contrary to the general case, that any real symmetric matrix is diagonalizable. We shall not prove here this property of real symmetric matrices, but pay attention to the properties of the diagonalizing matrix under the assumption that such a matrix exists. We introduce, first, a new concept.

Definition 16.4 A square matrix A is said to be orthogonal if $A^{-1} = A^{\mathrm{T}}$, that is, if its inverse is equal to its transpose.

Theorem 16.12 Let Q be the matrix which diagonalizes a real symmetric matrix A. Then Q is orthogonal.

Proof If Q diagonalizes A we have, from Theorem 16.8, $Q^{-1}AQ = \Lambda = \operatorname{diag}(\lambda_1, \ldots, \lambda_n)$. But Q is the matrix of characteristic column vectors of A. From Remark 16.9, $(x^i)^{\mathrm{T}}x^j = \delta_{ij}$. Thus $Q^{\mathrm{T}}Q = I$ and therefore $Q^{-1} = Q^{\mathrm{T}}$.

Triangularization of a Matrix

Even if it is not always possible to diagonalize a square matrix A, it is possible to transform it in a similar triangular matrix. Without proof we state here the following property as the counterpart of Theorem 16.7.

Theorem 16.13 Any linear transformation $t\colon V \to V$, $d(V) = n$, can be represented by a triangular matrix Δ.

We now study the triangularization theorem and shall afterward comment on some of the properties of the triangularizing matrix.

Theorem 16.14 Let A be any $n \times n$ matrix. Then there exists a matrix Q such that $Q^{-1}AQ = \Delta$, where Δ is triangular.

Proof Suppose λ_i, $i = 1, \ldots, n$, are the characteristic values of A. Consider the characteristic vector x^1 associated with λ_1 and define a matrix Q_1 such that the normalized vector $\hat{x}^1$ is its first column and the other columns are any vectors such that $\det(Q_1) \neq 0$. Now the matrix AQ_1 has first column $A\hat{x}^1 = \lambda_1\hat{x}^1$, while the matrix $Q_1^{-1}AQ_1$ has first column $Q_1^{-1}\lambda_1\hat{x}^1 = (\lambda_1, 0, \ldots, 0)^{\mathrm{T}}$. Note that the latter is the first column of $Q_1^{-1}AQ_1$. We have, therefore,

$$Q_1^{-1}AQ_1 = \begin{pmatrix} \lambda_1 & B_1 \\ Z & A_1 \end{pmatrix}.$$

Note that $Z = \theta$ is an $(n-1) \times 1$ submatrix and A_1 an $(n-1) \times (n-1)$ submatrix. Because they are similar the matrices A and $Q_1^{-1}AQ_1$ have the same characteristic values. But $|Q_1^{-1}AQ_1 - \lambda I| = (\lambda - \lambda_1)|A_1 - \lambda I|$; so $\lambda_2, \ldots, \lambda_n$ are obviously the characteristic values of the submatrix A_1. For a 2×2 matrix A, the theorem is proved since $A_1 = [\lambda_2]$ and thus

$$Q_1^{-1}AQ_1 = Q^{-1}AQ = \begin{pmatrix} \lambda_1 & b_{12} \\ 0 & \lambda_2 \end{pmatrix}.$$

If $n > 2$ we take a characteristic vector $\hat{x}^2$ associated with the value λ_2 of A_1 and define a matrix Q_2 with $\hat{x}^2$ as its first column and the remaining columns any vectors such that $\det(Q_2) \neq 0$. We get now

$$Q_2^{-1}A_1Q_2 = \begin{pmatrix} \lambda_2 & B_2 \\ Z & A_2 \end{pmatrix}.$$

A_2 is here an $(n-2) \times (n-2)$ matrix.

For a 3×3 matrix A, the theorem is proved since $A_2 = [\lambda_3]$ and thus

$$Q = Q_1 \cdot \begin{pmatrix} I_1 & 0 \\ 0 & Q_2 \end{pmatrix}.$$

If $n > 3$ we repeat the procedure. In general, for an $n \times n$ matrix A we apply the method $n - 1$ times until we arrive at

$$Q = Q_1 \cdot \begin{pmatrix} I_1 & Z \\ Z & Q_2 \end{pmatrix} \begin{pmatrix} I_2 & Z \\ Z & Q_3 \end{pmatrix} \cdots \begin{pmatrix} I_{n-2} & Z \\ Z & Q_{n-1} \end{pmatrix}.$$

This is the matrix which triangularizes A, that is, $Q^{-1}AQ = \Delta$.

Example From the earlier examples of this chapter, consider the matrix A with characteristic value $\lambda_1 = 3$ and corresponding characteristic vector $x^1 = (1, 1)^T$. The normalized version of x^1 is $\hat{x}^1 = (1/\sqrt{2}, 1/\sqrt{2})^T$. We can choose an appropriate triangularizing matrix Q by using the vector $(0, 1)^T$ as its second column. We thus have

$$A = \begin{pmatrix} 4 & -1 \\ 2 & 1 \end{pmatrix}, \qquad Q = \begin{pmatrix} 1/\sqrt{2} & 0 \\ 1/\sqrt{2} & 1 \end{pmatrix}, \qquad Q^{-1} = \begin{pmatrix} \sqrt{2} & 0 \\ -1 & 1 \end{pmatrix}.$$

The similar triangular matrix is

$$Q^{-1}AQ = \begin{pmatrix} \sqrt{2} & 0 \\ -1 & 1 \end{pmatrix} \begin{pmatrix} 4 & -1 \\ 2 & 1 \end{pmatrix} \begin{pmatrix} 1/\sqrt{2} & 0 \\ 1/\sqrt{2} & 1 \end{pmatrix} = \Delta = \begin{pmatrix} 3 & -\sqrt{2} \\ 0 & 2 \end{pmatrix}.$$

Note that Δ has its two characteristic values on its main diagonal.

The way in which Theorem 16.14 has been formulated is quite general, since it does not make any particular reference to possible properties of the triangularizing matrix Q. After the following definition, we shall refer to this point.

Definition 16.5 A square matrix A is called unitary if $A^{-1} = \bar{A}^T$, that is, if its inverse is equal to its conjugate transpose.

Orthogonality of real matrices, Definition 16.4, may be regarded now as a special case of the more general property described in Definition 16.5. Remember that a real matrix may be considered as a complex matrix, the elements of which have zero imaginary parts.

In the triangularization of the real matrix A of the last example, we did not use an orthogonal matrix though it would have been possible to use one. In the same sense, a complex matrix or a real matrix having complex eigenvalues can be triangularized by means of a unitary matrix. We shall not discuss this point any further but content ourselves with the following important remark.

Remark 16.10 Let A be a square matrix and Δ a similar triangular matrix. Then a matrix Q exists such that $Q^{-1}AQ = \Delta$ satisfying, additionally, the following properties: (1) If A is real and has only real eigenvalues, then Q is orthogonal; (2) if A is complex, or real with complex eigenvalues, then Q is unitary.

16.4 Matrices of Functions and Series of Matrices

We shall complete this chapter by referring briefly to a few types of infinite series of matrices and to matrices, the elements of which are real-valued functions rather than scalars. The notion of a polynomial of a matrix has already been studied in this chapter and the existence of powers of matrices has been referred to, in somewhat different contexts, more than once.

Matrix-Valued Functions, Sequences, and Series

The concept of a matrix-valued function has been suggested by our reference to the polynomial $p(A) = a_n A^n + \cdots + a_1 A + a_0 I$. Here we shall expand our views on the matter by means of two new concepts.

Definition 16.6 Let N be the set of natural numbers and $X = \{A^{(1)}, A^{(2)}, \ldots, A^{(k)}, \ldots\}$ a set of $m \times n$ matrices. A function from N to X, which assigns to each number $k \in N$ a matrix $A^{(k)} \in X$, is called a sequence of matrices and denoted $(A^{(k)} | k \in N)$. The sequence is said to be convergent to a matrix $A = [a_{ij}]$, denoted $A^{(k)} \to A$, if for every $i = 1, \ldots, m$ and $j = 1, \ldots, n$, the sequence of elements $(a_{ij}^{(k)} | k \in N)$ converges to the element a_{ij} of A.

A property of great importance and the proof of which is quite simple follows next.

Theorem 16.15 If a sequence $(A^{(k)} | k \in N)$ converges to A, then for fixed product-conformable matrices P and Q, the sequence $(PA^{(k)}Q | k \in N)$ converges to PAQ.

Proof We write $B^{(k)} = PA^{(k)}Q$ so that $b_{ij}^{(k)} = \sum_{r=1}^{m} \sum_{s=1}^{n} p_{ir} a_{rs}^{(k)} q_{sj}$. Since every sequence $(a_{rs}^{(k)} | k \in N)$ converges to a_{rs} of A we must have, for constant elements p_{ir}. and q_{sj}, that every sequence $(p_{ir} a_{rs}^{(k)} q_{sj} | k \in N)$ converges to $p_{ir} a_{rs} q_{sj}$ and thus every sequence of linear combinations

$$\left(\sum_{r=1}^{m} \sum_{s=1}^{n} p_{ir} a_{rs}^{(k)} q_{sj} \middle| k \in N \right) = (b_{ij}^{(k)} | k \in N)$$

converges to $b_{ij} = \sum_{r=1}^{m} \sum_{s=1}^{n} p_{ir} a_{rs} q_{sj}$. Therefore $(PA^{(k)}Q | k \in N)$ converges to PAQ.

The result of the theorem can also be formulated as follows: Convergence of matrices is preserved under matrix equivalence and, in particular, under matrix similarity.

Now, we can extend the convergence idea contained in Definition 16.6. to refer to an infinite series.

Definition 16.7 Let q_k, $k = 0, 1, \ldots, m, \ldots$ be scalars and let B^k, $k = 0, 1, \ldots, m, \ldots$ represent the kth power of an $n \times n$ matrix B, with the convention that $B^0 = I$. An expression $\sum_{k=0}^{\infty} a_k B^k = \sum_{k=0}^{\infty} A^{(k)}$, where $A^{(k)} = a_k B^k$, is called an infinite power series of matrices. The series is said to converge to an $n \times n$ matrix $A = [a_{ij}]$ if for every (i, j), the series of elements $\sum_{k=0}^{\infty} a_{ij}^{(k)}$ converges to the element a_{ij} of A.

In the same way as a scalar-valued function f is defined by power series of scalars, we may define a power series of matrices as a matrix-valued function.

Definition 16.8 Let $\sum_{k=0}^{\infty} a_k B^k$ be an infinite power series. Then $f(A) = \sum_{k=0}^{\infty} A^{(k)} = \sum_{k=0}^{\infty} a_k B^k$ defines a matrix-valued function for the matrices B for which the series is convergent.

Theorem 16.15 and Definition 16.7 together lead to following property:

Theorem 16.16 Let $f(A)$, defined as in Definition 16.8, converge and consider the product-conformable matrix Q. Then $f(Q^{-1}AQ)$ converges to $Q^{-1}f(A)Q$.

Using the Cayley–Hamilton theorem, we can show that if A is an nth-order square matrix with characteristic values λ_i, $i = 1, \ldots, n$ and $p(A)$ a polynomial on A, then $p(\lambda_i)$, $i = 1, \ldots, n$ are the characteristic values of $p(A)$. Both A and $p(A)$ have thus the same characteristic vectors.

When we refer, now more generally, to the matrix-valued function $f(A)$, we cannot answer the same question without paying due attention to the convergence of $f(A)$.

Theorem 16.17 Let λ_i, $i = 1, \ldots, n$ be a characteristic value of an $n \times n$ matrix A and $f(A)$ a matrix-valued function. If $f(A)$ converges, then $f(\lambda_i)$, $i = 1, \ldots, n$ is a characteristic value of $f(A)$.

Proof We assume here that the matrix A is diagonalizable (a more general proof could be given without this assumption, for example, by means of the Jordan canonical form, which we did not study here). We have $f(A) = I + A + \cdots + A^k + \cdots$. Diagonalizing, we have thus $Q^{-1}f(A)Q = I + Q^{-1}AQ + \cdots + Q^{-1}A^kQ + \cdots = I + \Lambda + \cdots + \Lambda^k + \cdots = f(\Lambda)$ [from property (4) of Remark 16.6 and the diagonalization theorem]. The ith diagonal element of $Q^{-1}f(A)Q$ is now a series $f(\lambda_i) = \sum_{k=0}^{\infty} \lambda_i^{\,k}$, which converges to

$(1 - \lambda_i)^{-1}$ provided $|\lambda_i| < 1$. But $Q^{-1}f(A)Q$ is the diagonal matrix similar to $f(A)$. Therefore $f(\lambda_i)$, $i = 1, \ldots, n$ is a characteristic value of $f(A)$.

Two matrix-valued functions of particular interest in economics are the algebraic function $(I - A)^{-1} = \sum_{k=0}^{\infty} A^k$ and the matrix exponential $e^{tA} = \sum_{k=0}^{\infty} (t^k/k!)A^k$. The first appears in the context of input–output economics. The second in dynamic problems involving differential equations.

Example The open Leontief model of Chapter 15, presented there as a production model, may be also interpreted as a multisectoral income generation model in the following way.

Suppose the total expenditure of the economy is represented by the sum of vectors

$$y = x + z,$$

where $y = (y_1, \ldots, y_n)^{\mathrm{T}}$, $x = (x_1, \ldots, x_n)^{\mathrm{T}}$, and $z = (z_1, \ldots, z_n)^{\mathrm{T}}$. The ith component of y represents total expenditure in goods of (that is, total income of) the ith sector. It is equal to "private" or induced expenditure x_i plus the "government" or autonomous expenditure z_i.

Assume "private" expenditure x_i in goods of the ith sector is the total amount of income that all sectors $j = 1, \ldots, n$ spend in the ith sector. If each of the $j = 1, \ldots, n$ sectors spends a constant fraction a_{ij} of its total income in goods of the ith sector, we can write $y_i = \sum_{j=1}^{n} a_{ij} y_j + z_i$. In matrix notation we have

$$y = Ay + z,$$

where the matrix $A = [a_{ij}]$ contains the $n \times n$ "propensities to spend." In other words, a_{ij} represents the "propensity to spend" income of the jth sector in goods of the ith sector.

We are interested in knowing about the effects of a given expenditure z of the "government" or autonomous sector upon the income-levels vector y. From

$$y = Ay + z,$$

we have

$$y - Ay = (I - A)y = z,$$

so that

$$y = (I - A)^{-1}z$$

or, as we know,

$$y = (I + A + A^2 + \cdots + A^k + \cdots)z.$$

The latter expression suggests that the income level determined by the government expenditure z is equal to the same expenditure Iz plus the first, second, ..., kth, ... round of induced expenditures Az, $A^2z, \ldots, A^kz, \ldots$. The process converges to a determinate level $(1 - \Lambda)^{-1}$, if each λ_i on the diagonal of Λ has $|\lambda_i| < 1$ (see Theorem 16.17). Clearly λ_i, $i = 1, \ldots, n$ are the characteristic values of A. The matrix $(I - A)^{-1}$ is called the matrix income multiplier. It determines the extent to which an autonomous expenditure vector z is multiplied up to an income vector y.

Matrices Whose Elements Are Functions

Our attention was focused so far on matrices, the elements of which are scalars. If we imagine now that the element $a_{ij} = a_{ij}(t)$, that is, a_{ij} is a real-valued function defined for $t \in [a, b]$, then we have extended our concept of a matrix to include matrices with real-valued functions as elements. We write $A = A(t) = [a_{ij}(t)]$.

It is easy to prove that the following rules of calculus apply.

Remark 16.11 Let A be an $m \times n$ matrix of real-valued functions $a_{ij}(t)$ for $t \in [a, b]$. Then

(1) $\dot{A} = [\dot{a}_{ij}(t)] = [da_{ij}(t)/dt] = DA$,
(2) $\int_\alpha^\beta A = [\int_\alpha^\beta a_{ij}(t)\, dt]$,
(3) $D(A + B) = DA + DB$,
(4) $D(A \cdot B) = A \cdot (DB) + (DA) \cdot B$,
(5) $D(A^k) = \sum_{i=1}^k A^{k-i}(DA)A^{i-1}$, for $k \in N$,
(6) $D(A^{-1}) = -A^{-1}(DA)A^{-1}$,
(7) $D(e^{At}) = Ae^{At} = e^{At} \cdot A$.

All these results remind us of the well-known rules of calculus. Observe that in (1) we have merely provided alternative symbolic expression for the derivative of a matrix.

Example Let us consider a simple case of dynamic price adjustment.

Regarding prices as functions of time, we write $p(t) = p$ to represent the n-component vector of normalized prices. Additionally, we let $z(p) = (z_1(p), \ldots, z_n(p))^T$ be the n-component vector of continuous single-valued excess-demand functions (see Chapter 9, Section 9.4).

We now describe an elementary price adjustment mechanism by means of the vector differential equation,

(1) $$Dp = \dot{p} = \operatorname{diag}(k_1, \ldots, k_n)\, z(p),$$

where $Dp = \dot{p}$ is the vector of first-order time derivatives of p, and $\operatorname{diag}(k_1, \ldots, k_n)$ the matrix of "speeds of responses" k_i.

Let $u = p - p^*$ be the vector of the price deviations from the equilibrium price vector p^*, and assume, for simplicity, that $k_i = 1$ for all i. Under circumstances, and for small deviations from the equilibrium price, Eq. (1) can be satisfactorily approximated by the linear vector differential equation,

$$(2) \qquad Du = Au,$$

where A is the $n \times n$ matrix of partial derivatives $a_{ij} = \partial z_i(p)/\partial p_j$.

Such an equation allows us to study the local stability of the adjustment process, that is, the stability in a small neighborhood of the equilibrium.

Let us try as a solution $u(t) = e^{At}u(0)$, where $u(0)$ is the vector of price deviation at the initial instant $t = 0$, and as such constant. From Remark 16.11, property (7), $D(e^{At}u(0)) = Ae^{At}u(0)$. Substituting in Eq. (2), $D(e^{At}u(0)) = Ae^{At}u(0)$; that is, $Ae^{At}u(0) - Ae^{At}u(0) = \theta$.

The expression $u(t) = e^{At}u(0)$ is thus a solution of Eq. (2).

Note that if e^{At} can be diagonalized into $e^{\Lambda t}$, then the latter has as diagonal elements the characteristic values

$$e^{\lambda_i} = 1 + \lambda_i + \lambda_i^2/2! + \cdots + \lambda_i^k/k! + \cdots, \qquad i = 1, \ldots, n.$$

Local asymptotic stability of the equilibrium, that is, convergence of $u(t) \to \theta$ as $t \to \infty$, requires that the real part of λ_i, $i = 1, \ldots, n$ be less than zero.

EXERCISES

Characteristic Polynomials, Values, and Vectors

E.16.1 Given the matrices below, find (a) their characteristic values; (b) a basis for each "eigenspace" (associated invariant space):

$$A = \begin{pmatrix} -6 & 6 & -4 \\ 4 & -8 & 6 \\ 7 & -9 & 9 \end{pmatrix}, \qquad B = \begin{pmatrix} 3 & 2 & 2 \\ -1 & 0 & -2 \\ 1 & 2 & 1 \end{pmatrix}.$$

Hint: The concept of "eigenspace" (associated invariant space) follows immediately from Theorem 16.1. See also E.16.2.

E.16.2 Find the characteristic values of the following matrices, and compute the normalized associated characteristic vectors:

$$A = \begin{pmatrix} 2 & 0 & 5 & -4 \\ 0 & -3 & -9 & 3 \\ -2 & 0 & -6 & 4 \\ -3 & -1 & -11 & 7 \end{pmatrix}, \qquad B = \begin{pmatrix} 4 & 4 & 1 \\ 2 & 2 & 0 \\ 2 & 2 & 1 \end{pmatrix}.$$

E.16.3 Prove that for two square matrices A and B, the product matrices AB and BA have the same eigenvalues.

Hint: Theorem 14.12.

E.16.4 Prove proposition (2) of Remark 16.6.

Hint: See Theorems 14.13 and 14.1.

E.16.5 Let S be the matrix defined below, where A and C are square matrices. Prove that the characteristic polynomial of S is the product of the characteristic polynomials of A and C.

$$S = \begin{pmatrix} A & B \\ Z & C \end{pmatrix}.$$

Hint: First solve E.14.6.

E.16.6 Prove proposition (3) of Remark 16.6.

E.16.7 Give examples of 2×2 and 3×3 matrices to illustrate Theorem 16.6.

Diagonalization, Triangularization, Matrix-Series

E.16.8 Diagonalize the following matrices and give the corresponding modal matrices:

$$A = \begin{pmatrix} 1 & 1 & 1 & 1 \\ 1 & 3 & 3 & 1 \\ 1 & 3 & 6 & 1 \\ 1 & 3 & 6 & 10 \end{pmatrix}, \qquad B = \begin{pmatrix} 2 & -4 & 2 \\ -4 & 2 & -2 \\ 2 & -2 & -1 \end{pmatrix}.$$

Hint: Observe that B is real symmetric.

E.16.9 Triangularize the following matrices, and give the triangularizing matrices:

$$A = \begin{pmatrix} 1 & 1 & 1 & 1 \\ 1 & 3 & 2 & 2 \\ 1 & 3 & 5 & 3 \\ 1 & 3 & 5 & 7 \end{pmatrix}, \qquad B = \begin{pmatrix} 1 & 1 & 2 \\ 1 & 3 & 5 \\ 1 & 3 & 8 \end{pmatrix}.$$

E.16.10 Prove that a necessary and sufficient condition for the series $\sum_{k=0}^{\infty} A^{(k)}$ to be convergent is that the sequence $(A^{(k)} \mid k \in N)$ be convergent.

PROBLEMS

P.16.1 Prove that if A is an orthogonal matrix with eigenvalue λ, then λ^{-1} is also eigenvalue of A.

Hint: $A^{-1} = A^{\mathrm{T}}$.

P.16.2 Prove that every characteristic value of an orthogonal matrix A has modulus unity, that is, for every λ_i, $(\lambda_i \cdot \bar{\lambda}_i)^{1/2} = 1$.

P.16.3 Prove that an orthogonal matrix of odd order has, at least, one eigenvalue equal to ± 1.

Hint: Begin by studying the case 3×3.

P.16.4 Prove that if A is $n \times n$, then there exists an orthogonal matrix Q which triangularizes A.

P.16.5 Let the $n \times n$ matrix $A = [a_{ij}]$ have $a_{ij} > 0$ for every (i, j). Prove that if A satisfies the Hawkins–Simon conditions, then $\lim_{k \to \infty} A^k = Z$.

Hint: See Morishima [1].

NOTES

See the work of Bellman [1], Gantmacher [1], and also Pease [1]. For a very interesting treatment of the subject with reference to economics see Karlin [1]. Gantmacher [1] contains a discussion of some stability topics. More advanced books on stability are those by Hahn [1] and Halanay [1]. Fundamental articles on the stability theory of general economic equilibrium are those of Arrow and Hurwicz [1], and Arrow *et al.* [1]. Here the relevant literature is very extensive but excellent; comprehensive discussions are presented by Arrow and Hahn [1], Morishima [1], and Nikaido [4]. Karlin's work [1] is most useful. See also Rader [2].

CHAPTER 17

REAL QUADRATIC FORMS

The subject of this chapter may be regarded as an application of some properties of real symmetric matrices to the study of a particular type of nonlinear mapping.

The type of nonlinear mapping we shall be concerned with may be written $q\colon V \to F$ and more precisely defined by $q(x) = f(x, x) = k \in F$, where $f\colon V \times V \to F$ is an inner product. Here x is a vector of the finite space V and k an element of the field F. We shall confine ourselves to studying some of the essential properties of the real-valued subclass of such mappings.

Mappings of the above type arise very frequently in optimization problems. As such they occupy an important place among the mathematical tools of theoretical economics.

Our main concern will be the behavior of $q(x)$ as the vector x varies over the space V or a subset of it. We shall pay particular attention to the conditions under which the positivity (or nonnegativity) and negativity (or nonpositivity) of the function q is guaranteed.

Nonlinear mappings such as q are described by real symmetric matrices. As we shall see, the positivity or negativity of q will completely depend on some specific properties of the real symmetric matrix A associated with q. Henceforth, the study of the properties of q will be reduced to the study of the properties of A.

We have stated above that the study of the variations of $q(x)$ may be accomplished by allowing the vector x to range over the whole space V or by confining its values to a specified subset of V. The first problem is dealt with in Section 17.1. There we try to answer some rather general questions. The second problem, that of the constrained variation of $q(x)$, will have our attention in Section 17.2.

To simplify matters, throughout this chapter x and y will always represent column vectors; their transposes x^T and y^T will, accordingly, represent row vectors; $(x_i)^2$ stands for the square of the ith component of x.

17.1 Unconstrained Quadratic Forms

Definition 17.1 Let A be a real symmetric matrix and x a vector of the real space V. Then a real-valued function q: $V \to R$ defined by $q(x) = x^T A x = \sum_{i=1}^{n} \sum_{j=1}^{n} x_i a_{ij} x_j$ is called a quadratic form over the real field.

Definition 17.1 can of course be extended to include the complex case. However, as announced before, we shall only pay attention to real forms.

Example

$$q(x) = (x_1)^2 - 5(x_2)^2 + 6(x_3)^2 + 4x_1x_2 + 6x_1x_3 + 8x_2x_3.$$

We can also write

$$\begin{aligned} q(x) = (x_1)^2 &+ 2x_2x_1 + 3x_3x_1 + 2x_1x_2 \\ &- 5(x_2)^2 + 4x_3x_2 + 3x_1x_3 + 4x_2x_3 + 6(x_3)^2. \end{aligned}$$

Or, finally,

$$q(x) = (x_1, x_2, x_3)\begin{pmatrix} 1 & 2 & 3 \\ 2 & -5 & 4 \\ 3 & 4 & 6 \end{pmatrix}\begin{pmatrix} x_1 \\ x_2 \\ x_3 \end{pmatrix} = x^T A x.$$

In general, one refers to the rank of the matrix A as the rank of the quadratic form $q(x) = x^T A x$. With this in mind, we distinguish between nonsingular and singular forms.

Definite, Semidefinite, and Indefinite Forms

Since our interest concerns the variations of $q(x)$ as x takes different values, it will be convenient to possess an appropriate classification of quadratic forms. This is given in the following definition.

Definition 17.2 Let $q(x) = x^{\mathrm{T}}Ax$ be a quadratic form. Then (1) q is said to be positive (negative) definite if $q(x) > 0$ (<0) for all vectors $x \neq \theta$; (2) q is called positive (negative) semidefinite if $q(x) \geq 0$ (≤ 0) for $x \neq \theta$ with $q(x) = 0$ for some $x \neq \theta$; (3) q is said to be nonnegative (nonpositive) definite if $q(x) \geq 0$ (≤ 0) for all $x \neq \theta$; (4) q is called an indefinite form if it is positive for some vectors x and negative for others.

Example 1 Replace different vectors x in

$$q(x) = (x_1, x_2)\begin{pmatrix} 4 & \sqrt{6} \\ \sqrt{6} & 3 \end{pmatrix}\begin{pmatrix} x_1 \\ x_2 \end{pmatrix}$$

and observe that the quadratic form is positive definite.

Example 2 Replace different vectors x in

$$q(x) = (x_1, x_2)\begin{pmatrix} 4 & 2 \\ 2 & 1 \end{pmatrix}\begin{pmatrix} x_1 \\ x_2 \end{pmatrix}$$

and observe that the quadratic form is positive semidefinite.

Example 3 Using different vectors x show that

$$q(x) = (x_1, x_2)\begin{pmatrix} -4 & -\sqrt{6} \\ -\sqrt{6} & -3 \end{pmatrix}\begin{pmatrix} x_1 \\ x_2 \end{pmatrix}$$

is a negative definite quadratic form.

Example 4 Using different vectors x show that

$$q(x) = (x_1, x_2)\begin{pmatrix} -4 & -2 \\ -2 & -1 \end{pmatrix}\begin{pmatrix} x_1 \\ x_2 \end{pmatrix}$$

is a negative semidefinite quadratic form.

Example 5 Observe that a case of indefiniteness is shown by

$$q(x) = (x_1, x_2)\begin{pmatrix} 5 & \sqrt{12} \\ \sqrt{12} & 1 \end{pmatrix}\begin{pmatrix} x_1 \\ x_2 \end{pmatrix}.$$

From all the cases considered in Definition 17.2, those to which we shall pay more attention are positive and negative definite and positive and negative semidefinite. This provides us with a corresponding way of classifying symmetric matrices.

Definition 17.3 A symmetric matrix A is said to be positive (negative) definite if the quadratic form $x^T Ax$ is positive (negative) definite. It is said to be positive (negative) semidefinite if $x^T Ax$ is positive (negative) semidefinite.

We have classified the matrices by referring to the property of the quadratic forms associated to them. It will become obvious soon that it is the matrix which determines the character of the form.

Congruence and Diagonalization

Quadratic forms can be simplified and reduced to canonical forms.

The fact that the matrix A of a real quadratic form $x^T Ax$ is symmetric has important consequences for the choice of the transformation matrix.

Definition 17.4 Let A and B be two $n \times n$ matrices. Then B is said to be congruent to A if there is a nonsingular matrix Q such that $B = Q^T AQ$.

It is clear that a congruence transformation is a type of similarity transformation.

In the case of a quadratic form we may make the following transformation. Suppose we are given $q(x) = x^T Ax$. We choose a nonsingular matrix Q such that $y = Q^{-1}x$, that is, $x = Qy$. Now replacing the expressions $x^T = (Qy)^T$ and $x = Qy$ in $x^T Ax$, we get $y^T Q^T AQy$. Since A is symmetric, so also is $Q^T AQ$. Writing $Q^T AQ = B$ it results that the original quadratic form $x^T Ax$ has been transformed into the new one $y^T By$, where B and A are congruent matrices.

The process of reduction can of course be carried further. From the preceding chapter we know that every real symmetric matrix is diagonalizable and precisely by means of an orthogonal matrix Q, that is, a matrix with $Q^{-1} = Q^T$.

Suppose again we are given $x^T Ax$. We can construct a nonsingular matrix Q using an orthonormal set of characteristic vectors of A. As before, we have a new vector y such that $y = Q^{-1}x$; that is, $x = Qy$. Replacing in $x^T Ax$, we get $y^T Q^T AQy$. In the way in which Q has been defined, $Q^T = Q^{-1}$ and Q is the matrix which diagonalizes A. We have thus $y^T Q^T AQy = y^T Q^{-1} AQy = y^T \Lambda y$, where $\Lambda = \text{diag}(\lambda_1, \ldots, \lambda_n)$. The λ_i, $i = 1, \ldots, n$ are, of course, the characteristic values of A.

The results of our inquiry can be summarized as follows:

Theorem 17.1 Any real quadratic form $q(x) = x^T Ax$ can be reduced to a diagonal canonical form $q(y) = y^T \Lambda y$ by means of an orthogonal matrix Q, with $x = Qy$.

Observe that if $x^{\mathrm{T}}Ax = \sum_{i=1}^{n} \sum_{j=1}^{n} x_i a_{ij} x_j$, its diagonal form is simply $y^{\mathrm{T}}\Lambda y = \sum_{j=1}^{n} \lambda_j (y_j)^2$, that is, it contains only quadratic terms.

Example By finding the eigenvalues of the corresponding matrix A, it is easy to reduce the quadratic forms of the preceding Examples 1–5 to the diagonal canonical form. Following the examples in the same order we get, respectively,

$$y^{\mathrm{T}}\begin{pmatrix} 6 & 0 \\ 0 & 1 \end{pmatrix} y = 6(y_1)^2 + (y_2)^2, \qquad y^{\mathrm{T}}\begin{pmatrix} 0 & 0 \\ 0 & 5 \end{pmatrix} y = 5(y_2)^2,$$

$$y^{\mathrm{T}}\begin{pmatrix} -6 & 0 \\ 0 & -1 \end{pmatrix} y = -6(y_1)^2 - (y_2)^2, \qquad y^{\mathrm{T}}\begin{pmatrix} 0 & 0 \\ 0 & -5 \end{pmatrix} y = -5(y_2)^2,$$

and

$$y^{\mathrm{T}}\begin{pmatrix} 7 & 0 \\ 0 & -1 \end{pmatrix} y = 7(y_1)^2 - (y_2)^2.$$

Note again that the first and the third case show a positive definite and a negative definite form, respectively. That the second and fourth case illustrate a positive semidefinite and a negative semidefinite form, respectively, can easily be shown by choosing any vector of the form $y = (y_1, 0)$. The fifth is, again, an indefinite form.

Positive Definite and Semidefinite Matrices

The properties of a quadratic form are condensed in the symmetric matrix representing it. The conditions for the definiteness or semidefiniteness of a quadratic form may thus be studied as the conditions for the definiteness or semidefiniteness of the corresponding matrix. We analyze here the essential elements of the problem.

Theorem 17.2 Let A be an $n \times n$ real symmetric matrix. Then A is positive definite if and only if all characteristic values of A are positive.

Proof Let $x^{\mathrm{T}}Ax$ be a quadratic form. Assume Q is an orthogonal matrix which diagonalizes A. We define $y = Q^{\mathrm{T}}x$ and get $x^{\mathrm{T}}Ax = (Qy)^{\mathrm{T}}AQy = y^{\mathrm{T}}Q^{\mathrm{T}}AQy = y^{\mathrm{T}}\Lambda y$. Thus $x^{\mathrm{T}}Ax = y^{\mathrm{T}}\Lambda y > 0$ if for all $i = 1, \ldots, n$, $\lambda_i > 0$. Conversely, assume A is positive definite and has a nonpositive eigenvalue, say $\lambda_1 \leq 0$. Define y' as the column vector with 1 in its first place and zeros elsewhere, and consider $x' = Qy'$. Since $x' \neq \theta$, we must have $(x')^{\mathrm{T}}Ax' = (y')^{\mathrm{T}}\Lambda y' \leq 0$. Thus, if $\lambda_1 \leq 0$, A is not positive definite violating our earlier assumption.

Using the same method, we can prove the following theorem.

Theorem 17.3 Let A be an $n \times n$ real symmetric matrix. Then A is positive semidefinite if and only if we have for the characteristic values λ_i, $i = 1, \ldots, n$, that all $\lambda_i \geq 0$ with at least one $\lambda_i = 0$.

Other necessary and sufficient conditions for a matrix to be positive definite can be immediately derived from Theorem 17.2. Let us introduce first the following definition:

Definition 17.5 Let A be an $n \times n$ symmetric matrix. Then the principal submatrix A_k of A is the $k \times k$ submatrix obtained from A by deleting the last $n - k$ rows and columns. The determinant $\det(A_k)$ is called a principal minor of A.

Example Given the 4×4 matrix A below,

$$A = \begin{pmatrix} a_{11} & a_{12} & a_{13} & a_{14} \\ a_{21} & a_{22} & a_{23} & a_{24} \\ a_{31} & a_{32} & a_{33} & a_{34} \\ a_{41} & a_{42} & a_{43} & a_{44} \end{pmatrix},$$

we have the principal submatrices,

$$A_3 = \begin{pmatrix} a_{11} & a_{12} & a_{13} \\ a_{21} & a_{22} & a_{23} \\ a_{31} & a_{32} & a_{33} \end{pmatrix}; \qquad A_2 = \begin{pmatrix} a_{11} & a_{12} \\ a_{21} & a_{22} \end{pmatrix}; \qquad A_1 = (a_{11}).$$

The corresponding principal minors of A are $\det(A_3)$, $\det(A_2)$, and $\det(A_1)$. They are sometimes called the "nested" principal minors of A.

We can return now to the study of the necessary and sufficient conditions for positive definiteness.

Theorem 17.4 An $n \times n$ real symmetric matrix A is positive definite if and only if $\det(A) > 0$ and all other principal minors of A are positive.

Proof Let $x^T Ax = y^T \Lambda y$. If A is positive definite then, by Theorem 17.2, $\lambda_i > 0$ for all $i = 1, \ldots, n$. Clearly then $\det(A) = \lambda_1 \cdot \lambda_2 \cdots \lambda_{n-1} \cdot \lambda_n > 0$, $\det(A_{n-1}) = \lambda_1 \cdot \lambda_2 \cdots \lambda_{n-1} > 0, \ldots,$ $\det(A_2) = \lambda_1 \cdot \lambda_2 > 0$, $\det(A_1) = \lambda_1 > 0$. Conversely, suppose A is positive definite with one principal minor, say $\det(A_k) \leq 0$. Then we must have at least one λ_i, $i = 1, \ldots, k$ which is not positive, thus violating our assumption about the positive definiteness of A.

From Theorem 17.4 it follows that a necessary condition for A to be positive definite is that $\det(A) > 0$. That this is not a sufficient condition follows from the fact that $\det(A)$ may have an even number of negative roots and be still positive.

A result closely related to Theorem 17.4 is given next. The proof can immediately be obtained using Theorem 17.4 and is left as an exercise.

Theorem 17.5 An $n \times n$ real symmetric matrix A is positive definite if and only if every principal submatrix of A is positive definite.

It is easy now to derive the corresponding conditions for the positive semidefiniteness of a real symmetric matrix.

Theorem 17.6 An $n \times n$ real symmetric matrix A is positive semidefinite if and only if $\det(A) = 0$ and all other principal minors of A are either positive or zero.

Proof Suppose $x^{\mathrm{T}}Ax = y^{\mathrm{T}}\Lambda y$. If A is positive semidefinite it must have at least one zero characteristic value. Assume $\lambda_i > 0$ for $i = 1, \ldots, n-1$ and $\lambda_n = 0$. Then clearly $\det(A) = \lambda_1 \cdot \lambda_2 \cdots \lambda_{n-1} \cdot 0 = 0$ and all other $\det(A_k) \geq 0$. Conversely, suppose the matrix A has $\det(A) = 0$ and all other $\det(A_k) \geq 0$. Since $\det(A) = \lambda_1 \cdot \lambda_2 \cdots \lambda_{n-1} \cdot \lambda_n$ and $\det(A_k) = \lambda_1 \cdot \lambda_2 \cdots \lambda_{k-1} \cdot \lambda_k$, it follows that $\lambda_n = 0$ and $\lambda_i \geq 0$ for $i = 1, \ldots, n-1$. Therefore A is positive semidefinite.

Another interesting result is the following:

Theorem 17.7 Let A be a positive definite matrix. Then its inverse A^{-1} is positive definite.

Proof Assume we have $Q^{\mathrm{T}}AQ = \Lambda$. Taking inverses $(Q^{\mathrm{T}}AQ)^{-1} = (Q^{-1}AQ)^{-1} = QA^{-1}Q^{-1} = QA^{-1}Q^{\mathrm{T}} = \Lambda^{-1}$. Thus if Q is an orthogonal matrix which diagonalizes A into $\Lambda = \operatorname{diag}(\lambda_1, \ldots, \lambda_n)$, Q^{T} is an orthogonal matrix which diagonalizes A^{-1} into $\Lambda^{-1} = \operatorname{diag}(\lambda_1^{-1}, \ldots, \lambda_n^{-1})$. Since every characteristic value λ_i of A is positive, we must have every characteristic value λ_i^{-1} of A^{-1} positive too. Therefore A^{-1} is positive definite.

So far we have been exclusively concerned with positive definite and semidefinite matrices. We can weaken some of our preceding results by considering nonnegative definite instead of positive semidefinite matrices. Proofs of the following properties are simple and suggested as exercises.

Remark 17.1 A symmetric matrix A is nonnegative definite if and only if each of its eigenvalues is nonnegative.

Remark 17.2 Every principal minor of a nonnegative definite matrix A is nonnegative.

Remark 17.3 Every principal submatrix of a nonnegative definite matrix A is nonnegative definite.

Negative Definite and Semidefinite Matrices

Our first investigation is into the signs of the characteristic values of a negative definite matrix. Afterward we shall follow a similar path as before. However, our task will be greatly simplified by our knowledge about positive definite matrices.

Theorem 17.8 Let A be an $n \times n$ real symmetric matrix. Then A is negative definite if and only if all characteristic values of A are negative.

Proof Let Q be an orthogonal matrix that diagonalizes the quadratic form $x^{\mathrm{T}}Ax$. After defining $y = Q^{\mathrm{T}}x$ and replacing, we get $x^{\mathrm{T}}Ax = y^{\mathrm{T}}Q^{\mathrm{T}}AQy = y^{\mathrm{T}}\Lambda y$. Thus $x^{\mathrm{T}}Ax = y^{\mathrm{T}}\Lambda y < 0$. Therefore $\lambda_i < 0$ for all $i = 1, \ldots, n$. Conversely, assume A is negative definite and has a nonnegative eigenvalue, say $\lambda_1 \geq 0$. Define y' as the column vector with 1 on its first place and zeros elsewhere, and consider $x' = Qy'$. Since $x' \neq \theta$, we have $(x')^{\mathrm{T}}Ax' = (y')^{\mathrm{T}}\Lambda y' \geq 0$. Thus, if $\lambda_1 \geq 0$, A is not negative definite contradicting our assumption.

In a similar way we can prove the following theorem.

Theorem 17.9 Let A be an $n \times n$ real symmetric matrix. Then A is negative semidefinite if and only if all characteristic values $\lambda_i \leq 0$ with at least one $\lambda_i = 0$.

We can examine now some alternative conditions for negative definiteness.

Theorem 17.10 An $n \times n$ real symmetric matrix A is negative definite if and only if $\det(A)$ has sign $(-1)^n$ and all other principal minors $\det(A_k)$ have sign $(-1)^k$.

Proof Let $x^{\mathrm{T}}Ax = y^{\mathrm{T}}\Lambda y$ and assume A is negative definite. By Theorem 17.8, $\lambda_i < 0$ for all $i = 1, \ldots, n$. Therefore $\det(A) = \lambda_1 \cdot \lambda_2 \cdots \lambda_{n-1} \cdot \lambda_n = (-1)^n|\lambda_1 \cdot \lambda_2 \cdots \lambda_{n-1} \cdot \lambda_n|$; $\det(A_{n-1}) = \lambda_1 \cdot \lambda_2 \cdots \lambda_{n-1} = (-1)^{n-1}|\lambda_1 \cdot \lambda_2 \cdots \lambda_{n-1}|$; $\ldots$; $\det(A_2) = \lambda_1 \cdot \lambda_2 = (-1)^2|\lambda_1 \cdot \lambda_2| = \lambda_1 \cdot \lambda_2 > 0$; and $\det(A_1) = \lambda_1 < 0$. Assume now A is negative definite and $\det(A_k) = (-1)^k|\lambda_1 \cdot \lambda_2 \cdots \lambda_k|$. Suppose $\det(A_{k+1})$ has the same sign as $\det(A_k)$, that is, $\det(A_{k+1}) = (-1)^k \times |\lambda_1 \cdot \lambda_2 \cdots \lambda_k \cdot \lambda_{k+1}|$. It follows that $\lambda_{k+1} > 0$ thus violating our assumption about the negative definiteness of A.

The result of Theorem 17.10 is usually formulated in the following terms:

Remark 17.4 If A is a negative definite matrix, then its successive principal minors of increasing order alternate in sign.

Using Theorem 17.10, we can immediately prove the following property.

Theorem 17.11 An $n \times n$ real symmetric matrix A is negative definite if and only if every principal submatrix of A is negative definite.

An important property which follows from both Theorems 17.4 and 17.5, and which is often given as a definition of negative definiteness is stated next.

Theorem 17.12 If A is positive definite, then $-A$ is negative definite and conversely.

Since a negative semidefinite matrix has at least one $\lambda_i = 0$, the next theorem is easily proved.

Theorem 17.13 Let the matrix A be negative semidefinite. Then $\det(A) = 0$.

Now, using Theorem 17.7 and Theorem 17.12, we immediately arrive at the following.

Theorem 17.14 Let A be a negative definite matrix. Then its inverse A^{-1} is negative definite.

As we did for nonnegative definite matrices, we can weaken some of our previous results and obtain properties of nonpositive definite matrices.

Remark 17.5 A symmetric matrix A is nonpositive definite if and only if each of its eigenvalues is nonpositive.

Remark 17.6 Every principal submatrix of a nonpositive definite matrix A is nonpositive definite.

Most of the properties studied up to this point can be examined by means of the examples given earlier in this chapter.

Example Quadratic forms provide a concise and elegant way to formulate the necessary and sufficient conditions for a local optimum in the classical optimization problem.

Let $f(x)$ be a real-valued function continuous over a set S with continuous first- and second-order partial derivatives $f_i(x)$ and $f_{ij}(x)$. For a sufficiently small variation h, the Taylor's expansion around $x = (x_1, \ldots, x_n)^T$ and up to the second-order term is

$$f(x + h) = f(x + dx) = f(x) + \sum_{i=1}^{n} f_i(x)\, dx + 1/2! \sum_{i=1}^{n} \sum_{j=1}^{n} f_{ij}(x)\, dx_i\, dx_j.$$

The well-known necessary and sufficient conditions for x^* to be an optimizer of $f(x)$ can be summarized as follows.

For a maximum

$$f_i(x^*) = 0, \quad \text{for} \quad i = 1, \ldots, n; \qquad \sum_{i=1}^{n} \sum_{j=1}^{n} f_{ij}(x^*)\, dx_i\, dx_j < 0.$$

For a minimum

$$f_i(x^*) = 0, \quad \text{for} \quad i = 1, \ldots, n; \qquad \sum_{i=1}^{n} \sum_{j=1}^{n} f_{ij}(x^*)\, dx_i\, dx_j > 0.$$

We can simplify these expressions using matrix notation. Let us introduce $\nabla f(x) = (f_1(x), \ldots, f_n(x))$, $dx = (dx_1, \ldots, dx_n)^{\mathrm{T}}$, and $H = [f_{ij}(x)]$. The latter, called the Hessian matrix, is an $n \times n$ real symmetric matrix with $f_{ij}(x) = f_{ji}(x)$.

The first three terms of the above Taylor's expansion can now be written,

$$f(x + h) = f(x + dx) = f(x) + \nabla f(x)\, dx + (1/2!)\, dx^{\mathrm{T}}\, H\, dx.$$

The second-order term is clearly a quadratic form and we can express the corresponding conditions by means of the matrix H. We summarize as follows.

For a maximum

$$\nabla f(x^*) = \theta, \qquad H = [f_{ij}(x^*)] \quad \text{negative definite.}$$

For a minimum

$$\nabla f(x^*) = \theta, \qquad H = [f_{ij}(x^*)] \quad \text{positive definite.}$$

Observe that we have considered here only the case of unconstrained optimization. An economic example which is formulated in the framework of the classical constrained optimization problem is given at the end of Section 17.2.

17.2 Constrained Quadratic Forms

In theoretical economics, in particular in the so-called neoclassical theory, many problems can be formulated by means of the classical constrained-optimization model. Individual demand functions, for instance, can be specified from a model in which the consumer is supposed to maximize a nonlinear utility function, by given prices, subject to a linear (budget) constraint. We shall show the usefulness of the approach after discussing the mathematical background of it.

Essentially, we are going to extend some of the results previously achieved. Our concern is still the variations of the quadratic form $q(x)$ and, in particular, like the preceding section, the conditions under which the positivity (or negativity) of $q(x)$ can be postulated. However, instead of allowing the vector x to vary over the whole space V, we constrain its possible values to a subset of V. To be specific we shall study the quadratic form $q(x) = x^T Ax$ with the condition that the n variables x_i satisfy the homogeneous linear system of m equations in n unknowns $Bx = \theta$. Before reaching a solution to this problem we have to get acquainted with some additional results on quadratic forms.

Simultaneous Diagonalization of Two Quadratic Forms and Finsler's Theorem

We begin our investigation by studying an important property of positive definite matrices.

Theorem 17.15 Let A be a positive definite matrix. Then a nonsingular matrix P exists such that $A = P^T P$.

Proof Let Q be an orthogonal matrix that diagonalizes A, that is, $Q^T AQ = \Lambda = \operatorname{diag}(\lambda_1, \ldots, \lambda_n)$. Consider the matrix $D = \operatorname{diag}(\lambda_1^{-\frac{1}{2}}, \ldots, \lambda_n^{-\frac{1}{2}})$, and define $S = QD$. Then $S^T AS = D^T Q^T AQD = D^T \Lambda D$. Now, obviously, $S^T AS = I$. Premultiplying by S and postmultiplying by S^T we obtain $S(S^T AS)S^T = SIS^T = SS^T$. But $S(S^T AS)S^T = (SS^T A)SS^T = SS^T$. Thus $SS^T A = I$. From the latter $SS^T = A^{-1}$. Taking inverses $(SS^T)^{-1} = (S^T)^{-1} S^{-1} = (S^{-1})^T S^{-1} = A$. Calling $(S^{-1})^T = P^T$ and $S^{-1} = P$, we finally get $A = P^T P$.

In the previous section we considered the problem of the diagonalization of a quadratic form. We shall see now that in general a matrix R exists such that it is possible to simultaneously diagonalize two quadratic forms of which one is positive definite. This result cannot be regarded as a generalization of what we know about diagonalizing a single quadratic form for, as we shall also see, the diagonalizing matrix R will be here orthogonal only in a very special case.

Theorem 17.16 Let $q_1(x) = x^T Ax$ and $q_2(x) = x^T Bx$ be two quadratic forms, where $q_1(x)$ is positive definite. Then, there is a matrix R such that $q_1(x)$ and $q_2(x)$ can be reduced to the diagonal forms

$$q_1(y) = y^T Iy = \sum_{j=1}^{n} (y_j)^2 \qquad \text{and} \qquad q_2(y) = y^T \Gamma y = \sum_{j=1}^{n} \mu_j (y_j)^2,$$

respectively.

Proof We have to prove that a matrix R exists such that $A = (R^{-1})^{\mathrm{T}}R^{-1}$ and $B = (R^{-1})^{\mathrm{T}}\Gamma R^{-1}$, where $\Gamma = \operatorname{diag}(\mu_1, \ldots, \mu_n)$. If this is true, then clearly $R^{\mathrm{T}}AR = I$ and $R^{\mathrm{T}}BR = \Gamma$. Since A is positive definite, we can find a nonsingular matrix P such that $A = P^{\mathrm{T}}P$.

Define $C = (P^{\mathrm{T}})^{-1}BP^{-1}$. Observe that C is a symmetric matrix. Then, an orthogonal matrix Q exists which diagonalizes C. We have thus $Q^{\mathrm{T}}CQ = Q^{\mathrm{T}}(P^{\mathrm{T}})^{-1}BP^{-1}Q = \Gamma$. Taking $R = P^{-1}Q$, we get $R^{\mathrm{T}}BR = \Gamma = \operatorname{diag}(\mu_1, \ldots, \mu_n)$ and

$$\begin{aligned} R^{\mathrm{T}}AR &= R^{\mathrm{T}}P^{\mathrm{T}}PR = R^{\mathrm{T}}(QR^{-1})^{\mathrm{T}}QR^{-1}R \\ &= R^{\mathrm{T}}(R^{\mathrm{T}})^{-1}Q^{\mathrm{T}}QR^{-1}R = R^{\mathrm{T}}(R^{\mathrm{T}})^{-1}R^{-1}R = I. \end{aligned}$$

We can write

$$|B - \mu A| = |B - \mu P^{\mathrm{T}}P| = |(P^{\mathrm{T}})^{-1}BP^{-1} - \mu I| = |C - \mu I|.$$

Observe that, in the preceding result, $\mu_1, \ldots, \mu_n$ are the characteristic values of the matrix $C = (P^{\mathrm{T}})^{-1}BP^{-1}$.

In general, the diagonalizing matrix R is not orthogonal. However, if the matrices A and B commute, that is, $AB = BA$, we can show that they have the same characteristic values and thus can be diagonalized by the same orthogonal matrix Q. We summarize this in the following remark.

Remark 17.7 Let A and B be two real symmetric and commutative matrices. Then there exists an orthogonal matrix Q such that $Q^{\mathrm{T}}AQ$ and $Q^{\mathrm{T}}BQ$ are diagonal matrices.

We study next an important property of quadratic forms usually known as Finsler's theorem. It is essential for the study of constrained forms which we shall undertake immediately afterward.

Theorem 17.17† Let $x^{\mathrm{T}}Ax$ and $x^{\mathrm{T}}Bx$ be two quadratic forms, where the latter is positive semidefinite. If $x^{\mathrm{T}}Ax > 0$ whenever $x^{\mathrm{T}}Bx = 0$, then there exists a number $r > 0$ such that $x^{\mathrm{T}}(A + rB)x$ is positive definite.

Proof Let Q be an orthogonal matrix which diagonalizes B. Since B has rank k and is positive semidefinite we have, for $x = Qy$,

$$x^{\mathrm{T}}Bx = y^{\mathrm{T}}Q^{\mathrm{T}}BQy = y^{\mathrm{T}}\Lambda y = \sum_{i=1}^{n} \lambda_i (y_i)^2 = \sum_{i=1}^{k} \lambda_i (y_i)^2.$$

Applying the same transformation to $x^{\mathrm{T}}Ax$, we have

$$x^{\mathrm{T}}Ax = y^{\mathrm{T}}Q^{\mathrm{T}}AQy = y^{\mathrm{T}}Cy = \sum_{i=1}^{n}\sum_{j=1}^{n} c_{ij}y_i y_j.$$

† Adapted from Bellman [1].

Now, we must have

$$\sum_{i=1}^{n}\sum_{j=1}^{n} c_{ij} y_i y_j > 0 \qquad \text{whenever} \qquad \sum_{i=1}^{n} \lambda_i (y_i)^2 = \sum_{i=1}^{k} \lambda_i (y_i)^2 = 0.$$

The latter certainly happens for $y = (0, \ldots, 0, y_{k+1}, \ldots, y_n)$. Thus, for all nonzero y_i, $i = k+1, \ldots, n$ we must have

$$\sum_{i=k+1}^{n}\sum_{j=k+1}^{n} c_{ij} y_i > 0.$$

There is, of course, an orthogonal $(n-k) \times (n-k)$ matrix which diagonalizes the latter quadratic form. Letting y_i, $i = 1, \ldots, k$ remain unchanged, we incorporate this transformation into the $n \times n$ orthogonal matrix P and define $y = Pw$. We have now

$$\begin{aligned} x^{\mathrm{T}}Ax &= \sum_{i=1}^{n}\sum_{j=1}^{n} c_{ij} y_i y_j = w^{\mathrm{T}}P^{\mathrm{T}}APw = w^{\mathrm{T}}Rw \\ &= \sum_{i=k+1}^{n} \lambda_i (w_i)^2 + \sum_{i=k+1}^{n}\sum_{j=1}^{k} p_{ij} w_i w_j + \sum_{i=1}^{k}\sum_{j=1}^{k} p_{ij} w_i w_j = Q(w). \end{aligned}$$

Using P to transform $x^{\mathrm{T}}Bx$, we simply have

$$x^{\mathrm{T}}Bx = y^{\mathrm{T}}\Lambda y = w^{\mathrm{T}}P^{\mathrm{T}}\Lambda Pw = w^{\mathrm{T}}\Lambda w = \sum_{i=1}^{k} \lambda_i (y_i)^2 = \sum_{i=1}^{k} \lambda_i (w_i)^2,$$

for y_i, $i = 1, \ldots, k$ were left unchanged. We have now,

$$x^{\mathrm{T}}(A + rB)x = w^{\mathrm{T}}(R + r\Lambda)w = Q(w) + r \sum_{i=1}^{k} \lambda_i (w_i)^2.$$

It would be easy to show that, for $\lambda_i > 0$, $i = 1, \ldots, n$ there exists an $r > 0$ large enough as to make $x^{\mathrm{T}}(A + rB)x = w^{\mathrm{T}}(R + r\Lambda)w > 0$.

Sufficient Conditions

Sufficiency criteria for the positive or negative definiteness of a constrained quadratic form play an important role in relation to the second-order conditions for an optimum in the classical, constrained optimization problem. As such they are most relevant for theoretical economics, especially in the context of the neoclassical theory.

The derivation of sufficient conditions can be done on the basis of Finsler's theorem. Bellman [1] and especially Lancaster [1] have approached the question in a manner which is particularly suitable for economic analysis. Here, essentially we follow Lancaster's [1] treatment of the problem.

We concentrate our attention on a quadratic form $x^{\mathrm{T}}Ax$ subject to an equality constraint $Bx = \theta$. We shall agree on A and B being $n \times n$ and $m \times n$

matrices, respectively. Thus B^TB will represent an $n \times n$ matrix. As usual, Z will stand for the zero matrix.

Theorem 17.18† Let $Q(x) = x^TAx$ be a quadratic form subject to the constraint $Bx = \theta$. If $\det(A)$ and every principal minor of

$$M = \begin{pmatrix} Z & B \\ B^T & A \end{pmatrix}$$

has the sign $(-1)^m$, then $Q(x)$ is positive definite for any vector x such that $Bx = \theta$.

Proof Consider the matrix product

$$M_1 \cdot M_2 = M_3 = \begin{pmatrix} -I_m & B \\ kB^T & A \end{pmatrix}\begin{pmatrix} I_m & B \\ Z & I_n \end{pmatrix} = \begin{pmatrix} -I_m & Z \\ kB^T & A + kB^TB \end{pmatrix}.$$

Note that in M_3 the expression $A + kB^TB$ is analogous to the one we studied in Finsler's theorem, for B^TB is here an $n \times n$ matrix. Taking determinants we have $\det(M_1) \cdot \det(M_2) = \det(M_3)$. Observe that $\det(M_2) = 1$. Therefore $\det(M_1) = \det(M_3)$. Now

$$\begin{aligned} \det(M_3) &= (-1)\begin{vmatrix} -I_{m-1} & Z \\ kB_{m-1}^T & A + kB^TB \end{vmatrix} \\ &= (-1)^2\begin{vmatrix} -I_{m-2} & Z \\ kB_{m-2}^T & A + kB^TB \end{vmatrix} \\ &\vdots \\ &= (-1)^{m-1}\begin{vmatrix} -1 & 0 \\ kB_1{}^T & A + kB^TB \end{vmatrix} \\ &= (-1)^m \det(A + kB^TB). \end{aligned}$$

Hence, if $A + kB^TB$ is positive definite, $\det(A + kB^TB) > 0$ and therefore the sign of $\det(M_1)$ is $(-1)^m$. But

$$\det(M_1) = \begin{vmatrix} -I_m & B \\ kB^T & A \end{vmatrix} = k^m \begin{vmatrix} (-1/k)I_m & B \\ B^T & A \end{vmatrix}.$$

Since we can always choose k large enough so as to make the sign of $\det(M_1)$ the same as the sign of

$$\begin{vmatrix} Z & B \\ B^T & A \end{vmatrix} = \det(M),$$

we can reduce the problem to stating the condition in terms of the signs of $\det(M)$ and the principal minors obtained from it by successively deleting the last $1, 2, \ldots, n - m + 1$ rows and columns.

† Adapted from Lancaster [1].

Sufficient conditions for the negative definiteness of the constrained form $Q(x)$ can be easily obtained from the above theorem by replacing $-A$ for A in the matrix M. As we know, if A is negative definite, then $-A$ is positive definite and conversely. As in the previous theorem, it is necessary to study the sign of $\det(M)$ and of the principal minors obtained by deleting the last $1, 2, \ldots, n - m + 1$ rows and columns. The proof is left here as an exercise.

Theorem 17.19 Let $Q(x) = x^{\mathrm{T}}Ax$ be a quadratic form subject to the constraint $Bx = \theta$. A sufficient condition for A to be negative definite is that the sign of $\det(M)$ be $(-1)^n$ and that of the principal minor of order $(m + n - r)$ be $(-1)^{n-r}$.

Example Consider the demand theory referred to in Chapter 6 and discussed in Chapter 9. Assume the consumption set of a consumer is totally preordered by a strongly convex preference preordering, that is, for any two commodity bundles x^1, $x^2 \in X$ and $\lambda \in]0, 1[$, if $x^2 \sim x^1$ then $\lambda x^2 + (1 - \lambda)x^1 \succ x^1$.

Let the vectors x, p and the scalar m represent, respectively, an n-commodity bundle, a given (fixed) price system and the consumer's income. Assume a real-valued, continuous, twice differentiable utility function $u(x)$ exists. If the consumer is a utility maximizer and p and m are given, then his choice problem can be fairly well represented by means of the classical constrained optimization model:

$$\text{Maximize} \quad u(x) \quad \text{subject to} \quad c(x) = px - m = 0.$$

As we know from calculus, the question reduces to finding the maximum of the Lagrangean function $l(x, \lambda) = u(x) - \lambda(px - m)$. First-order conditions are given as usual by $\nabla l(x^*, \lambda^*) = \theta$. The second-order conditions depend now on the behavior of the matrix M of the two preceding theorems, estimated at the point (x^*, λ^*).

Let u_i, c_i and u_{ij}, c_{ij} represent first-order and second-order partial derivatives, respectively. In our special case of a single constraint, the matrix B of Theorems 17.18 and 17.19 reduces to a vector. We thus have

$$\begin{pmatrix} 0 & c_1 & \cdots & c_n \\ c_1 & u_{11} - \lambda c_{11} & \cdots & u_{1n} - \lambda c_{1n} \\ \vdots & \vdots & & \vdots \\ c_n & u_{1n} - \lambda c_{1n} & \cdots & u_{nn} - \lambda c_{nn} \end{pmatrix} = M$$

or, equivalently,

$$\begin{pmatrix} 0 & p_1 & \cdots & p_n \\ p_1 & u_{11} & \cdots & u_{1n} \\ \vdots & \vdots & & \vdots \\ p_n & u_{1n} & \cdots & u_{nn} \end{pmatrix} = M,$$

since $c_i(x) = p_i$, $i = 1, \ldots, n$ and $c_{ij}(x) = 0$, $i, j = 1, \ldots, n$.

From the first-order conditions $l_i = u_i - \lambda p_i$, $i = 1, \ldots, n$, we find $p_i = (1/\lambda)\, u_i = k u_i$. Replacing in M,

$$\begin{pmatrix} 0 & ku_1 & \cdots & ku_n \\ ku_1 & u_{11} & \cdots & u_{1n} \\ \vdots & \vdots & & \vdots \\ ku_n & u_{1n} & \cdots & u_{nn} \end{pmatrix} = M.$$

Taking determinants, observe that

$$\det(M) = k^2 \cdot \begin{vmatrix} 0 & u_1 & \cdots & u_n \\ u_1 & u_{11} & \cdots & u_{1n} \\ \vdots & \vdots & \cdots & \vdots \\ u_n & u_{1n} & \cdots & u_{nn} \end{vmatrix} = k^2 \det(U).$$

The matrix U is usually called the bordered Hessian of $u(x)$. Obviously in our problem it must be studied at (x^*, λ^*). Since now the determinant and principal minors of M and U behave in the same way in regard to their signs, it is clear that we shall prefer to state the second-order conditions for a maximum in terms of U rather than in terms of M. We can formulate the conditions as follows: Given the problem,

$$\text{maximize} \quad u(x) \quad \text{subject to} \quad px - m = 0,$$

it is sufficient for a critical point to be a maximum of $u(x)$ that the determinant of the bordered Hessian of the utility function and its principal minors of successively lower order alternate in sign.

It is suggested as an exercise to study the behavior of the determinants and principal minors of M and the bordered Hessian of the objective function when the problem is one of constrained minimization.

EXERCISES

E.17.1 Write in matrix form the following quadratic forms:

(a) $q(x) = (x_1)^2 - 2x_1x_2 + (x_2)^2$;
(b) $q(x) = 4(x_1)^2 + (x_2)^2 + 6x_2x_1$;
(c) $q(x) = (x_1)^2 + 5(x_2)^2 + 3(x_3)^2 - 2x_1x_2 + 4x_1x_3 + 8x_2x_3$;
(d) $q(x) = -(x_2)^2 + (x_3)^2 + 6x_1x_2 - 4x_1x_3 + 10x_2x_3$.

Hint: Note that the matrix A in $x^{\mathrm{T}}Ax$ has following properties: (1) a_{ii} is the coefficient of $(x_i)^2$; (2) $a_{ij} = a_{ji}$ is one-half the coefficient of $x_i x_j$.

E.17.2 Find the diagonal canonical form of each of the quadratic forms of the preceding exercise, and discover the corresponding diagonalizing matrix Q.

Hint: Form the block matrix $(A\,|\,I)$, and perform on it the necessary row and column operations to obtain the block matrix $(\Lambda\,|\,M)$, where Λ is diagonal. Observe that $M = Q^{\mathrm{T}}$, that is, $Q = M^{\mathrm{T}}$.

E.17.3 Using the propositions of the appropriate theorems of this chapter, construct examples of 3×3 real symmetric matrices of the following types: (a) positive definite; (b) positive semidefinite; (c) negative definite; (d) negative semidefinite.

E.17.4 Use the matrices of the preceding exercise to show that any quadratic form defined on them has the definiteness property of the associated matrix (see Bellman [1]).

E.17.5 Prove Theorem 17.3.

Hint: See Theorem 17.2.

E.17.6 Prove Theorem 17.12.

E.17.7 Using Theorem 17.7 and the result of E.17.6 prove Theorem 17.14.

Hint: Note that if A is positive definite so is A^{-1}.

E.17.8 Find the matrix R which simultaneously diagonalizes $q_1(x)$ and $q_2(x)$ in the following so that $q_1(y) = y^{\mathrm{T}}Iy$ and $q_2(y) = y^{\mathrm{T}}\Gamma y = \sum_{j=1}^{3} \mu_j (y_j)^2$:

$$q_1(x) = (x_1, x_2, x_3)\begin{pmatrix} 5 & 1 & 0 \\ 1 & 2 & 3 \\ 0 & 3 & 10 \end{pmatrix}\begin{pmatrix} x_1 \\ x_2 \\ x_3 \end{pmatrix},$$

$$q_2(x) = (x_1, x_2, x_3)\begin{pmatrix} 2 & 0 & 4 \\ 0 & 1 & -1 \\ 4 & -1 & 5 \end{pmatrix}\begin{pmatrix} x_1 \\ x_2 \\ x_3 \end{pmatrix}.$$

E.17.9 Construct an example with 2×2 matrices as illustration of Remark 17.7.

E.17.10 Assume an economy with only three commodities, x_1, x_2, and x_3; a consumer with a utility-index function defined on the nonnegative orthant of R^3 by $u(x) = x_1 x_2 x_3$, and having an income of 100; and let $p = (1, 2, 3)$ be the given price vector. Determine the constrained utility maximum, and study the bordered Hessian of $u(x)$, and its principal minors, at the maximum.

PROBLEMS

P.17.1 The economic problem discussed as last example of this chapter can be formulated, alternatively, as that of minimizing the money income the consumer needs to be able to attain a given utility level u^0, that is,

$$\text{minimize} \quad m = px \quad \text{subject to} \quad u(x) - u^0 = 0.$$

Study the second-order conditions for a minimum.

Hint: Lancaster [1].

P.17.2 Prove that the price adjustment process $Du = Au$ is locally asymptotically stable if A is negative definite.

Hint: See Example in Chapter 16.

P.17.3 Let $f(y) = 0$ be a twice-differentiable production function, and $y = (y_1, \ldots, y_n)^T$ a commodity bundle (inputs and outputs). For a given price vector $p = (p_1, \ldots, p_n)$, profit is thus defined $\pi = py$. Study the necessary and sufficient conditions for a maximum of the problem,

$$\text{maximize} \quad \pi = py \quad \text{subject to} \quad f(y) = 0.$$

and analyze the relevant bordered Hessian.

P.17.4 Consider the hyperplane $H(p, 0) = \{x \mid x \in R^n, px = 0\}$ (see Definition 18.1, Chapter 18), and the quadratic surface represented by the form $x^T Ax = 0$ where A is $n \times n$. Determine the minimum distance between surface and hyperplane. (Source: Bellman [1].)

P.17.5 Prove that a real symmetric matrix A is positive definite if and only if a real nonsingular matrix P exists such that $P^T P = A$.

Hint: See Theorem 17.15.

NOTES

This chapter is mainly inspired by Bellman [1] and Lancaster [1]. An advanced discussion of the same topic is presented by Gantmacher [1]. Intriligator [1] discusses extensively the use of classic and modern optimization methods in economic theory, particularly in neoclassical and growth economics.

CHAPTER 18

LINEAR INEQUALITIES, HYPERPLANES, AND CONVEX CONES

Modern linear economic analysis is based considerably upon the concepts and tools of the theory of linear inequalities. To a certain extent, it seems reasonable to say that a distinct trait of modern theory when compared to the classical approach is manifested in the study of inequality systems as opposed to equation models. Some basic problems of the theory of linear inequalities are discussed in Section 18.1.

Section 18.2 introduces a few definitions and theorems on separating and supporting hyperplanes, tools which are standard by now in analytical economics.

Finally, Section 18.3 is confined to the study of a special class of convex sets, namely, convex cones. The theory of convex cones is of the utmost significance in production theory.

At first sight, it may seem that we are dealing in this chapter with three types of problems which have very little in common. However, that this is not the case will soon be seen. The idea of convexity, as discussed in Chapter 4, will be present all through our study and will become particularly relevant in the last two sections. The inequalities discussed in Section 18.1 will become standard ingredients of the analysis of Sections 18.2 and 18.3. Finally, in a way, the discussion of the last section may be regarded as an alternative approach to the subject studied in Section 18.1.

We conclude these preliminary remarks by recalling notation agreed upon in Chapter 3. For two vectors x and x', we write $x \gg x'$ if $x_i > x_i'$ for all i, $x > x'$ if $x_i \geq x_i'$ with at least one $x_i > x_i'$, and $x \geq x'$ if $x_i \geq x_i'$. In particular, we say that a vector x is positive, semipositive, or nonnegative if $x \gg \theta$, $x > \theta$, or $x \geq \theta$, respectively.

18.1 Linear Inequality Systems

In this first section we shall be concerned with the existence of solutions for systems of the type $Ax \geq b$ or $Ax \leq b$, and $Ax \leq \theta$. In particular, we shall pay some attention to some additional conditions we might require from a solution, namely, that it be say positive or nonnegative. Our frame of reference will be the real space.

The question we have to discuss is at first sight more general than the one tackled in Chapter 15. We are no longer asking about the existence of a vector x, mapped by the matrix A into the constant vector b. In $Ax \geq b$ ($Ax \leq b$), we are obviously concerned with vectors which are mapped by A not only into b, but also into any constant vector b' such that $b' > b$ ($b' < b$).

In economics, the question has usually a straightforward interpretation. Let A and x be the technology matrix and an output vector of a Leontief closed economy. The input requirements, as seen in another place, are immediately determined by the product Ax. Now, since inputs are generated only within the system, we must have $Ax \leq x$. Any nonnegative solution vector would seem to make some sense in the present context, though our real interest will only be satisfied by vectors $x > \theta$, and especially $x \gg \theta$. A strictly positive vector x^*, which gives $Ax^* = x^*$, is usually called a strong or interior equilibrium of the system. But note that this problem is only a special case of the more general question about the "feasibility" of the system. To answer the latter, we have to be able to decide whether a vector $x \geq \theta$, $x > \theta$, or $x \gg \theta$ (in order of increasing economic relevance) exists, such that $Ax \leq x$, that is, such that the quantitative input requirements for a given production is not greater than the actual production from which they can be satisfied.

In spite of obvious differences, there is a common ground to the theories of linear equations and linear inequalities. We begin our study here with two results on linear equations systems.

Linear Equations: Nonnegative Solutions†

We study first the question of the solvability of a nonhomogeneous equations system in a manner which is appropriate to our main subject.

Theorem 18.1 Consider the systems (1) $Ax = b$ and (2) $yA = \theta$, $yb = 1$. Then only one of the following two alternatives is true: Either (1) has a solution or (2) has a solution.

Proof We show first that (1) and (2) are incompatible systems. Premultiplying (1) by y and postmultiplying $yA = \theta$ by x, we get $yAx = yb$ and $yAx = 0$. This is inconsistent with $yb = 1$. Thus (1) and (2) cannot both have solutions. We finally show that if (1) has no solution, then (2) has a solution. Assume $Ax = b$ has no solution and let $r(A) = r$. Thus, $\{A_1, \ldots, A_r\}$ forms a linearly independent set of column vectors of A and therefore a basis. By the same assumption $\{A_1, \ldots, A_r, b\}$ is linearly independent too and the matrix $\hat{A} = (A_1, \ldots, A_r, b)$ has $r(\hat{A}) = r + 1 \leq m$. Hence, the system $y\hat{A} = \hat{c}$ has a solution for every possible vector $\hat{c}$. For instance, we may choose $\hat{c} = (0, 0, \ldots, 0, 1)$. Now, the equations $yA_j = 0$, $j = 1, \ldots, r$ and $yb = 1$ form an equivalent system to $y\hat{A} = \hat{c}$ and have therefore a solution. But every column of A is a linear combination of $A_1, \ldots, A_r$. For all k we can thus write $A_k = \sum_{j=1}^{r} h_j A_j$ so that $yA_k = 0$. Therefore y is a solution of $yA = \theta$, $yb = 1$ and the theorem is proved.

An indirect way to test the solvability of a system of type (1) is, thus, to study the existence of a solution for the corresponding system of type (2). The criterion is both simple and straightforward.

Of particular interest is the case in which our question concerns not only the existence of a solution but rather a solution satisfying additional conditions. We confine ourselves to the case of a nonnegative solution.

Theorem 18.2 Consider the systems (1) $Ax = b$ and (2) $yA \geq \theta$, $yb < 0$. Then only one of the following two alternatives is true: Either (1) has a nonnegative solution or (2) has a solution.

Proof We prove first that (1) and (2) cannot both have a solution. Proceeding as in the previous theorem, we obtain $yAx = yb$ and $yAx \geq 0$, which is obviously inconsistent with $yb < 0$.

† Theorems 18.1 and 18.2 in this subsection adapted from Gale [3].

Next, we show that if no nonnegative solution for (1) exists, then (2) has a solution. Note, by the way, that if (1) has no solution at all, then there will exist a vector y such that $yA = \theta$ and $yb = -1$ (Theorem 18.1).

Suppose (1) has a solution, but not a nonnegative one, and let us proceed by induction on n, the number of columns of the $m \times n$ matrix A.

(a) Let $n = 1$. We have $Ax = b$, where $A = (a_{11}, \ldots, a_{m1})^{\mathrm{T}}$ and x is a scalar. Putting $y = -b^{\mathrm{T}}$ we obtain $yb = -b^{\mathrm{T}}b$, $yA = yb/x = -b^{\mathrm{T}}b/x > 0$. Hence y is a solution of (2).

(b) Assume the theorem holds for $n - 1$ columns. Note that if (1) has no nonnegative solution, neither will $\sum_{j=1}^{n-1} A_j x_j = b$ have one. By hypothesis, there must exist y' such that $y'A_j \geq 0$, $j = 1, \ldots, n-1$, and $y'b < 0$. Thus, the theorem is proved for $j = n$ if $y'A_n \geq 0$. It may occur, though, that $y'A_n < 0$. If so, then we define $\hat{A}_j = (y'A_j)A_n - (y'A_n)A_j$, for $j < n$, and $\hat{b} = (y'b)A_n - (y'A_n)b$. Let us denote these expressions by (3), and consider (4) $\sum_{j=1}^{n-1} \hat{A}_j \hat{x}_j = \hat{b}$. That the latter does not admit a nonnegative solution can be shown by substituting (3) into (4). We get

$$\begin{aligned} &\hat{x}_1(y'A_1)A_n - \hat{x}_1(y'A_n)A_1 + \hat{x}_2(y'A_2)A_n - \hat{x}_2(y'A_n)A_2 \\ &\quad + \cdots + \hat{x}_{n-1}(y'A_{n-1})A_n - \hat{x}_{n-1}(y'A_n)A_{n-1} = y'bA_n - y'A_nb \end{aligned}$$

or, rearranging terms,

$$(-y'A_n)^{-1}\left(\sum_{j=1}^{n-1} \hat{x}(y'A_j) - y'b\right)A_n + \sum_{j=1}^{n-1} \hat{x}_j A_j = b.$$

A nonnegative solution for this expression means a nonnegative solution for $Ax = b$, contrary to assumption. We can, thus, have a vector $\hat{y}$ such that $\hat{y}A_j \geq 0$ for $j < n$ and $\hat{y}\hat{b} < 0$. Define $y = (\hat{y}A_n)y' - (y'A_n)\hat{y}$. Since for $j < n$, $yA_j = \hat{y}\hat{A}_j \geq 0$, $yb = \hat{y}\hat{b} < 0$, and $yA_n = 0$, it follows that the vector y is a solution of (2), thus proving the theorem.

Linear Inequalities: Nonnegative Solutions†

We begin by considering a nonhomogeneous system, and before approaching the question of the existence of nonnegative solutions, we concentrate on the more general problem of solvability.

Theorem 18.3 Consider the systems (1) $Ax \geq b$ and (2) $yA = \theta$, $yb = 1$. Then only one of the following alternatives is true: Either (1) has a solution or (2) has a nonnegative solution.

Proof The proof that (1) and (2) cannot both have a solution follows along the identical lines of the preceding theorems. We can dispense with it and concentrate on the second aspect.

† Theorems 18.3–18.5 in this subsection adapted from Gale [3].

Given $yA = \theta$ and $yb = 1$, note that we can write

$$yA = (y_1, \ldots, y_m)(A^1, \ldots, A^m)^{\mathrm{T}} = \sum_{i=1}^{m} y_i A^i = \theta$$

and

$$yb = \sum_{i=1}^{m} y_i b_i = 1.$$

Observe that if this system has no nonnegative solution, we could invoke Theorem 18.2 and show that a vector x and a scalar γ exist, such that $A^i x + \gamma b_i \geq 0$, $i = 1, \ldots, m$, and also $0 \cdot y + \gamma = \gamma < 0$. This means $A^i(-x/\gamma) \geq b_i$. Therefore $(-x/\gamma)$ is the solution.

We look now into the question of the nonnegativity of the solution.

Theorem 18.4 Consider the systems (1) $Ax \leq b$ and (2) $yA \geq \theta$, $yb < 0$. Then only one of the following two alternatives is true: Either (1) has a nonnegative solution or (2) has a nonnegative solution.

Proof By the same method of previous theorems it can be shown that systems (1) and (2) cannot both have solutions. Let us look at the second part of the proof.

Assume (1) has no nonnegative solution. Then neither has the system $\hat{A}\hat{x} = b$ a nonnegative solution, where $\hat{A} = (A \mid I)$ is an $m \times (n + m)$ matrix (here I is the identity matrix of mth order), and $x = (x_1, \ldots, x_n, z_1, \ldots, z_m)^{\mathrm{T}}$, an $(n + m)$ vector containing the "slack variable" $z = (z_1, \ldots, z_m)^{\mathrm{T}}$. If this is the case, Theorem 18.2 states that $y\hat{A} \geq \theta$, $yb < 0$ has a solution. But $y\hat{A} \geq \theta$ consists of the subsystems $yA \geq \theta$ and $yI = y \geq \theta$. From the latter, it results that $yA \geq \theta$, $yb < 0$ has a solution $y \geq \theta$, therefore proving the theorem.

A useful result on semipositive solutions can be easily obtained for homogeneous inequality systems. We give it in the next theorem.

Theorem 18.5 Let there be the systems (1) $Ax \leq \theta$ and (2) $yA > \theta$. Then only one of the following two alternatives is true: Either (1) has a semipositive solution or (2) has a nonnegative solution.

Proof The incompatibility of systems (1) and (2) can be proved as in previous theorems.

Regarding the second aspect of the proof, let us begin with the assumption that $Ax \leq \theta$ has no semipositive solution.

$$Ax = (A_1, \ldots, A_n)(x_1, \ldots, x_n)^{\mathrm{T}} = \sum_{j=1}^{n} A_j x_j \leq \theta \qquad \text{and} \qquad -\sum_{j=1}^{n} x_j \leq -1,$$

have consequently no nonnegative solution. We can now invoke Theorem 18.4, to state that there must exist a vector $y \geq \theta$ and a scalar $\gamma \geq 0$, such that $yA_j - \gamma \geq 0$ and $-\gamma < 0$. Therefore, $yA_j \geq \gamma > 0$ and $yA > \theta$.

18.2 Bounding Hyperplanes

The subject of this section is, in a way, a continuation of our discussion on convexity initiated in Chapter 4. It is, thus, advisable to keep in mind the basic properties of convex sets, and of mappings on convex sets, discussed earlier.

The notion of a hyperplane is well known to the reader, and we have often referred to it in previous chapters. Nevertheless, it seems appropriate to introduce here formal definitions of this and other closely related concepts.

Definition 18.1 Consider the nonzero point $p \in R^n$ and the number $c \in R$. Then the set $H(p, c) = \{x \mid x \in R^n, px = c\}$ is called a hyperplane with normal p.

For any vector $x' \in H(p, c)$ we have $px' = c$, that is, for $x \in H(p, c)$, $p(x - x') = 0$. Thus for given x', $H(p, c)$ contains all those points of R^n satisfying this equation. Since $x - x'$ and p are orthogonal, we also say that the point p is orthogonal to the hyperplane $H(p, c)$, or simply that p and $H(p, c)$ are orthogonal.

Definition 18.2 Let $H(p, c)$ be a hyperplane with normal p. Then the sets

$$H^+ = \{x \mid x \in R^n, px > c\} \quad \text{and} \quad H^- = \{x \mid x \in R^n, px < c\}$$

are called the open half-spaces of R^n and their closures, $\overline{H}^+$ and $\overline{H}^-$, are the closed half-spaces of R^n.

The half-spaces H^+ and H^- are said to be "above" or "below" $H(p, c)$, respectively. In the same sense, one says that a vector x lies "above" or "below" the hyperplane if $x \in H^+$ or $x \in H^-$, respectively.

Definition 18.3 An intersection of hyperplanes in R^n is called a linear manifold.

The following property is obvious and does not require proof.

Remark 18.1 Consider the closed half-spaces $\overline{H}^+ = \{x \mid x \in R^n, \ px \geq c\}$ and $\overline{H}^- = \{x \mid x \in R^n, px \leq c\}$. Then $\overline{H}^+ \cap \overline{H}^- = H(p, c)$.

A closed half-space is clearly a closed convex set. From the properties of the intersection of closed convex sets we have immediately the following result.

Remark 18.2 A closed half-space, a hyperplane, and a linear manifold are all closed, convex sets.

Minkowski's Theorem

Definition 18.4 Let S be a subset of R^n. Then a hyperplane is said to be bounding for S, if S is entirely contained in one of the closed half-spaces determined by the hyperplane.

The idea of a bounding hyperplane is quite general and straightforward. Every time that a set S is completely contained in one of the two closed half-spaces into which R^n has been divided by $H(p, c)$, we say that the latter is bounding for S. The definition obviously includes two special cases. The case in which S has no common points with $H(p, c)$, and the one in which S has, at least, one common point with the hyperplane. Observe that, in the latter case, the common points can only be on the boundary of S.

If the hyperplane is bounding for two sets, say S and S', then three cases may happen. First, S and S' have no common points with $H(p, c)$, and each lies completely in a different open half-space. Second, only one of the sets, say S, has common (boundary) points with $H(p, c)$, whereas S' is entirely included in the opposite open half-space. Finally, the two sets have some (boundary) points in common. Clearly, these points lie on $H(p, c)$. The diagrams in Fig. 18.1 may help to clarify these different situations.

It should be pointed out that the terminology we are following here is not necessarily that most commonly found in books on the subject. We think, however, it is consistent and appropriate.

To complete our list of concepts, let us give names to the two special cases of bounding hyperplanes distinguished above.

Definition 18.5 Let S be a closed subset of R^n. A bounding hyperplane $H(p, c)$ is said to be separating for S, if S is entirely contained in one of the open half-spaces determined by $H(p, c)$. It is said to be supporting for S, if S is wholly contained in one of the closed half-spaces determined by $H(p, c)$, with at least one of its points lying on $H(p, c)$.

Summarizing, separating and supporting hyperplanes are two different varieties of bounding hyperplanes.

Assume now S is a closed, convex subset of R^n. It is intuitive that through any point $z \in R^n$, noninterior to S, a hyperplane passes which is bounding

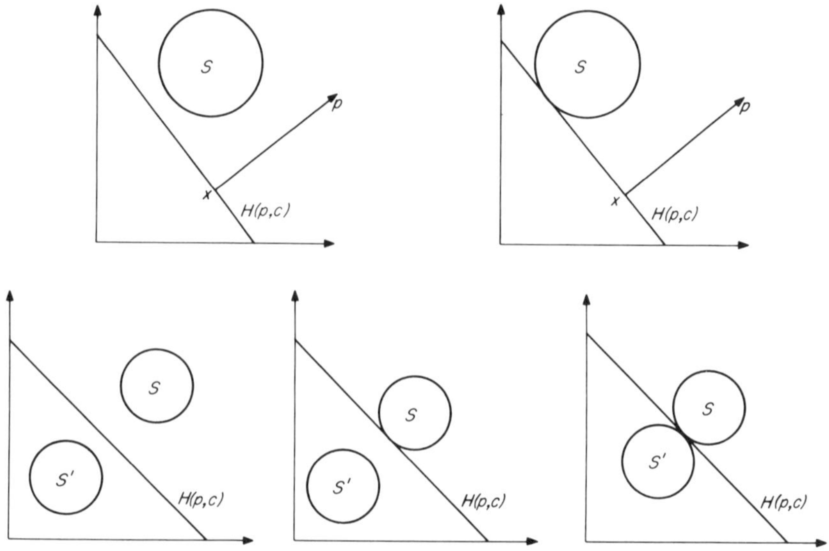

Figure 18.1

for S. If the point is exterior, the hyperplane is clearly separating for S; if it is a boundary point, then the hyperplane is supporting S at z.

If we drop the assumption about the convexity of S, the situation becomes less clear. Figure 18.2 shows a separating and a supporting hyperplane for a convex set S. In the figure on the right-hand side we see a nonconvex set with an exterior point through which no bounding (separating or supporting) hyperplane can pass.

Let us return to the case of a convex set $S \subset R^n$. Simple as it might appear, the question of the existence of a bounding hyperplane, passing through a noninterior point to S, is not without difficulties when, approached from a formal point of view. The results which we shall examine next, generally

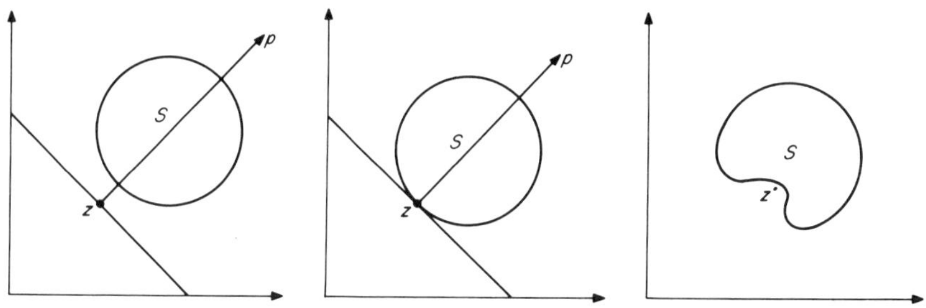

Figure 18.2

presented in varying versions as the Minkowski theorem, play an important role in linear programming and game theory. We shall study, in this order, the question of the existence of a separating and of a suppporting hyperplane. In this way, we will have proved in two steps our third theorem in the series, which following Debreu, we shall call the Minkowski theorem.†

Let us begin with a theorem on the existence of a separating hyperplane.

Theorem 18.6 Let S be a closed convex subset of R^n and z a point of R^n. Then either $z \in S$ or there exists a hyperplane $H(p, c)$ such that $z \in H(p, c)$ and S is contained in one of the open half-spaces determined by the hyperplane.

Proof Assume $z \notin S$ and define the mapping $f\colon S \to R$ such that, for every $x \in S$, $f(x) = (x - z)^{\mathrm{T}}(x - z)$. That f has a minimum follows from the fact that, for any $x \in S$, $(x - z)^{\mathrm{T}}(x - z) \geq 0$. Now, suppose $x^* \in S$ is such that $(x^* - z)^{\mathrm{T}}(x^* - z) \leq (x - z)^{\mathrm{T}}(x - z)$ for all $x \in S$. Since $z \notin S$, we must have $p = (x^* - z)^{\mathrm{T}} \neq \theta$. Consider any $x \in S$ different from x^*. For $\lambda \in [0, 1]$, we clearly have by convexity $x^* + \lambda(x - x^*) \in S$. Then

$$(x^* - z)^{\mathrm{T}}(x^* - z) \leq [x^* + \lambda(x - x^*) - z]^{\mathrm{T}}[x^* + \lambda(x - x^*) - z].$$

Expanding and rearranging terms in the latter expression, we obtain

$$0 \leq 2(x^* - z)^{\mathrm{T}}(x^* - x) + \lambda(x - x^*)^{\mathrm{T}}(x - x^*).$$

Now, when $\lambda \to 0$ we get

$$\begin{aligned} 0 \leq (x^* - z)^{\mathrm{T}}(x - x^*) &= p(x - x^*) = p[(x - z) - (x^* - z)] \\ &= p[(x - z) - p^{\mathrm{T}}] = p(x - z) - pp^{\mathrm{T}}. \end{aligned}$$

Since $pp^{\mathrm{T}} > 0$, it must be that $p(x - z) > 0$, that is, $px > pz$. On the other hand, we have $p(x - x^*) \geq 0$, that is, $px \geq px^*$, and hence $px \geq px^* > pz$. Therefore, there is a hyperplane $H(p, c) = \{y \mid y \in R^n, py = pz = c\}$ such that S is contained entirely in the open half-space $H^+ = \{x \mid px > pz = c\}$.

The preceding theorem has been proved for a hyperplane "below" the set S. Clearly, a similar proof follows for the case in which $H(p, c)$ separates S from "above."

We examine next the question of the existence of a supporting hyperplane.

Theorem 18.7 Let z be a boundary point of the closed convex set $S \subset R^n$. Then, there exists a supporting hyperplane of S at z.

Proof Consider a vector $z \in S$, and for each integer $k = 2, 3, \ldots,$ a point $z^k \notin S$ such that $d(z^k, z) = 1/k$. Furthermore, let $x^k \in S$ be the point with the

† Theorems 18.6–18.8 have been adapted from Lang [1].

smallest distance $d(z^k, x^k)$, and define $p^k = (x^k - z^k)^T$. We shall call $\hat{p}^k$ the corresponding normalized expression for p^k and consider the sequence $(\hat{p}^k)$. It is clear that $(\hat{p}^k)$ converges to the limit point $\hat{p}$ on the unit sphere. Invoking the preceding theorem, we have for all $x \in S$ and every k, $p^k x > p^k z^k$ and thus $\hat{p}^k x > \hat{p}^k z^k$. Since $\lim_{k \to \infty} z^k = z$, and $\hat{p}$ is a limit point of $(\hat{p}^k)$, we have $\hat{p}x \geq \hat{p}z$ for all $x \in S$ ($=$ for $x = z$). Therefore, S has a supporting hyperplane $H(p, c) = \{x \mid \hat{p}x = \hat{p}z = c\}$ through its boundary point z.

The results obtained for a separating and a supporting hyperplane can be joined now in a more general version (Minkowski theorem).

Theorem 18.8 Let $S \subset R^n$ be a closed, convex set and z a point of R^n. Then either z is interior to S or there exists a bounding hyperplane for S through z.

Observe that the latter theorem establishes a necessary and sufficient condition for the existence of the bounding hyperplane. In other words, the bounding hyperplane at z will exist if and only if z is not an interior point of $S \subset R^n$. Theorems 18.6 and 18.7 are obviously special cases of Theorem 18.8, and consider the two distinct situations of $z \in R^n$ as exterior point and boundary point of S, respectively.

An interesting result, which immediately follows from Theorem 18.7, is the following:

Remark 18.3 Any closed, convex set in R^n is the intersection of all the closed half-spaces of its supporting hyperplanes.

Example An important economic theorem (Debreu [4]) concerning one side of the relationships between equilibrium and optimal states will serve to illustrate a possible application of Minkowski's theorem.

Using the notation of Chapter 9, define an economy as a system $\{\{X_i, \precsim_i\}, \{Y_j\}, r\}$ (see discussion in Problem P.9.1), and suppose it satisfies all the relevant assumptions regarding the private ownership economy of Theorem 9.3.

We shall call A the set of all attainable states of the economy. Formally, a state $((x_i), (y_j))$ is in A if (1) $x_i \in X_i$ for every i; (2) $y_j \in Y_j$ for every j; (3) $x - y = r$.

Observe that the preorderings on X_i, $i = 1, \ldots, m$ induce a preordering on A. We say that a state $((x_i), (y_j))$ is at least as desired as a state $((x_i'), (y_j'))$, denoted $((x_i'), (y_j')) \precsim ((x_i), (y_j))$, if for every consumer his consumption in the latter state is at least as desired as his consumption in the former state, that is, if $x_i' \precsim_i x_i$ for every i.

Now, movements from one state to another may have for different consumers, quite different results. Noncomparable states are thus admissible: A is only partially preordered under $\precsim$.

A maximal element of A for $\precsim$ will be called an optimum state (Pareto optimum). If $((x_i^*), (y_j^*))$ is such a state, then no movement to another state could be made which would improve the satisfaction level of an individual consumer without, simultaneously, worsening that of, at least, another consumer.

Let us define now certain sets, some of which will be useful tools in the proof of the theorem which follows.

For the ith consumer

$$X_i^* = \{x_i \mid x_i \in X_i, x_i \succsim_i x_i^*\} \qquad \text{and} \qquad X_i^{**} = \{x_i \mid x_i \in X_i, x_i \succ_i x_i^*\}$$

represent the set of all consumptions which are at least as desired as, and strictly preferred to x_i^*, respectively.

Accordingly, the sets of total resources of an economy, which can attain the consumption levels $(x_i) \succsim_i (x_i^*)$ and $(x_i) \succ_i (x_i^*)$, may be denoted

$$G^* = \sum_{i=1}^{m} X_i^* - \sum_{j=1}^{n} Y_j \qquad \text{and} \qquad G^{**} = \sum_{i=1}^{m} X_i^{**} - \sum_{j=1}^{n} Y_j,$$

respectively. However, without loss of generality, let us concentrate our attention on the mth consumer. If we consider his set $X_m^{**} = \{x_m \mid x_m \in X_m, x_m \succ_m x_m^*\}$, then $G_m^{**} = X_m^{**} + \sum_{i=1}^{m-1} X_i^* - \sum_{j=1}^{n} Y_j$ is the set of total resources of an economy which can attain the state $((x_i), (y_j))$ with $x_i \succsim_i x_i^*$ for $i = 1, \ldots, m-1$ and $x_m \succ_m x_m^*$. The sets G^* and G_m^{**} will be used in the proof.

Theorem 18.9† For the economy $\{\{X_i, \precsim_i\}, \{Y_j\}, r\}$, an optimum $((x_i^*), (y_j^*))$ is an equilibrium relative to some price system $p^* > \theta$ provided for each i, $p^*x_i^* \neq \min p^*X_i$.

Proof We note that $((x_i^*), (y_j^*))$ is an optimum state. Being attainable, $x^* - y^* = r$, but, being an optimum, $r \notin G_m^{**}$. From the standard convexity assumption, it follows that X_m^{**}, X_i^* for every i, and Y_j for every j, are convex sets. Thus, so are their sums, and in particular, G_m^{**} is a convex set. Hence, by Minkowski's theorem, there is a bounding hyperplane for G_m^{**} which passes through the resources vector r. This means that a nonzero price system $p^* \in R^k$ exists such that, for any point $\rho \in G_m^{**}$, $p^* \cdot \rho \geq p^* \cdot r$.

Note that $\bar{X}_m^{**} = \{x_m \mid x_m \in X_m, x_m \succsim_m x_m^*\}$. We can thus write $X_m^* \subset \bar{X}_m^{**}$, and it is easy to see that, by sum of closures,

$$G^* \subset \bar{X}_m^{**} + \sum_{i=1}^{m-1} \bar{X}_i^* - \sum_{j=1}^{n} \bar{Y}_j \subset \bar{G}_m^{**}.$$

Thus G^* is contained in the closed half-space above the hyperplane through r. But $r \in G^*$ and is a minimizer of $p^* \cdot \rho$ over G^*. In other words, it is on the

† Adapted from Debreu [4]

boundary of G^*. Since $r = x^* - y^*$, it can be seen that x_i^* is a minimizer of $p^* \cdot \rho$ on X_i^* for every i, while $-y_j^*$ is a minimizer of $p^* \cdot \rho$ on $-Y_j$ for every j. Since $p^*x_i^* \neq \min p^*X_i$, for each i, x_i^* is also a largest element of $\{x_i | x_i \in X_i, p^*x_i \leq p^*x_i^*\}$, for every i. Clearly, for every j, y_j^* is a profit maximizer on Y_j relative to p^*; and $x^* - y^* = r$. Therefore, $((x_i^*), (y_j^*))$ is an equilibrium of the economy relative to p^*.

Krein–Milman Theorem

We shall study next two theorems which play an essential role in the theory of linear programming.

Suppose x^r and x^s are two points of a convex set X. For a number $\lambda \in [0, 1]$, we know, by definition of a convex set, the convex combination $\lambda x^r + (1 - \lambda) x^s$ defines points which also belong to X. These points lie precisely on the closed segment which connects x^r and x^s.

If we now confine the scalar λ to values of the open unit interval, that is, $\lambda \in]0, 1[$, then it is clear that $\lambda x^r + (1 - \lambda)x^s$ can only represent points which are interior to the closed segment from x^r to x^s. Some points which belong to the closed segment can not be represented in this way.

Definition 18.6 Let $S \subset R^n$ be a convex set. Then $x^w \in S$ is said to be an extreme point of S if it cannot be written $x^w = \lambda x^u + (1 - \lambda)x^v$ for $\lambda \in]0, 1[$, for any $x^u, x^v \in S$.

It is intuitive that a convex set may have a finite or infinite number of extreme points.

Example Consider the boundaries of S, S', and S'' in Fig. 18.3. S has an infinite number of extreme points, and its boundary consists exclusively of such points. S' also, has an infinite number of extreme points, but only on the arc $\widehat{x^1x^2}$ of its boundary. Finally, S'' has a boundary with only four extreme points, x^1, x^2, x^3, x^4.

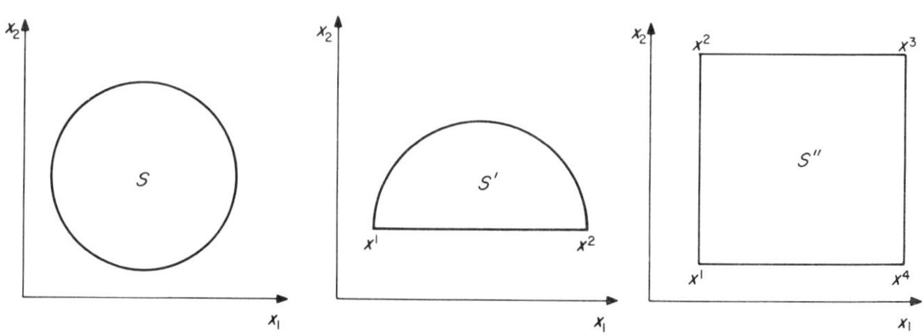

Figure 18.3

With the concept of an extreme point in mind, we can study now an important property of supporting hyperplanes.

Theorem 18.10† Let the set $S \subset R^n$ be closed, convex, and bounded from below, and $H(p, c)$ a supporting hyperplane for S at the point x^0. Then $H(p, c)$ contains an extreme point.

Proof We have, by hypothesis, $H(p, c) = \{x \mid px = px^0 = c\}$. Thus, for every $x \in S$, $px \geq px^0$. Now consider the set $S \cap H(p, c)$. From the intersection properties, $S \cap H(p, c)$ is clearly closed, convex, and bounded from below. Assume, first, that $z \in S \cap H(p, c)$ is an extreme point of $S \cap H(p, c)$. Suppose, moreover, that for some x^1, $x^2 \in S$ and $\lambda \in]0, 1[$ we can write $z = \lambda x^1 + (1 - \lambda)x^2$. Hence, $pz = \lambda px^1 + (1 - \lambda)px^2$. Now, $pz = px^0$, so that $px^0 = \lambda px^1 + (1 - \lambda)px^2$. However, since x^1, $x^2 \in S$, it is also true that $px^1, px^2 \geq px^0$. Should it be that $px^1 > px^0$ or $px^2 > px^0$, then we would have $\lambda px^1 + (1 - \lambda)px^2 > \lambda px^0 + (1 - \lambda)px^0 = px^0$, which is a contradiction. Thus, we must have $x^1, x^2 \in H(p, c)$; so $x^1, x^2 \in S \cap H(p, c)$. But this violates the assumption about $z \in S \cap H(p, c)$ being an extreme point. So if z is an extreme point of $S \cap H(p, c)$, it is an extreme point of S.

Next, consider all points $x \in S \cap H(p, c)$. Because of the closedness and lower boundedness of this set, we can find subsets $Z', Z'', \ldots, Z^{(n)}$ of $S \cap H(p, c)$, $Z^{(n)} \subset \cdots \subset Z'' \subset Z'$, such that Z' contains all vectors of $S \cap H(p, c)$ with smallest first component, Z'' all vectors of Z' with smallest second component, $\ldots, Z^{(n)}$ all vectors of $Z^{(n-1)}$ with smallest nth component. It is true that each of these subsets is closed and bounded from below, and that $Z^{(n)}$ contains only one member. Let us call this vector z. Suppose now, that for some y^1, $y^2 \in S \cap H(p, c)$ and $\lambda \in]0, 1[$, we can write $z = \lambda y^1 + (1 - \lambda)y^2$. For the first component, we have ${y_1}^1, {y_1}^2 \geq z_1$, and $z_1 = \lambda {y_1}^1 + (1 - \lambda){y_1}^2$. Note, however, that ${y_1}^1 > z_1$ or ${y_1}^2 > z_1$ is excluded for, in such case, we would have $\lambda {y_1}^1 + (1 - \lambda){y_1}^2 > \lambda z_1 + (1 - \lambda)z_1 = z_1$. Therefore, ${y_1}^1 = {y_1}^2 = z_1$. By induction, if we assume that this is true for all components $i = 1, \ldots, n - 1$, it follows that $\lambda {y_n}^1 + (1 - \lambda){y_n}^2 = z_n$ and, hence, $y^1 = y^2 = z$. Therefore, z is an extreme point of $S \cap H(p, c)$ and hence an extreme point of S contained in $H(p, c)$.

Example Several interesting economic problems can be formulated as that of finding the optimum (maximum or minimum) of a linear function under a given set of linear constraints (usually, a system of linear inequalities) and some direct constraints on the values of the variables involved. The formulation of a problem in this manner is called a linear program.

Suppose a multiproduct firm with a linear technology wants to maximize its revenue derived from the production of n commodities, the quantity of

† Adapted from Lang [1].

which are given by the vector $x^T = (x_1, \ldots, x_n)$. If $p = (p_1, \ldots, p_n)$ is the vector of given prices, the total revenue is represented by the "objective function" $r = px$.

If the firm's technology is given by the $m \times n$ matrix $A = [a_{ij}]$—there are m inputs entering into the production of the n outputs—and the total quantity of available resources is represented by the vector $b^T = (b_1, \ldots, b_m)$, then the "functional" constraints can be written $Ax \leq b$. Besides, since only nonnegative production will make economic sense, it will also be required that $x \geq \theta$.

The complete linear programming problem can, thus, be written:

$$\text{maximize} \quad r = px \quad \text{subject to} \quad Ax \leq b, \ x \geq \theta.$$

The set of all nonnegative vectors x which satisfy the constraints $Ax \leq b$ is called the "set of feasible solutions." It is among these that we have to look for the maximizer of $r = px$.

Let us interpret the problem in terms of Theorem 18.10. For simplicity, we reduce our quantity vector to $x^T = (x_1, x_2)$ and our resources vector to $b^T = (b_1, b_2, b_3)$. Clearly, A is now a 3×2 matrix. We can refer so to the diagram of Fig. 18.4.

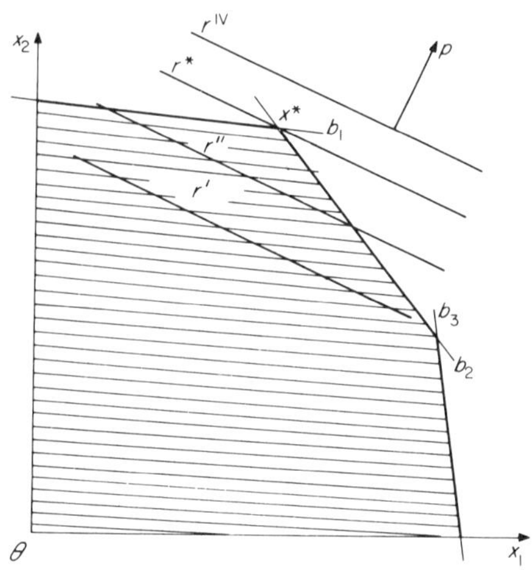

Figure 18.4

Observe that the set of feasible solutions is closed, convex, and bounded from below (by $x \geq \theta$), and that the objective function $r = px$ is a hyperplane. We can easily see from Fig. 18.4 that a vector x which maximizes $r = px$ must be both a point belonging to the feasible solutions set and a boundary point to it. In other words, $r = px$ is a supporting hyperplane for the feasible set.

Moving the hyperplane over the feasible set we obtain the optimum. Thus the optimal hyperplane is supporting for the set determined by $Ax \leq b$, $x \geq \theta$. And, provided a feasible solution exists, at least one of the extreme points of $\{x | Ax \leq b, x \geq \theta\}$ will be optimal.

The convex hull or convex closure of a set X, denoted $\dot{X}$, has been defined in Chapter 4 as the intersection of all convex sets containing X. The convex hull $\dot{X}$ is, thus, the smallest convex set containing X.

With the above notion in mind, and the idea of a simplex as defined in Chapter 7, we should not find difficulties in understanding the definition which follows.

Definition 18.7 Let $S \subset R^n$ have the extreme points $x^1, \ldots, x^r$. Then the convex hull or closure of the extreme points of S is the set

$$\ddot{S} = \{x | x = \sum_{i=1}^{r} \lambda_i x^i, \lambda_i \in R, \lambda_i \geq 0, \sum_{i=1}^{r} \lambda_i = 1, i = 1, \ldots, r\}.$$

A closed simplex in R^n is the set of all linear combinations of $n + 1$ linearly independent points in R^n. If the points are the extreme points of a set X, then the simplex is the convex closure or hull of X.

An interesting relation exists between the *convex closure* and the *convex closure of extreme points* of a set.

Every set in R^n has a convex closure, but not necessarily a convex closure of extreme points. This is the case, of course, of open or unbounded sets. If the set is not bounded, then we would not have all the required extreme points. If it is open, then extreme points might be lacking.

A special situation occurs when S is itself a convex set. That the above exceptions regarding unbounded or open sets still hold, is clear. Sets with these properties will not have a convex hull of extreme points. However, when a convex set is neither unbounded nor open, the following result (Krein–Milman) is valid.

Theorem 18.11† Let $S \subset R^n$ be a closed, bounded, convex set. Then $S = \ddot{S}$, that is, S is the convex hull of its extreme points.

Proof We called $\ddot{S}$ the convex hull or closure of the extreme points of S. Suppose we have a point $z \in S$, $z \notin \ddot{S}$. By Theorem 18.6, there exists a hyperplane $H(p, c) = \{x | px = pz = c\}$ such that $px > c$ for all $x \in \ddot{S}$. Consider a linear mapping f: $R^n \to R$ defined by $f(x) = px$. Since S is closed, bounded, and convex, so is the set $\{f(x) | x \in S\}$ which, being a closed interval of R, we shall denote $[a, b]$. Hence, $a \leq c \leq b$. Let $H(p, a)$ be the hyperplane $\{x | px = a\}$.

† Adapted from Lang [1].

Since $px \geq a$ for all $x \in S$, it follows that $H(p, a)$ is supporting for S. Moreover, by Theorem 18.10, $H(p, a)$ contains an extreme point, say x^k, of S. But, then, $x^k \in \ddot{S}$. We have earlier shown that, for all $x \in \ddot{S}$, $px > c$. It follows, thus, that $px^k > c$. But, now, we have $x^k \in \ddot{S}$ such that $px^k = a \leq c$. This and the preceding statement are in contradiction. Therefore, we conclude that there can be no point in S which does not belong to $\ddot{S}$, and no point in $\ddot{S}$ which is not in S. That is, $S = \ddot{S}$.

Amongst the class of convex sets, we can distinguish sets with a finite and sets with an infinite number of extreme points. We conclude this section with a definition.

Definition 18.8 A convex set $S \subset R^n$ having a finite number of extreme points is called a polytope or convex polyhedral set.

An illustration of a convex polyhedral set is the "feasible set" of our preceding linear programming example.

18.3 Convex Cones

In this last section of the chapter, we shall briefly examine a few properties of a type of convex set which plays a nonnegligible role in economic analysis, especially, though not only, in production theory. We introduce first the basic concept.

Definition 18.9 Let $C \subset R^n$ and $\lambda_r, \lambda_s \in [0, \infty[$. Then C is said to be a convex cone if for every $x^r, x^s \in C$ we have $\lambda_r x^r + \lambda_s x^s \in C$.

A convex cone is, thus, a set closed under the operations vector addition and vector multiplication by a nonnegative scalar. From the definition it is obvious that the sum of two vectors in C is a vector in C, and, for any $x \in C$ and $\lambda \geq 0$, λx belongs to C. The definition also implies $\theta \in C$.

Example Figure 18.5 illustrates a convex cone in R^2. However, note that any of the following sets are convex cones in the sense of Definition 18.9: (a) for any vector $v \in R^n$, the half-line in R^n defined by $(v) = \{x \mid x = \lambda v, \lambda \geq 0\}$; (b) the half-space $\bar{H}^- = \{x \mid x \in R^n, ax \leq 0\}$, which may be regarded as the solution set of the homogeneous inequality $ax \leq 0$; (c) as a generalization of the preceding case, the solution set of the linear homogeneous system $Ax \leq \theta$; (d) the nonnegative orthant of R^n; (e) in general, any linear subspace of R^n.

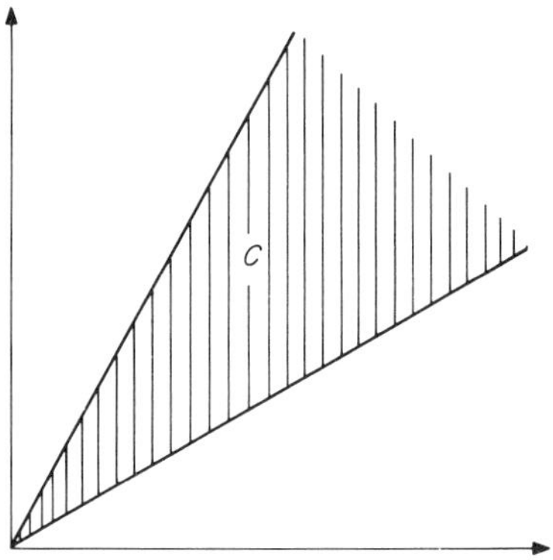

Figure 18.5

A careful consideration of Definition 18.9 would suggest to us that, with the exception of perhaps θ, no point of a convex cone C can be an extreme point. The origin θ itself may be an extreme point, but only provided $C \cap (-C) = \{\theta\}$. If this is not the case, that is, $C \cap (-C) \neq \{\theta\}$, then some point x must exist in C, with the property that $-x$ is also in C. The origin θ can then be represented as the convex combination $\frac{1}{2}x + \frac{1}{2}(-x)$, and it is clear that it is not an extreme point. A linear subspace, hence, is a convex cone without an extreme point.

Definition 18.10 A convex cone is said to be pointed if it has θ as its extreme point.

The next remark follows from Definition 18.10 and Minkowski's theorem.

Remark 18.4 If C is a pointed convex cone in R^n, then there is a hyperplane through θ which is supporting for C and $-C$.

Operations with Convex Cones

The first two operations we shall deal with are well known to us. We mainly adapt to our current special situation, concepts already studied in earlier chapters. The third operation will be a matter of particular concern.

Definition 18.11 Let $C_1, \ldots, C_m$ be a finite number of convex cones in R^n. Then the sum $\sum_{j=1}^{m} C_j$ is the set

$$\{x \mid x = \sum_{j=1}^{m} x_j, \ x_j \in C_j \text{ for } j = 1, \ldots, m\}.$$

The next important result follows immediately from the corresponding property of the sum of convex sets and the definition of a convex cone.

Remark 18.5 The sum of m convex cones in R^n is a convex cone.

Example Figure 18.6 illustrates the sum of two convex cones in the space R^2.

Figure 18.6

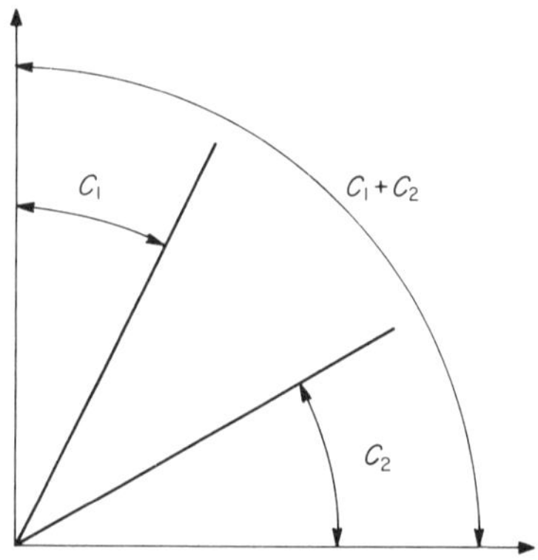

Now let us consider the intersection. That the union of convex cones does not need to be a convex cone is obvious, and does not deserve further examination.

Definition 18.12 Let $C_1, \ldots, C_m$ represent m convex cones in R^n. Then the intersection $\bigcap_{j=1}^{m} C_j$ is the set $\{x \mid x \in C_j, \text{ for all } j = 1, \ldots, m\}$.

Arguments similar to those supporting Remark 18.5, except that we refer to intersection instead of sum of sets, can be invoked to back up the following property.

Remark 18.6 The intersection of m convex cones in R^n is a convex cone.

Example Figure 18.7 shows the intersection of two convex cones in R^2.

Figure 18.7

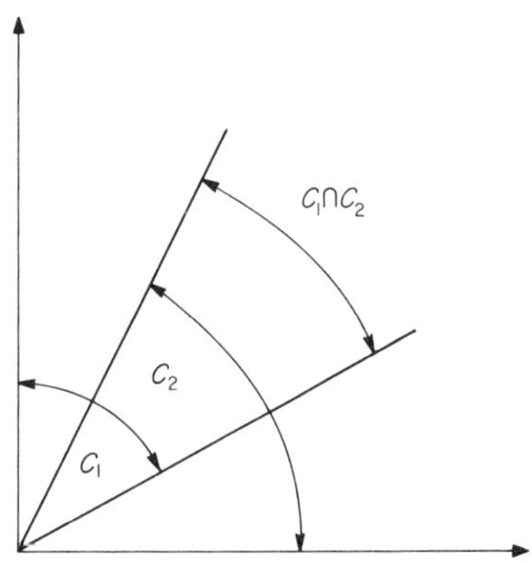

Our last operation, dualization, consists in assigning to a convex cone C in R^n, another cone in R^n called the dual or polar of C and denoted C^*. To properly define this operation, it is sufficient, thus, to describe the dual or polar cone of C.

Definition 18.13 Let C be a convex cone in R^n. Then the dual or polar of C, denoted C^*, is the set in R^n $\{y \mid yx \leq 0, \text{ all } x \in C\}$.

From the definitions of a convex cone and of its dual Remark 18.7 easily follows.

Remark 18.7 If C is a convex cone in R^n, then its dual or polar C^* is a convex cone.

Other significant features of the polar C^* of C will be examined very soon.

Example Figure 18.8 shows the polar cone of C in three different situations. It should help to convince us that C^* can only be a convex cone.

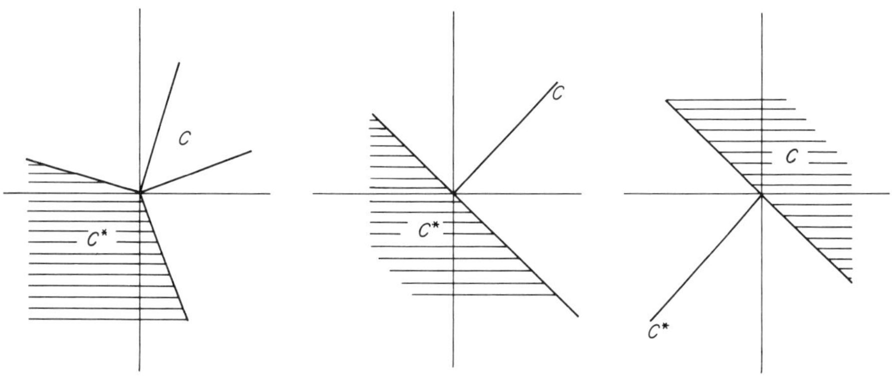

Figure 18.8

Definition 18.14 Let $C_1, \ldots, C_m$ be pointed convex cones in R^n. Then, they are said to be positively semi-independent if $\sum_{j=1}^{m} x_j = \theta$ with $x_j \in C_j$ for $j = 1, \ldots, m$ implies $x_j = \theta$ for $j = 1, \ldots, m$.

Definition 18.14 can be interpreted as follows: Suppose, for simplicity, we have only two cones, C_1 and C_2. The cones are positively semi-independent only if for any pair of vectors $x_1 \in C_1$ and $x_2 \in C_2$, the sum $x_1 + x_2 \neq \theta$ every time, with exception of the trivial case $x_1, x_2 = \theta$. Further, it can readily be shown that a necessary and sufficient condition for the positive semi-independence of C_1 and C_2 is that $C_1 \cap (-C_2) = \{\theta\}$.

Example Observe that C_1 and C_2 in Fig. 18.9 are two positively semi-independent cones.

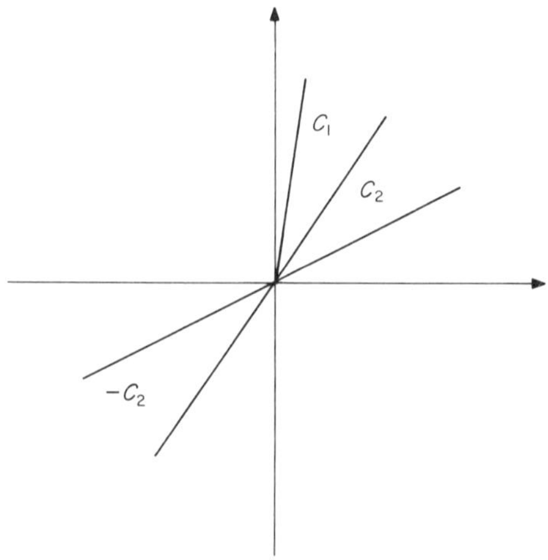

Figure 18.9

Finite Convex Cones

Within the class of convex cones, there is a subclass which is of special interest as an analytical tool frequently used in linear economic theory.

Definition 18.15 A cone $C \subset R^n$ is said to be finite, if there is a finite number of vectors v^k such that $x = \sum_k \lambda_k v^k$, for any $x \in C$ and $\lambda_k \in [0, \infty[$. The finite number of half-lines (v^k) are called the generators of C.

Half-lines, as seen in an earlier example, are a quite elementary type of cone. A finite cone has the property that it can be represented as the sum of a finite number of half-lines. The notation $C = \sum_k (v^k)$ to express the finite cone generated by the half-lines (v^k) is most frequent.

From the properties of the sum and intersection of convex cones and the definition of a finite cone we easily derive the following properties.

Remark 18.8 Let $C_1, \ldots, C_m$ be finite cones. Then the sum $\sum_{j=1}^m C_j$ and the intersection $\bigcap_{j=1}^m C_j$ are also finite cones.

Example A linear subspace of R^n is easily seen to be a finite cone. So also is the solution set of the homogeneous system $Ax \leq \theta$.

If A is an $m \times n$ matrix, then $Ax \leq \theta$ can be thought as a set of m inequalities $A^i x \leq 0$, where, for each $i = 1, \ldots, m$ there is defined a closed half-space $\overline{H}_i^- = \{x | A^i x \leq 0\}$. Each of these is a finite cone, and thus, so is the intersection $\bigcap_{i=1}^{m} \overline{H}_i^-$. But, obviously, $\bigcap_{i=1}^{m} \overline{H}_i^-$ is the solution set of $Ax \leq \theta$.

Finite convex cones are often called convex polyhedral cones.

Example The set of net outputs of an economy may be called a production set. If T represents an $m \times n$ "technology" matrix (n activities to produce m commodities) with $m < n$, y an n-component vector of "levels of activity operation," and x an m-component vector of net outputs, then the production set is defined as $X = \{x \mid x = Ty,\ y \geq \theta\}$. The ith component of x may be positive, null, or negative, according to whether the final result of the corresponding operation is a "final output," a "pure intermediate good," or a "primary input" (that is, labor), respectively. The vector y, on the other hand, is nonnegative, since it does not make sense to speak of an activity operated at a negative level. Observe that X is a finite convex cone, for it is the result of a finite number of possible activities. In Fig. 18.10 we

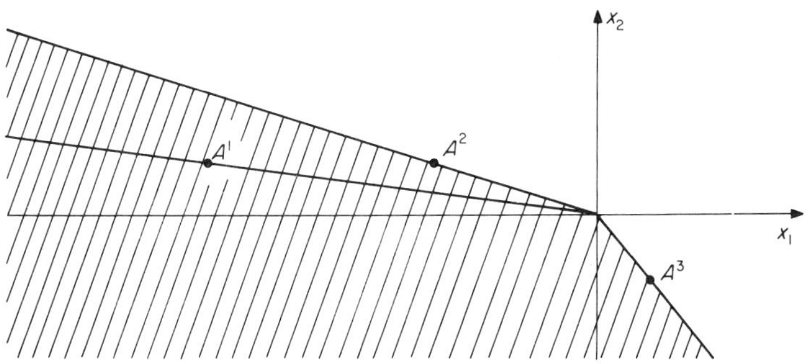

Figure 18.10

have represented a two-commodity production set. A^1, A^2, and A^3 represent three activities. Observe also the following assumptions implied in the form of the production set: (1) impossibility of producing without use of, at least, one input ($X \cap \Omega = \{\theta\}$); (2) irreversibility of the production process ($X \cap (-X) = \{\theta\}$); (3) possibility of free disposal, that is, "wasteful" production ($-\Omega \subset X$); (4) nonincreasing returns to scale and linear technology.

The Dual or Polar Cone

We return to the study of the dual or polar cone C^* of C, already introduced in Definition 18.13. Our concern now will be some of the properties of C^*.

We shall first focus our attention on certain features of a dual "in general," and thereafter examine the particularities of a dual when C is a finite cone.

The first property we shall point out is a direct consequence of Definition 18.13 (definition of the polar cone) and does not require a special proof. We state it in the following theorem:

Theorem 18.12 Let C_1 and C_2 be two convex cones in R^n. Then $C_1 \subset C_2$ implies $C_2^* \subset C_1^*$.

Example Figure 18.11 illustrates the content of Theorem 18.12 for two cones and their polars in R^2.

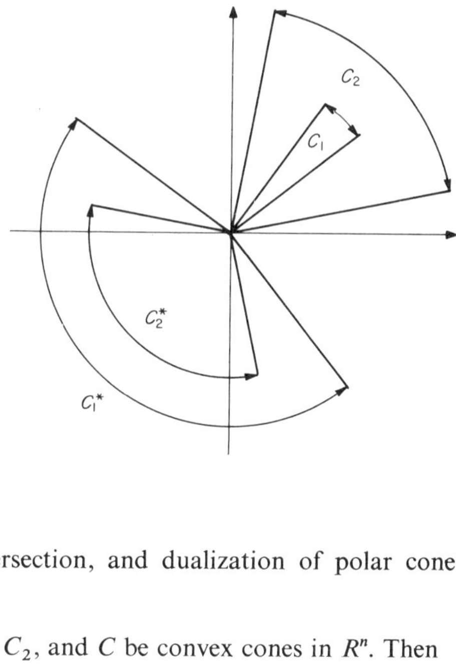

Figure 18.11

Regarding sum, intersection, and dualization of polar cones we have the following results.

Theorem 18.13 Let C_1, C_2, and C be convex cones in R^n. Then

(a) $(C_1 + C_2)^* = C_1^* \cap C_2^*$;
(b) $C_1^* + C_2^* \subset (C_1 \cap C_2)^*$;
(c) $C \subset (C^*)^*$.

Proof We confine ourselves to the statements (a) and (c) and leave the proof of (b) as an exercise. Let us begin with $(C_1 + C_2)^* = C_1^* \cap C_2^*$. Assume $x \in C_1^* \cap C_2^*$. Then, for all $y_1 \in C_1$ and all $y_2 \in C_2$ we have $y_1 \cdot x \leq 0$ and $y_2 \cdot x \leq 0$, respectively. Consider now $y \in C_1 + C_2$. It must be $y = y_1 + y_2$ for some $y_1 \in C_1$ and $y_2 \in C_2$. Hence, $yx = y_1 \cdot x + y_2 \cdot x \leq 0$. Thus, $x \in (C_1 + C_2)^*$.

Conversely, assume $x \in (C_1 + C_2)^*$. We have, then, $yx \leq 0$ for $y = y_1 + y_2$ with $y_1 \in C_1$ and $y_2 \in C_2$. In particular, if $y_2 = \theta$, we have $y_1 \cdot x \leq 0$ for all $y_1 \in C_1$; if $y_1 = \theta$, we have $y_2 \cdot x \leq 0$ for all $y_2 \in C_2$. Therefore $x \in C_1{}^*$ and $x \in C_2{}^*$. This obviously means $x \in C_1{}^* \cap C_2{}^*$.

Finally, let us show that $C \subset (C^*)^*$. Assume $z \in (C^*)^*$. It follows that $yz \leq 0$ for $y \in C^*$. But now $y \in C^*$. We have thus $yx \leq 0$ for all $x \in C$. Therefore $C \subset (C^*)^*$.

Finite cones, as said before, are of special interest as analytical tools. We have referred to their role in linear economic theory, and the reader will have no difficulty in understanding the reasons for their importance, if he thinks of the close relation existing between the idea of a finite cone and that of a system of linear inequalities.

A finite cone, according to Definition 18.15, can be represented as the sum of a finite number of half-lines (v^k). Of course, that is equivalent to the statement that a finite cone is a linear combination of a finite number of vectors v^k in R^n.

Theorem 18.14 Let C be a finite convex cone in R^n. Then its dual C^* is also a finite cone.

Proof Let us write C has a sum of half-lines, namely, $C = \sum_k (v^k)$. But then, by definition, the dual C^* is the solution set of the inequalities system $yv^k \leq 0$ and is, therefore, a finite cone.

The property which we study next is usually known as the duality theorem.

Theorem 18.15 Let C be a finite convex cone in R^n. Then $C = (C^*)^*$.

Proof We have already shown in Theorem 18.13 that for any convex cone $C \subset (C^*)^*$. We strengthen this result, by proving that $(C^*)^*$ cannot contain other vectors than those belonging to C, that is, if $b \notin C$ then $b \notin (C^*)^*$.

Using matrix notation, let us assemble all generators of C in the matrix A. We can then define the finite cone C by $C = \{y \mid Ax = y,\ x \geq \theta\}$. Now choose some vector $b \notin C$. $Ax = b$ has, therefore, no nonnegative solution, and there is some vector y such that $yA \leq \theta$ and $yb > 0$ (from Theorem 18.2). Since $yAx \leq 0$ for all $x \geq \theta$, it follows that $y \in C^*$. On the other hand, $yb > 0$ so that $b \notin (C^*)^*$. Summarizing, from the fact that $C \subset (C^*)^*$ and that if $b \notin C$, then $b \notin (C^*)^*$, it follows $C = (C^*)^*$.

In Theorem 18.13 we have seen that, for general convex cones, $C_1{}^* + C_2{}^* \subset (C_1 \cap C_2)^*$. This inclusion relation can be strengthened to equality in the case of finite cones, as shown next.

Theorem 18.16 Let C_1 and C_2 be two finite convex cones in R^n. Then $C_1{}^* + C_2{}^* = (C_1 \cap C_2)^*$.

Proof From the fact that $C = (C^*)^*$, we can write $C_1^* + C_2^* = ((C_1^* + C_2^*)^*)^*$ and $((C_1^*)^* \cap (C_2^*)^*)^* = (C_1 \cap C_2)^*$. Now, from result (a) of Theorem 18.13, $((C_1^* + C_2^*)^*)^* = ((C_1^*)^* \cap (C_2^*)^*)^*$. Therefore, $C_1^* + C_2^* = (C_1 \cap C_2)^*$.

Example Consider the production set $X \subset R^2$ defined $X = \{x \mid x = Ty, y \geq \theta\}$ in a previous example (see Fig. 18.12). It is a closed convex cone

Figure 18.12

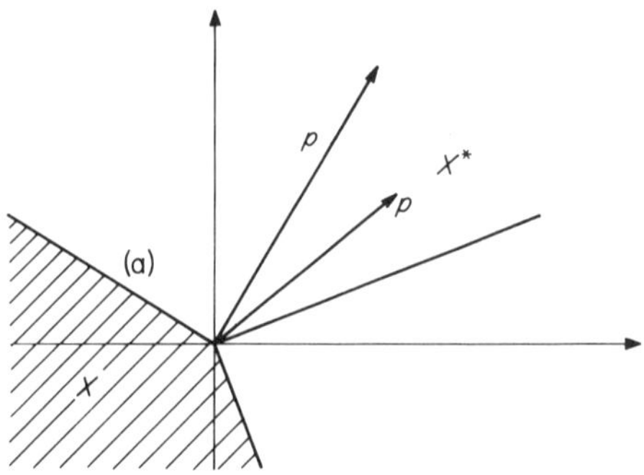

with polar X^*. Consider a price vector $p \in X^*$, and observe that when p is an interior point of X^*, the only profit maximizer is θ, whereas when p is on the boundary of X^*, the maximizers are the boundary points on the half-line (a).

EXERCISES

Inequality Systems

E.18.1 Let there be given the system,

$$x_1 + x_2 \leq 1, \qquad -x_1 - 8x_2 \leq 2, \qquad -2x_1 + x_2 \leq 2.$$

(a) Use Theorem 18.3 to show that the system has a solution. (b) Draw the solution set in R^2. (c) What kind of solution does this system admit?

E.18.2 Show that the following system has a solution:

$$10x_1 - 8x_2 \leq -32, \qquad -15x_1 + 20x_2 \leq 40, \qquad 10x_1 - 20x_2 \leq -20.$$

Does it have a nonnegative solution?

Hint: Use Theorems 18.3 and 18.4.

E.18.3 Does the system below have a solution? If the answer to this question is affirmative, what kind of solution?

$$-5x_1 + 4x_2 \leq 4, \qquad x_1 - x_2 \leq -1, \qquad -x_1 + 2x_2 \leq 2.$$

E.18.4 Show that the following system has no solution:

$$-x_1 + x_2 \leq 1, \qquad 2x_1 + 3x_2 \leq 6, \qquad x_1 - 3x_2 \leq 9.$$

Hint: Use Theorem 18.3.

E.18.5 Using Theorem 18.4 show that the following system has a nonnegative solution:

$$-2x_1 + 4x_2 \leq 4, \qquad x_1 - x_2 \leq -2, \qquad 2x_1 - x_2 \leq 2.$$

Hyperplanes

E.18.6 Show that the solution set of the linear system $Ax = b$ is convex.

Hint: Consider convex combinations of solutions.

E.18.7 Let the set $S \subset R^2$ be defined by $S = \{x \mid x \in R^2,\ x^{\mathrm{T}}x \leq 25\}$. Give (a) the formula of one separating hyperplane for S; (b) the formula of the supporting hyperplane at the point (3, 4); (c) the general formula for a supporting hyperplane at any point on $\{x \mid x \in R^2,\ x^{\mathrm{T}}x = 25\}$.

E.18.8 Study the effects of replacing the closed set S by an open one in Theorems 18.6, 18.7, and 18.8. Illustrate this by means of the set $\{x \mid x \in R^2,\ x^{\mathrm{T}}x < 25\}$.

E.18.9 Illustrate Remark 18.3 with the set $S = \{x \mid x \in R^2,\ x^{\mathrm{T}}x \leq 25\}$. What is the geometric meaning of the supporting hyperplane of S?

Convex Cones

E.18.10 Prove Remark 18.4.

E.18.11 Prove Remarks 18.5–18.7.

E.18.12 Show with an example that if C_1 and C_2 are convex cones in R^2, $C_1 \cup C_2$ does not need to be a convex cone.

E.18.13 Prove proposition (b) of Theorem 18.13.

E.18.14 Consider a cone C and its polar C^* in R^n. Show that $C + C^* = R^n$.

PROBLEMS

P.18.1 Let there be given the homogeneous system:

$$-x_1 + x_2 \leq 0, \qquad 2x_1 + x_2 \leq 0.$$

(a) Use theorem 18.5 and show that this system has no semipositive solution.

(b) Show that it has a nonzero solution and that, in general, any homogeneous system of two inequalities in two unknowns has a nonzero solution.

(c) Generalize the result of (b) to homogeneous systems of n inequalities in n unknowns.

P.18.2 Let S be a compact, convex subset of R^n. Prove that if S has two points on the same supporting hyperplane $H(p, c)$, then it has, at least, two extreme points.

Hint: Consider the points $x', x'' \in S \cap H(p, c)$ such that (1) x' has the largest distance to θ; (2) x'' has the largest distance to x'.

P.18.3 Let t: $R^n \to R^n$ be a bijective linear mapping, and $S \subset R^n$ a convex set. Show that if x^w is an extreme point of S, then $t(x^w)$ is an extreme point of $t(S)$.

Hint: You may compare this proposition with the concepts of Definitions 7.12 and 7.13 in Chapter 7.

P.18.4 Consider the two-traders, two-commodities pure exchange economy of the Edgeworth box in Fig. 18.13. (a) List some properties which would

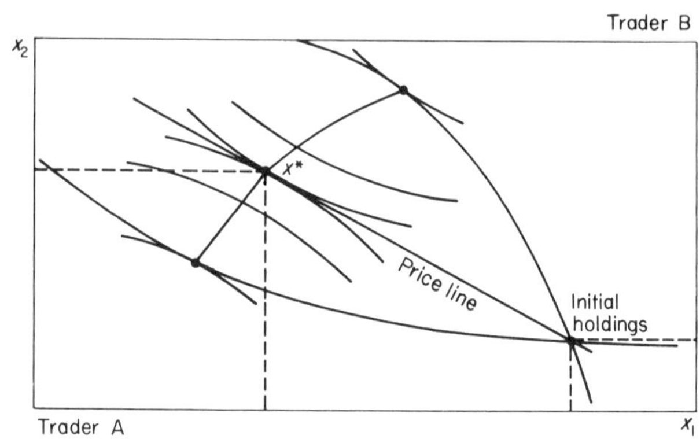

Figure 18.13

guarantee the existence of an equilibrium (x^*, p^*). (b) Use your knowledge on bounding hyperplanes to prove that x^* is an equilibrium allocation.

P.18.5 Assume that the production set $X \subset R^n$ is closed, convex, and contains θ. Furthermore, assume that free production is impossible (nothing can be produced without using inputs), production processes are irreversible (if x is possible $-x$ is not), and the possibility of free disposal exists. Prove that (a) $-\Omega \subset X$; (b) X cannot contain an entire hyperplane passing through θ; (c) there is a hyperplane through θ which is bounding for X, and such that X lies entirely on one of the half-spaces thus determined; (d) X is smaller than a half-space.

NOTES

This chapter borrows a great deal of material from Gale [3] and Lang [1]. For examples of how the ideas discussed here have become essential tools of economic analysis, the reader should consult Gale [1, 3], Koopmans [2], Nikaido [3], and Morgenstern [1]. The theorem on the relationship between competitive equilibrium and Pareto optima is from Debreu [4]. For this latter subject the reader is specially referred to Nikaido [4].

CHAPTER 19

NONNEGATIVE SQUARE MATRICES

One characteristic which, one is tempted to say, distinguishes economic variables from those studied in other disciplines, is that they are constrained to nonnegative values. This is not, of course, a rule without exceptions. Nevertheless, a glance to any book dealing with economic problems would convince us of the essential role which nonnegativity constraints play in this field. It is obvious that for variables like production, supply, demand, income and, in most of the cases, prices, negative magnitudes do not make very much sense.

The nonnegative character of economic variables has induced the interest of mathematical economists in the properties of nonnegative mappings. The problem can be studied, of course, in quite general terms, but we shall confine ourselves to the discussion of certain essential features of nonnegative linear mappings on R^n. And this, as we know, implies that we will have to focus our attention on the n-square matrices associated with them.

In Section 19.1 we introduce the notion of matrix decomposability.

Sections 19.2 and 19.3 will show some classical results on nonnegative indecomposable and nonnegative decomposable matrices, respectively. Properties of the inverse matrix $(\mu I - A)^{-1}$, of great importance in linear economic models, will be considered in Section 19.4. But, first of all, we should introduce some basic definitions and refer to some elementary properties which follow from them.

According to an earlier convention, we agreed that a vector, say $a = (a_1, \ldots, a_n)$, would be called positive, semipositive, or nonnegative, denoted, respectively, $a \gg \theta$, $a > \theta$, $a \geq \theta$, if we have that $a_i > 0$ for all i, $a_i \geq 0$ for all i but $a \neq \theta$, and $a_i \geq 0$ for all i, respectively. The idea can easily be extended to a square matrix $A = [a_{ij}]$. A positive matrix A, denoted $A \gg Z$, is one with $a_{ij} > 0$ for all (i, j). On the other hand, semipositivity of A, denoted $A > Z$, will imply $a_{ij} \geq 0$ for all (i, j) but $A \neq Z$. Finally, $A \geq Z$ will stand for a nonnegative matrix, that is, a matrix with $a_{ij} \geq 0$ for all index pairs (i, j).

In the present chapter we only consider real n-square matrices and real vectors. The following properties are obvious.

Remark 19.1 If $x \geq \theta$, $A \geq Z$, and k is any nonnegative real number, then (1) $kx \geq \theta$; (2) $kA \geq Z$.

Remark 19.2 If x, $y \geq \theta$ and A, $B \geq Z$, then (1) $x + y \geq \theta$; (2) $x \cdot y \geq 0$; (3) $A + B \geq Z$; (4) $A \cdot B \geq Z$; (5) $Ax \geq \theta$.

Remark 19.3 Let x and y be two vectors. Then $x - y \gg \theta$, $x - y > \theta$, and $x - y \geq \theta$ imply $x \gg y$, $x > y$, and $x \geq y$, respectively.

Remark 19.4 Let A and B be n-matrices. Then $A - B \gg Z$, $A - B > Z$, and $A - B \geq Z$ imply $A \gg B$, $A > B$, and $A \geq B$, respectively.

19.1 Indecomposable and Decomposable Matrices

We remind the reader that a collection $\{I, J\}$ is a partition of the set K, if the nonempty sets I, $J \subset K$ are such that $I \cup J = K$ and $I \cap J = \emptyset$. The notion of partition is useful in the following definition.

Definition 19.1 Let $A = [a_{ij}]$ be $n \times n$ and $K = \{1, 2, \ldots, n\}$. A is said to be indecomposable if no partition $\{I, J\}$ of K exists such that $a_{ij} = 0$ for $i \in I$ and $j \in J$; A is said to be decomposable if such a partition exists; finally, A is said to be completely decomposable if it is decomposable and, additionally, $a_{ji} = 0$ for $j \in J$ and $i \in I$.

Example Consider the matrices A, B, and C below. It is easy to see that A is indecomposable, since no partition of K would satisfy the requirements of Definition 19.1. On the contrary, B is decomposable. We have the partition $\{I, J\}$ of K, where $I = \{3, 4, 5\}$ and $J = \{1, 2\}$ for which $a_{ij} = 0$ every time $i \in I$ and $j \in J$. It also follows immediately that C is completely decomposable. For the same partition of K we have $a_{ij} = 0$ for $i \in I$ and $j \in J$, but we also have $a_{ji} = 0$ when $j \in J$ and $i \in I$:

$$A = \begin{pmatrix} a_{11} & a_{12} & a_{13} & a_{14} & a_{15} \\ a_{21} & a_{22} & a_{23} & a_{24} & a_{25} \\ a_{31} & a_{32} & a_{33} & a_{34} & a_{35} \\ a_{41} & a_{42} & a_{43} & a_{44} & a_{45} \\ 0 & a_{52} & a_{53} & a_{54} & a_{55} \end{pmatrix}; \qquad B = \begin{pmatrix} b_{11} & b_{12} & b_{13} & b_{14} & b_{15} \\ b_{21} & b_{22} & b_{23} & b_{24} & b_{25} \\ 0 & 0 & b_{33} & b_{34} & b_{35} \\ 0 & 0 & b_{43} & b_{44} & b_{45} \\ 0 & 0 & b_{53} & b_{54} & b_{55} \end{pmatrix};$$

$$C = \begin{pmatrix} c_{11} & c_{12} & 0 & 0 & 0 \\ c_{21} & c_{22} & 0 & 0 & 0 \\ 0 & 0 & c_{33} & c_{34} & c_{35} \\ 0 & 0 & c_{43} & c_{44} & c_{45} \\ 0 & 0 & c_{53} & c_{54} & c_{55} \end{pmatrix}.$$

The preceding example suggests that decomposable matrix can be written

$$\begin{pmatrix} A_{11} & A_{12} \\ Z & A_{22} \end{pmatrix}, \tag{1}$$

where A_{11} and A_{22} are square submatrices, though not necessarily of the same order. A completely decomposable matrix, on the other hand, is an array expressible as

$$\begin{pmatrix} A_{11} & Z & \cdots & Z \\ Z & A_{22} & \cdots & Z \\ \vdots & \vdots & \vdots & \vdots \\ Z & Z & \cdots & A_{kk} \end{pmatrix}, \tag{2}$$

where $A_{11}, A_{22}, \ldots, A_{kk}$ are square matrices, again, not necessarily of the same order.

A matrix A need not be originally given in form (1) or (2) to be decomposable. The question is rather, whether A can be transformed into a matrix with form (1) or (2) by performing suitable and identical permutations on the rows and columns. This can be done, as we know, by premultiplying and postmultiplying A by the permutation matrix P and its inverse P^{-1} (note that here $P^{-1} = P^{\mathrm{T}}$), respectively.

Remembering forms (1) and (2), we can summarize as follows:

Remark 19.5 If an $n \times n$ matrix A is decomposable or completely decomposable, then a permutation matrix P exists such that PAP^{-1} has form (1) or (2), respectively.

Example 1 Observe that A following is completely decomposable:

$$A = \begin{pmatrix} 1 & 0 & 2 & 0 \\ 0 & 8 & 0 & 7 \\ 3 & 0 & 4 & 0 \\ 0 & 6 & 0 & 5 \end{pmatrix}; \quad P = \begin{pmatrix} 1 & 0 & 0 & 0 \\ 0 & 0 & 1 & 0 \\ 0 & 0 & 0 & 1 \\ 0 & 1 & 0 & 0 \end{pmatrix}; \quad PAP^{-1} = \begin{pmatrix} 1 & 2 & 0 & 0 \\ 3 & 4 & 0 & 0 \\ 0 & 0 & 5 & 6 \\ 0 & 0 & 7 & 8 \end{pmatrix}.$$

Example 2 Note that B is a decomposable matrix:

$$B = \begin{pmatrix} 1 & 4 & 2 & 3 \\ 0 & 12 & 0 & 11 \\ 5 & 8 & 6 & 7 \\ 0 & 10 & 0 & 9 \end{pmatrix}; \quad P = \begin{pmatrix} 1 & 0 & 0 & 0 \\ 0 & 0 & 1 & 0 \\ 0 & 0 & 0 & 1 \\ 0 & 1 & 0 & 0 \end{pmatrix};$$

$$PAP^{-1} = \begin{pmatrix} 1 & 2 & 3 & 4 \\ 5 & 6 & 7 & 8 \\ 0 & 0 & 9 & 10 \\ 0 & 0 & 11 & 12 \end{pmatrix}.$$

The notion of matrix decomposability has a profound meaning in terms of systems theory, and it may be useful here to dedicate some thoughts to this point.

Technically, a linear system is a set of objects interconnected in a determinate way and satisfying the linearity principle. Mathematically, a linear system is a linear system of equations. It reduces, essentially, to a linear mapping which transforms a set of vectors ("inputs") into a set of vectors ("outputs"). If a "closed system" is representable by the equation $Ax = \lambda x$, then the role played by the matrix A is intuitive. Of course, it determines the way in which x is transformed into λx. But, beyond that, it summarizes the structure of the system, that is, in systems-theory jargon, the structural properties of the "net" of coupled elements. Three basic cases can be distinguished.

Independence

Suppose the matrix A is completely decomposable, so that $Ax = \lambda x$ can be written, with an appropriate partition of x (in this chapter x_1, x_2 denote *subvectors* rather than *components*),

$$\begin{pmatrix} A_{11} & Z \\ Z & A_{22} \end{pmatrix} \begin{pmatrix} x_1 \\ x_2 \end{pmatrix} = \lambda \begin{pmatrix} x_1 \\ x_2 \end{pmatrix}.$$

The "system" consists now of the two "subsystems,"

$$A_{11} x_1 = \lambda x_1, \qquad A_{22} x_2 = \lambda x_2,$$

which are, obviously, totally independent of each other. There is no relation whatsoever between them, and the use of the term "system" to refer to what is in reality a set of two autonomous units is rather formal.

If $Ax = \lambda x$ stands for a closed Leontief economy and A is completely decomposable, then we can separate the industries of the economy into two sets having no relations between them. We might think as well of two autonomous economies rather than in a single economic system.

One-Way Dependence

If A is only decomposable, then $Ax = \lambda x$ becomes

$$\begin{pmatrix} A_{11} & A_{12} \\ Z & A_{22} \end{pmatrix} \begin{pmatrix} x_1 \\ x_2 \end{pmatrix} = \lambda \begin{pmatrix} x_1 \\ x_2 \end{pmatrix},$$

or, equivalently

$$A_{11}x_1 + A_{12}x_2 = \lambda x_1, \qquad A_{22}x_2 = \lambda x_2.$$

We have a typical case of one-way dependence, since the system described by the first equation depends on the system $A_{22}x_2 = \lambda x_2$, but the converse is not true.

A decomposable closed Leontief economy $Ax = \lambda x$ would thus imply the possibility of separating industries into two sets such that industries in one set would sell to, but not buy from, or buy from, but not sell to, industries in the other set.

Two-Way Dependence

When the matrix A is indecomposable, $Ax = \lambda x$ is

$$\begin{pmatrix} A_{11} & A_{12} \\ A_{21} & A_{22} \end{pmatrix} \begin{pmatrix} x_1 \\ x_2 \end{pmatrix} = \lambda \begin{pmatrix} x_1 \\ x_2 \end{pmatrix},$$

that is,

$$A_{11}x_1 + A_{12}x_2 = \lambda x_1, \qquad A_{21}x_1 + A_{22}x_2 = \lambda x_2.$$

The two-way dependence is obvious since each subsystem depends not only on itself but on the other as well.

If a closed Leontief economy is indecomposable, then it is impossible to separate industries into two sets without relations or even with only one-way relations.

The structural importance of a matrix is thus evident for any linear system.

From our concepts of nonnegative and positive matrices and the definition of indecomposability, the next remark immediately follows.

Remark 19.6 A positive matrix is indecomposable. A nonnegative indecomposable matrix is either positive or semipositive.

19.2 Nonnegative Indecomposable Matrices

The important results we are going to study next, together with other properties we shall not discuss here, form the content of what is known in the literature as the Frobenius theorem. The Frobenius theorem provides a set of analytical tools whose role in theoretical economics can hardly be overestimated.

According to Remark 19.6, a positive matrix is always indecomposable. On the other hand, a nonnegative indecomposable matrix is either positive or semipositive. The Frobenius theorem, customarily formulated to cover nonnegative indecomposable matrices, holds properly for positive and semipositive indecomposable matrices. Terminology differences, thus, should not produce confusion.

Theorem 19.1† Let A be an $n \times n$ nonnegative indecomposable matrix. Then

(1) A has a characteristic value $\lambda^* > 0$;
(2) if x^* is a characteristic vector corresponding to λ^*, we have $x^* \gg \theta$;
(3) for any other characteristic value λ of A, we have $\lambda^* \geq |\lambda|$;
(4) the value of λ^* increases with the increase of any element a_{ij} of A.

Proof Let us consider the four points separately.

(1) A has an eigenvalue $\lambda^* > 0$. Our concern is here with vectors $x > \theta$, so that the set of all normalized vectors satisfying this property is the simplex

$$S = \{x \mid x \in R^n, \ x > \theta, \ \sum_{i=1}^{n} x_i = 1\}.$$

We define now the mapping f: $S \to S$ by $f(x) = (1/\rho(x))Ax$, where $\rho(x) > 0$, and such that for every $x \in S$ also $f(x) \in S$. Since there exists such a $\rho(x)$ for every $x \in S$, the continuity of f is easily seen. By Theorem 7.6 (Brouwer) the mapping f has a fixed point. Let us call this fixed point x^* and so $x^* = f(x^*)$. By definition, we have $x^* = (1/\rho(x^*))Ax^*$, that is, $x^* \cdot \rho(x^*) = Ax^*$. Writing $\rho(x^*) = \lambda^*$ we get $Ax^* = \lambda^* x^*$. Since $A > Z$ and $x^* > \theta$ by assumption, clearly $Ax^* > \theta$. Therefore, $\lambda^* > 0$.

† Adapted from Debreu and Herstein [1].

(2) Next we show that associated with λ^* there is an $x^* \gg \theta$. Suppose we write x^* in partitioned form $x^* = (x_1^* \mid x_2^*)^T$ where x_1^*, x_2^* are subvectors. Doing the same with A, we have

$$\left(\begin{array}{c|c} A_{11} & A_{12} \\ \hline A_{21} & A_{22} \end{array}\right)\left(\begin{array}{c} x_1^* \\ \hline x_2^* \end{array}\right) = \lambda^*\left(\begin{array}{c} x_1^* \\ \hline x_2^* \end{array}\right).$$

Assume, furthermore, $x_2^* = \theta$. Then

$$\left(\begin{array}{c|c} A_{11} & A_{12} \\ \hline A_{21} & A_{22} \end{array}\right)\left(\begin{array}{c} x_1^* \\ \hline \theta \end{array}\right) = \lambda^*\left(\begin{array}{c} x_1^* \\ \hline \theta \end{array}\right).$$

$A_{11}x_1^* = \lambda^* x_1^*$ and $A_{21}x_1^* = \theta$. It follows that $A_{21} = Z$. But this violates our assumption about the indecomposability of the matrix A. Therefore $x^* \gg \theta$.

(3), (4) Any other λ of A is $|\lambda| \leq \lambda^*$, and λ^* increases with the increase of any element of A. We prove these two points together. Let us have a matrix B such that $Z \leq B \leq A$, with characteristic value k and associated eigenvector y. We show that $|\lambda| \leq \lambda^*$, indirectly, by proving that $|k| \leq \lambda^*$. Then, setting $B = A$, the proof of (4) is completed.

Take the transpose of A. Since it is semipositive indecomposable, it has an eigenvalue $\lambda^0 > 0$ and vector $x^0 \gg \theta$ such that $A^T x^0 = \lambda^0 x^0$. We also have $By = ky$. Take $|k|$, define the vector $\bar{y} = (|y_1|, \ldots, |y_n|)$ and note that the matrix $A = [a_{ij}] = [|a_{ij}|]$. From operations with absolute values we know $|a| \cdot |b| \leq |a \cdot b|$ for two numbers a, b. Thus, $|k|\bar{y} \leq B\bar{y}$. Since, on the other hand, $B\bar{y} \leq A\bar{y}$, it follows that (i) $|k|\bar{y} \leq B\bar{y} \leq A\bar{y}$ and also (ii) $|k|(x^0)^T\bar{y} \leq (x^0)^T A\bar{y} = \lambda^0 (x^0)^T \bar{y}$. From $x^0 \gg \theta$ it follows that $(x^0)^T\bar{y} > 0$. Hence, we have for λ^0 of A^T and every k of B, $|k| \leq \lambda^0$.

Now set $B = A$. Writing λ for k we get (iii) $|\lambda|\bar{y} \leq A\bar{y}$ and also (iv) $|\lambda|(x^0)^T\bar{y} \leq (x^0)^T A\bar{y} = \lambda^0(x^0)^T\bar{y}$. From $x^0 \gg \theta$ it follows that $(x^0)^T\bar{y} > 0$. Hence, for λ^0 of A^T and every λ of A we have $|\lambda| \leq \lambda^0$.

Since we have shown $|\lambda| \leq \lambda^0$, it must be that $\lambda^* \leq \lambda^0$. Similarly, proceeding in the reverse order, we can show that $\lambda^0 \leq \lambda^*$. Thus $\lambda^0 = \lambda^*$. Also, $|\lambda| \leq \lambda^*$. Consider again the eigenvalue k of B, and assume $|k| = \lambda^*$. We would have $\lambda^*\bar{y} = B\bar{y} = A\bar{y}$. Now, since $\lambda^*\bar{y} = A\bar{y}$ it follows $\bar{y} \gg \theta$. Furthermore, from $B\bar{y} = A\bar{y}$ and $B \leq A$ it follows $B = A$.

19.3 Nonnegative Decomposable Matrices

The results summarized in the Frobenius theorem can be substantially weakened, if we drop the assumption about the matrix A being indecomposable. This we do in the theorem which follows.

Theorem 19.2† Let A be an $n \times n$ nonnegative matrix. Then the following properties hold:

(1) A has a characteristic value $\lambda^* \geq 0$.
(2) If x^* is a characteristic value corresponding to λ^*, we have $x^* > \theta$.
(3) For any other characteristic value λ of A, we have $\lambda^* \geq |\lambda|$.
(4) The value of λ^* does not decrease with the increase of any element a_{ij} of A.

Proof First we prove points (1), (3), and (4) together and then proceed with (2).

(1), (3), (4) A has a $\lambda^* \geq 0$ with an $x^* > \theta$, and such that for any other λ, $|\lambda| \leq \lambda^*$. When any a_{ij} is increased, λ^* does not decrease. Assume A has been decomposed into a number of matrices from which $A_{11}, \ldots, A_{rr}$ are indecomposable. Call A_R the submatrix with largest eigenvalue. Since this eigenvalue is also an eigenvalue of A, points (1), (3), and (4) are immediately proved by using Theorem 19.1.

(2) Let $(A^t \mid t \in N)$ be a sequence of matrices $A^t \gg Z$ converging to A. From Theorem 19.1 it follows that every A^t has an eigenvalue $\lambda_t{}^* > 0$ and an eigenvector $x^{*,t} \in S = \{x \mid x \in R^n,\ x > \theta,\ \sum_{i=1}^n x_i = 1\}$. Consider now the sequence $(x^{*,t} \mid t \in N)$.

Since S is compact, the sequence contains a subsequence $(x^{*,t'} \mid t' \in N)$ which converges to $x^* > \theta$. Clearly, λ^* is the limit of the sequence $(\lambda_{t'}^* \mid t' \in N)$. For every t' we have $A^{t'}x^{*,t'} = \lambda_{t'}^* \cdot x^{*,t'}$. Therefore, in the limit, $Ax^* = \lambda^* x^*$.

19.4 The Matrices $\mu I - A$ and $(\mu I - A)^{-1}$

In the two previous sections we dealt with some properties of a nonnegative matrix A, both with and without the assumption of indecomposability. We can derive now some useful results for a matrix $(\mu I - A)^{-1}$, where $\mu > 0$ is a real number and A belongs to one of the two classes mentioned above.

A matrix $(\mu I - A)^{-1}$ plays a very important role in the analysis of linear economic models, and we shall see a simple illustrative example at the end of this chapter. The study of the matrix $(\mu I - A)^{-1}$ depends on our knowledge about its inverse $\mu I - A$ so that we will have to give some time to the study of the latter.

Definition 19.2 Let A be a nonnegative square matrix. Then the eigenvalue λ^* satisfying the properties of either Theorems 19.1 or 19.2 is called the dominant or maximal eigenvalue of A.

† Adapted from Debreu and Herstein [1].

The Matrix $(\mu I - A)^{-1}$ When A is Nonnegative

In the proofs of the theorems given below we follow closely the approach of Debreu and Herstein's [1] fundamental article on the subject. The main theorems, Theorem 19.4 in this subsection and 19.6 in the next, are preceded by the paving theorems, Theorems 19.3 and 19.5, respectively.

Theorem 19.3 Let $A \geq Z$ have the dominant characteristic value λ^*. It holds that

(1) if $Ax \leq \mu x$ for $x \gg \theta$, then $\lambda^* \leq \mu$;
(2) if $Ax \geq \mu x$ for $x \gg \theta$, then $\lambda^* \geq \mu$;
(3) if $Ax \ll \mu x$ for $x \geq \theta$, then $\lambda^* < \mu$;
(4) if $Ax \gg \mu x$ for $x \geq \theta$, then $\lambda^* > \mu$.

Proof The proofs for the four statements are similar. Thus we only give proof for the first of them. Consider A^{T} with dominant value λ^* and associated vector x^0. From Theorem 19.2 (2) we have $A^{\mathrm{T}}x^0 = \lambda^* x^0$. We have now $Ax \leq \mu x$, $x \gg \theta$. Thus $(x^0)^{\mathrm{T}}Ax \leq (x^0)^{\mathrm{T}}\mu x$, that is, $\lambda^*(x^0)^{\mathrm{T}}x \leq (x^0)^{\mathrm{T}}\mu x$. Since $x \gg \theta$ and $x^0 > \theta$ it follows $(x^0)^{\mathrm{T}}x > 0$. Therefore $\lambda^* \leq \mu$.

We can refer now to the matrix $(\mu I - A)^{-1}$. The meaning of $(\mu I - A)^{-1}$ can easily be seen. Suppose, for instance, we are given the system $y = (\mu I - A)x$. The relation between such an equation and a Leontief open economy will be touched in an example later. It is obvious, however, that $(\mu I - A)^{-1}$ is the matrix of the solution $x = (\mu I - A)^{-1}y$.

Theorem 19.4 Let A be nonnegative. Then the matrix $(\mu I - A)^{-1}$ is nonnegative if and only if $\mu > \lambda^*$.

Proof Assume first $\mu > \lambda^*$. It follows that, for each y, the solution x is unique. We must show that $y \geq \theta$ implies $x \geq \theta$. Suppose, on the contrary, x has negative components. By proper permutations and a partition it would be possible to write

$$\left(\begin{array}{c|c} \mu I - A_{11} & -A_{12} \\ \hline -A_{21} & \mu I - A_{22} \end{array}\right)\left(\begin{array}{c} -x_1 \\ \hline x_2 \end{array}\right) = y,$$

where $x_1 \gg \theta$ and $x_2 \geq \theta$ are subvectors of x, I the appropriate identity matrix, and $y \geq \theta$. We have $-(\mu I - A_{11})x_1 - A_{12}x_2 \geq \theta$. In consequence, $-(\mu I - A_{11})x_1 \geq \theta$. It follows $A_{11}x_1 \geq \mu x_1$. From statement (2) of Theorem 19.3 we get $\lambda_1^* \geq \mu$. But λ_1^* is only the dominant eigenvalue of A_{11}. For the matrix A it must hold that $\lambda^* \geq \lambda_1^*$. Now, this contradicts our initial assumption $\mu > \lambda^*$. Therefore, $y \geq \theta$ implies $x \geq \theta$, which was to be proved.

Conversely, assume $(\mu I - A)^{-1} \geq Z$. Then, to any $y > \theta$, there must correspond a vector $x \geq \theta$. Thus from $\mu x - Ax = y$ it follows that $Ax \ll \mu x$. From Theorem 19.3, statement (3), we finally get $\lambda^* < \mu$ and the proof is complete.

The Matrix $(\mu I - A)^{-1}$ When A Is Nonnegative Indecomposable

As in the previous subsection, we first study the relations between the dominant eigenvalue λ^* of A and the number μ. Unlike Theorem 19.3, in no case does Theorem 19.5 put such a strong requirement as the strict positiveness of the vector x. But here, the matrix A is assumed to be indecomposable. Proofs are not given since they are essentially similar to that of Theorem 19.3, statement (1).

Theorem 19.5 Let $A \geq Z$ be indecomposable with dominant eigenvalue λ^*. It holds that

(1) if $Ax \leq \mu x$ for $x > \theta$, then $\lambda^* \leq \mu$;
(2) if $Ax \geq \mu x$ for $x > \theta$, then $\lambda^* \geq \mu$;
(3) if $Ax < \mu x$ for $x \geq \theta$, then $\lambda^* < \mu$;
(4) if $Ax > \mu x$ for $x \geq \theta$, then $\lambda^* > \mu$.

Our object of attention is again the matrix $(\mu I - A)^{-1}$, the matrix of the solution $x = (\mu I - A)^{-1} y$ of the original system $y = (\mu I - A)x$.

Theorem 19.6 Let A be nonnegative and indecomposable. Then the matrix $(\mu I - A)^{-1}$ is positive if and only if $\mu > \lambda^*$.

Proof We have first to show that $y > \theta$ implies $x \gg \theta$. From Theorem 19.4, we already know that $x > \theta$. Suppose, now, x has some zero components. After appropriate permutations and a partition it would be possible to write

$$\left(\begin{array}{c|c} \mu I - A_{11} & -A_{12} \\ \hline -A_{21} & \mu I - A_{22} \end{array}\right)\left(\begin{array}{c} x_1 \\ \hline x_2 \end{array}\right) = y,$$

where $y > \theta$, $x_1 = \theta$, $x_2 \gg \theta$, and I is an identity matrix. Thus, $-A_{12}x_2 \geq \theta$. Since $x_2 \gg \theta$, it follows $A_{12} = Z$. This contradicts the assumption about the indecomposability of A. Thus, it is sufficient that $\mu > \lambda^*$ to have $(\mu I - A)^{-1} \gg Z$. The converse proof follows immediately. Observe that $(\mu I - A)^{-1} \gg Z$ implies $(\mu I - A)^{-1} \geq Z$. But, for $(\mu I - A)^{-1} \geq Z$, Theorem 19.4 has shown that a necessary condition is given by $\mu > \lambda^*$.

Example Suppose $y = x - Ax$ represents an open Leontief economy, where x is output, Ax the input requirements, and y a surplus of output over input. If y is given, then we call it "final demand." An important question to answer is whether a linear economy, like the one described here, is able to satisfy any y (primary factor constraints are not taken account of, here, although their introduction into the analysis would not change essentially the conclusions), that is any final demand, regardless of the relative composition of y.

The model can be written $y = (I - A)x$, and the solution is given by $x = (I - A)^{-1}y$. We are particularly interested in $(I - A)^{-1} \gg Z$ for we wish to ensure conditions under which y is satisfied, independently of its composition.

We can, of course, write $(I - A)^{-1} = (\mu I - A)^{-1}$, provided $\mu = 1$. Assume now λ^* is the dominant eigenvalue of A and, furthermore, assume A is indecomposable. It results that, under the given assumptions, $(I - A)^{-1} \gg Z$ if and only if $1 = \mu > \lambda^*$. In other words, the open Leontief economy $y = x - Ax$ (with indecomposable matrix A) is able to satisfy any demand vector y, whatever the proportions in which its components appear, provided A has a dominant characteristic value $\lambda^* < 1$.

EXERCISES

E.19.1 Find whether any of the following matrices is decomposable or totally decomposable. When appropriate, give the corresponding permutation matrix.

$$\begin{pmatrix} 1 & 2 & 3 & 0 & 4 \\ 0 & 5 & 6 & 0 & 7 \\ 8 & 9 & 0 & 10 & 11 \\ 12 & 0 & 13 & 14 & 15 \\ 16 & 17 & 18 & 19 & 0 \end{pmatrix}; \quad \begin{pmatrix} 1 & 2 & 3 & 4 & 5 \\ 6 & 7 & 8 & 9 & 10 \\ 11 & 0 & 12 & 0 & 13 \\ 14 & 0 & 15 & 0 & 16 \\ 17 & 18 & 19 & 20 & 21 \end{pmatrix};$$

$$\begin{pmatrix} 0 & 1 & 0 & 2 & 0 \\ 3 & 0 & 4 & 0 & 5 \\ 0 & 6 & 0 & 7 & 0 \\ 8 & 0 & 9 & 0 & 10 \\ 0 & 11 & 0 & 12 & 0 \end{pmatrix}.$$

E1.9.2 Show that a matrix A is indecomposable if and only if its transpose A^T is indecomposable.

E.19.3 Prove propositions (2)–(4) of Theorem 19.3

Hint: Essentially similar to proposition (1).

E.19.4 Prove Theorem 19.5.

Hint: The proof is almost a repetition of that of Theorem 19.3.

E.19.5 Let A be a nonnegative indecomposable (strictly speaking, positive) matrix. Show that (a) its dominant characteristic value λ^* is positive; (b) the characteristic vector x^* associated with λ^* is positive; (c) λ^* increases with an increase in any element a_{ij}.

$$A = \begin{pmatrix} 2 & 2 & 1 \\ 1 & 3 & 1 \\ 1 & 2 & 2 \end{pmatrix}.$$

Hint: (a) Check $\lambda^* = 5$; (b) consider $(1, 1, 1)^{\mathrm{T}}$; (c) add 1 to any element of A and compute a new λ^*.

E.19.6 Let A and B be the nonnegative decomposable matrices below. Show that their dominant roots are nonnegative, the associated characteristic vectors nonnegative, and that λ^* does not decrease with increase in any element,

$$A = \begin{pmatrix} 2 & 1 & 1 \\ 1 & 2 & 1 \\ 0 & 0 & 1 \end{pmatrix}, \qquad B = \begin{pmatrix} 2 & 2 & 0 \\ 2 & 2 & 0 \\ 0 & 0 & 1 \end{pmatrix}.$$

Hint: Same as in preceding exercise; check $\lambda_A{}^* = 3$, with $x^* = (1, 1, 0)^{\mathrm{T}}$, and $\lambda_B{}^* = 4$, with $x^* = (1, 1, 0)^{\mathrm{T}}$.

E.19.7 Let A be the same matrix as in the preceding exercise. Study the matrix $(\mu I - A)^{-1}$ under the following assumptions: (a) $\mu = 4$; (b) $\mu = 3$; (c) $\mu = 2$.

E.19.8 Let A be the matrix of exercise E.19.5. Study the matrix $(\mu I - A)^{-1}$ under the assumptions: (a) $\mu = 6$; (b) $\mu = 5$; (c) $\mu = 4$.

E.19.9 Consider the matrix A of E.19.5. (a) Show that A has more than one real characteristic value; (b) Show that the dominant characteristic value of any submatrix M_{ii}, obtained from A by deleting the ith row and the ith column, lie on the interval having as extremes the two largest characteristic values of A.

Hint: Study A, M_{11}, M_{22}, and M_{33}.

E.19.10 Generalize the result of (b) of the preceding exercise to cover the case of any $n \times n$ matrix $A \gg Z$ having at least two real characteristic values, and its $(n-1) \times (n-1)$ submatrices M_{ii}.

PROBLEMS

P.19.1 Let $Ax - x$ represent a closed Leontief economy. The economy is said to be viable if there exists $x' \geq \theta$ such that $Ax' \leq x'$. The vector x' is called an interior equilibrium if $x' \gg \theta$ and $Ax' = x'$. Prove that a necessary condition for the economy to have an interior equilibrium is the existence of an eigenvalue $\lambda' = 1$. What would happen if every eigenvalue is greater (smaller) than one?

P.19.2 Let A be nonnegative indecomposable. Prove that if λ^* is the dominant characteristic value, and x^* the associated characteristic vector, then x^* is the only nonnegative vector satisfying the inequality $Ax \leq x$.

P.19.3 For the closed economy of problem P.19.1 prove the following propositions: (a) if A is indecomposable, then $\lambda^* = 1$ is a sufficient condition for the existence of an interior equilibrium x^*; (b) if A is indecomposable, then x^* is the only viable solution of the system.

Hint: Use the results of P.19.1 and P.19.2.

P.19.4 Prove that a closed Leontief economy has no interior equilibrium (is not viable) if the dominant eigenvalue of A is less than one (greater than one).

P.19.5 In a closed Leontief economy, given a price vector $p = (p_1, \ldots, p_n)$, we may define the profit per unit of output in the jth sector by $p_j - \sum_{i=1}^{n} p_i a_{ij}$. Prove: (a) The condition for the existence of a dominant eigenvalue $\lambda^* = 1$ is the existence of a vector p' such that either $p'A \leq p'$ is satisfied with at least one equality, or $p'A \geq p'$ is satisfied with at least one equality. Give the economic meaning of this proposition. (b) If there is any p' such that $p'A \ll p'$ $(p'A \gg p')$, the model has no equilibrium (is not viable).

NOTES

Properties of positive matrices were first studied by Perron [1], and these results extended to the case of nonnegative matrices by Frobenius [1]. The discussion of this chapter follows Debreu and Herstein [1]. A different proof of Frobenius theorem, using only algebraic methods, has recently been given by Murata [1]. Comments on alternative proofs are presented by Bellman [1]. The subject of this chapter is discussed by Bellman [1], Karlin [1, 2], Lancaster [1], Nikaido [3, 4], and Gantmacher [1], to give only few references. The first four authors mentioned also deal with the application of the theory of nonnegative matrices to economics. For generalizations of the concept of indecomposability see the Notes to Chapter 20.

CHAPTER 20

MULTISECTORAL BALANCED GROWTH

In this chapter we show how some of the concepts developed in Part II can become tools of economic analysis. The example we have chosen to illustrate this is the classical von Neumann's [1] model of an expanding economy. It is clear that many other examples from the fields of linear economics would have served our purpose as well. However, the study of von Neumann's growth model put us in contact with several of the subjects previously discussed—linear equations and inequalities, nonnegative matrices, matrix indecomposability—and is, thus, especially well suited to our aims.

The growth system we are going to study may be considered as a sort of generalization of the closed Leontief economy considered in previous chapters. First, our model is dynamic in the sense that it is designed to describe a process over time. Looking at this question in terms of equilibrium theory, we may say that our concern here is not the existence of a (static) solution, represented by an output vector x, called an "interior equilibrium," but rather of a vector-valued function $x(t + \tau)$, $\tau = 0, 1, 2, \ldots$, called a "balanced growth mode," and which can be considered as a kind of dynamic equilibrium. Second, the model does not constrain the technology to a one-to-one correspondence between the number of productive activities and the number of produced commodities. In principle, on the contrary, the same activity may produce more than one commodity, and the same commodity may be produced by more than one activity.

Two main problems will occupy our attention here. In the first place, we shall answer the question of whether the model discussed has a "balanced growth mode," that is, we shall be concerned with the existence problem. In the second place, we shall consider the question of the uniqueness of such a solution. On the contrary, the study of the dynamic adjustment properties of the model, the typical stability problem, will not be analyzed here.

In Section 20.1 the model is described and its assumptions commented upon. Sections 20.2 and 20.3 are dedicated to the study of the so-called technological expansion and economic expansion, respectively. These two sections, like Section 20.4, where the two sides of the problem are related, contain the discussion on existence, whereas Section 20.5 deals with conditions for uniqueness of the desired solution.

Before going into the discussion of the model, it seems appropriate to make precise the meaning with which the expression "balanced growth mode" is used in this chapter.

Definition 20.1 A multisectoral economic model is said to have a balanced growth mode at a constant factor α, if it has a solution with the property that each sector grows at the same factor α.

In a balanced growth process, as it follows from the above definition, the scale of output is allowed to change over time, whereas the proportions among the different sectoral outputs remain constant.

20.1 Framework and Assumptions

We shall begin our study by stating the assumptions upon which the von Neumann's growth model is built. Here too, we make a distinction between the "technological" and the "economic" aspect of the system. The dynamic assumptions are discussed afterward.

Technological Assumptions

The economy is supposed to produce m commodities by means of n activities. Each activity may produce more than one commodity, and the same commodity may be produced by more than one activity. A technology like that described is said to be of the "activity analysis" type.

A brief comment might be useful here. If we assume, as in fact we do, that the technology is linear, then it is clear that we can describe it by means of $m \times n$ "technology matrices." The Leontief technology discussed in previous chapters is a special case of the technology discussed here. If $m = n$, then clearly the "technology matrices" reduce to square matrices. In linear economic theory it is shown, however, that even in the general case, if $m < n$, the efficient production of the m commodities can be obtained by the use of at most m different activities.

Let us return now to our main argument. The (linear) technology of our system will be condensed in the matrix pair (B, A) where both matrices are $m \times n$.

An activity j $(j = 1, \ldots, n)$ is completely described by an input vector $A_j = (a_{1j}, \ldots, a_{mj})^{\mathrm{T}} \geq \theta$ together with an output vector $B_j = (b_{1j}, \ldots, b_{mj})^{\mathrm{T}} \geq \theta$. The vector components a_{ij} and b_{ij} denote, respectively, the amount of the ith commodity which is used and produced when the jth activity is operated at the unit intensity level. The $m \times n$ matrices $A = [a_{ij}]$ and $B = [b_{ij}]$ are called the input and output matrix, respectively. Introducing the intensity vector $y = (y_1, \ldots, y_n)^{\mathrm{T}}$, we can determine the total inputs and outputs of the system by the products Ay and By, respectively. If we call $x = (x_1, \ldots, x_m)^{\mathrm{T}}$ the net-output vector, then the system, operating at intensities given by y, can be written $x = (B - A)y$. Note, however, that we shall assume our model is closed in the same sense as the closed Leontief system: Outputs are only consumed within the economy and inputs are only generated inside it.

The technology assumptions we now impose on the model are

(i) $A^i > \theta$, for all $i = 1, \ldots, m$;
(ii) $B_j > \theta$, for all $j = 1, \ldots, n$;
(iii) $Ay \leq By$.

The first assumption simply reminds us that production necessarily requires the use of inputs. The second, that every commodity is the product of some activity. The last one, without making reference here to possible dynamic relations specifies that, since the system is closed, the input requirements of each activity cannot exceed the available amounts, that is, the quantities of the corresponding outputs.

Economic Assumptions

If p_i represents the price of the ith commodity, we can describe a given price system by the price vector $p = (p_1, \ldots, p_m)$. We assume now

(iv) $pB \leq pA$.

This assumption might be interpreted in the usual sense that in a closed, competitive system, there is no positive profit in any activity. In other words, if an activity shows positive profit, the effects of competition will make it soon disappear. Here too, as in the technology assumption (iii), we have not specified dynamic characteristics. We come to this question next.

Dynamic Specifications

The model, as discussed so far, only shows its essential technological and economic traits, without relating these, however, to time.

In the study of von Neumann's model we make in this chapter, true dynamic analysis is left out of the discussion. Nevertheless, since our subject of analysis is a growth model, insight into its working can only be won after providing a dynamic specification of the relevant technological and economic assumptions.

We refer here to the discrete-time analysis method, representing the variable time by t and letting $t + \tau$, where $\tau = 0, 1, 2, \ldots$, symbolize a sequence of discrete time intervals or periods. Both the vectors y and p may be regarded now as sequences or discrete functions of time, if we write $y(t + \tau)$ and $p(t + \tau)$, respectively.

We now assume

(v) $Ay(t + 1) \leq By(t)$; $y(t + 1), y(t) \geq \theta$;
(vi) $p(t + 1)A \geq p(t)B$; $p(t + 1), p(t) \geq \theta$.

Notice that assumptions (v) and (vi) are, essentially, dynamic versions of assumptions (iii) and (iv), respectively. The requirement regarding the nonnegativity of the intensity and price vectors is straightforward.

Assumption (v) states that, in our closed model, the input requirements for period $t + 1$ cannot exceed the available amounts, that is, the outputs of the preceding period. Assumption (vi) merely takes account of the dynamic "way of operation" of the model, and accommodates to it the "competitive" assumption that no activity shall make a positive profit.

We have said before that our main concerns are the existence of a balanced growth mode and, furthermore, the conditions for its uniqueness. An answer to these questions requires first some knowledge about the technological and economic expansion feasibility of the system.

The analysis is done on the basis of inequalities (v) and (vi), respectively. To simplify the notation (the reader is advised to keep in mind the dynamic "way of operation" of the model) we drop the reference to time in the above-mentioned inequalities. We shall simply write, as in assumptions (iii) and (iv), $Ay \leq By$ and $pA \geq pB$.

A growth model with that feature, and satisfying the assumptions discussed so far, will be denoted by its technology matrix pair (B, A). In what follows we always refer to "the model (B, A)."

20.2 Technological Expansion

We start the analysis of the balanced growth properties of the model (B, A) by first focusing our attention on the problem of its technological expansion feasibility. Some basic concepts are defined next.

Definition 20.2 The model (B, A) is said to be expanding at a "technological expansion factor" at least equal to λ, if there exists a number $\lambda > 0$ and an intensity vector $y > \theta$ such that $By \geq \lambda Ay$.

Note, that in the preceding definition, it is only required that $\lambda > 0$. It is obvious that the definition also includes the case of a decaying economy, that is, the case $\lambda < 1$. From the economic point of view, however, it is easy to understand that the question is really relevant for expansion factors $\lambda > 1$.

Using the standard mathematical programming formulation, we can express the technological expansion problem as that of finding the largest factor λ compatible with the basic structural and nonnegativity constraints, that is,

$$\text{maximize} \quad \lambda \qquad \text{subject to} \quad (B - A)y \geq \theta \quad \text{and} \quad y \geq \theta.$$

The answer to whether (B, A) possesses an expansion factor and an associated vector with the desired properties is the object of the following existence theorem:

Theorem 20.1 (Existence)† For a model (B, A), there exists a positive technological expansion factor λ^*.

Proof Consider the inequality $(B - \lambda A)y \geq \theta$. For some $y \geq \theta$ we must have $By \geq \theta$. Thus, for some sufficiently small $\lambda > 0$, we have $(B - \lambda A)y \geq \theta$. On the other hand, taking a sufficiently large $\lambda > 0$, we must be able to obtain $(B - \lambda A)y < \theta$. The problem reduces, therefore, to choosing the least upper bound of the set $\{\lambda \mid \lambda > 0, (B - \lambda A)y \geq \theta\}$. Calling this number λ^* and y^* the associated intensity vector, the theorem is proved since, certainly, $(B - \lambda^* A)y^* \geq \theta$.

Observe the different economic meaning of a factor for which $(B - \lambda A)y > \theta$ and one for which $(B - \lambda A)y = \theta$. In the first case, we clearly have waste

† Adapted from Gale [3].

of some commodities, namely, of those for which production exceeds the amounts in which they are required as inputs. Only in the second case can we speak of a balanced technological expansion. Note, however, that no economically meaningful answer can be given to the problem before studying the behavior of prices.

In the literature, the vector y^* corresponding to the expansion factor λ^* is called an "optimal intensity vector."

20.3 Economic Expansion

We discuss next the counterpart of the problem previously analyzed. Let us define, once more, some essential concepts.

Definition 20.3 The model (B, A) is said to be expanding at an "economic expansion factor" at most equal to μ, if there exists a number $\mu > 0$ and a price vector $p > \theta$ such that $pB \leq \mu pA$.

As in the technological expansion problem, we can use here, too, a mathematical programming formulation to condense the object of our inquiry. The preceding definition clearly indicates that we confront here a minimization problem. Our problem is now,

$$\text{minimize} \quad \mu \qquad \text{subject to} \quad p(B - \mu A) \leq \theta \quad \text{and} \quad p \geq \theta.$$

The answer to whether (B, A) possesses an expansion factor and an associated price vector with the mentioned properties is given in Theorem 20.2. Since the proof is analogous to that given for Theorem 20.1, it is left here as an exercise.

Theorem 20.2 (Existence)† For a model (B, A) there exists a positive economic expansion factor μ^*.

It follows that μ^* is the greatest lower bound of $\{\mu \mid \mu > 0, p(B - \mu A) \leq \theta\}$. The associated price vector will be denoted p^*. For μ^* and p^* we clearly must have $p^*(B - \mu^* A) \leq \theta$. Note that p^* is the "optimal price vector."

20.4 Von Neumann's Theorem‡

We have already completed our separate analysis of the two sides of the problem. We know that a model (B, A) always has a positive technological expansion factor λ^* and a positive economic expansion factor μ^*. Associated

† Adapted from Gale [3].
‡ Theorems 20.3 and 20.4 in this section adapted from Gale [3].

with these, we have the semipositive intensity vector y^* and price vector p^*, respectively. But our main problem is not yet solved, and we wonder what is the relation between the technological and the economic aspects.

Given the model (B, A), it seems natural to expect that the connection between the two aspects will find its expression in the relations between the factors λ^* and μ^*, and the vectors y^* and p^*. The theorem which follows gives a preliminary answer to the first question.

Theorem 20.3 For a model (B, A) it always holds that $\mu^* \leq \lambda^*$.

Proof Consider first the inequality $(B - \lambda^* A) \gg \theta$. That it has no non-negative solution results from the following argument. Suppose y' is a non-negative solution. We would have $By' \gg \lambda^* Ay'$. And, for some $\lambda^{**} > \lambda^*$, we would obtain $By' \geq \lambda^{**} Ay'$. This, however, would violate the property of λ^* of being a maximum. Now, from Theorem 18.5, we know that a vector $p > \theta$ exists such that $p(B - \lambda^* A) \leq \theta$, that is, $pB \leq p\lambda^* A$. Therefore, μ^* being a minimum by Theorem 20.2, it follows $\mu^* \leq \lambda^*$.

We are now able to construct a complete theory for a model (B, A). This is precisely the content of von Neumann's theorem, which we study next.

Theorem 20.4 For a model (B, A), there exists a number $\gamma > 0$ and vectors $y, p > \theta$ such that (1) $(B - \gamma A)y \geq \theta$; (2) $p(B - \gamma A) \leq \theta$; (3) $p(B - \gamma A)y = 0$.

Proof The proof of (1) and (2) follows immediately. From Theorem 20.3 we know that $\mu^* \leq \lambda^*$. We can thus take $\gamma \in [\mu^*, \lambda^*]$. It follows that γ, together with the vectors y^* and p^* will satisfy (1) and (2), respectively. Let us concentrate now on point (3). Take $pB \leq \gamma pA$. If $y > \theta$, then clearly $pBy \leq \gamma pAy$. Analogously, consider $By \geq \gamma Ay$. For $p > \theta$, we obtain $pBy \geq \gamma pAy$. From the two weak inequalities, it follows that we can only have $pBy = \gamma pAy$, that is, $p(B - \gamma A)y = 0$.

The theoretical significance of the preceding theorem is remarkable.

First, we have shown the compatibility of the results previously obtained regarding the technological and the economic expansion of a model (B, A). This is done in points (1) and (2) of Theorem 20.4 by proving the existence of a factor $\gamma \in [\mu^*, \lambda^*]$ for which both the technological and the economic constraints are satisfied. It should be pointed out that, in the general case, the economic expansion factor is at most equal to the technological expansion factor. Thus, between the extremes of the interval $[\mu^*, \lambda^*]$, any value of γ will be a solution. This implies that there will be pairs of intensity and price vectors, other than the "optimal pair" y^*, p^*, compatible with the expansion of the two aspects of the model. In other words, in the general case the solution is not unique. We shall return to this point in the next section.

Point (3) of Theorem 20.4 has an especially deep economic meaning. The statement says that, for the chosen $\gamma \in [\mu^*, \lambda^*]$, we will have $p(B - \gamma A)y = 0$. This leads us to examine two situations of special interest.

Suppose, first, that for some i, we have $B^i y - \gamma A^i y > 0$. Since $p(B - \gamma A)y = 0$ must be satisfied, this implies that, for this i, we must have $p_i = 0$. A standard economic interpretation of this result is that, for all those commodities produced in excess, prices are zero.

Suppose, next, that for some j, we have $pB_j - \gamma pA_j < 0$. Here too, the equation $p(B - \gamma A)y = 0$ has to be satisfied. For our particular j, therefore, we shall have $y_j = 0$. We may interpret this as stating that all those activities which show losses are not used.

20.5 Indecomposability and Duality

The main theorem of the preceding section put together the two pieces into which the problem had been split. This was mainly done by showing the existence of an expansion factor γ which simultaneously satisfies the technological and the economic constraints. In the literature, especially when the whole question is approached from the mathematical programming point of view, the technological and the economic problem are referred to as dual problems.

Theorem 20.4, however, did not guarantee the uniqueness of the factor γ. On the contrary, as we have seen, there is no unique γ in the general case. Any value between μ^* and λ^* is a solution of the system.

A most important special case occurs, on the other hand, when we include a stronger assumption regarding the technology of (B, A), namely, that it is indecomposable. An indecomposable technology means here that there exists no proper subset of commodities which can be produced using exclusively inputs from the same subset, that is, without using at least one input which is not in the subset.

The indecomposability of our system depends, of course, on the technology matrix pair (B, A). That B and A are semipositive matrices follows immediately from the initial assumption. The indecomposability of the system will be, therefore, a consequence of the relations between the zeros in the input matrix A and the output matrix B. In formal terms, we have the following definition.

Definition 20.4† Let $I = \{1, 2, \ldots, m\}$ and $J = \{1, 2, \ldots, n\}$ be the sets of row and column indices, respectively, of the matrices of the model (B, A),

† The content of this definition is essentially due to Weil [1].

and define the matrix $C = A + B$. The model (B, A) is said to be indecomposable if there is no partitions $\{I_1, I_2\}$ of I and $\{J_1, J_2\}$ of J such that $c_{ij} = 0$ for $i \in I_2$ and $j \in J_1$; it is said to be decomposable if such partitions exist; finally, it is said to be completely decomposable if it is decomposable and, additionally, $c_{ij} = 0$ for $i \in I_1$ and $j \in J_2$.

The preceding definition is a generalization of the concepts of Definition 19.1. The notion of decomposability is complicated here by the fact that we have two rectangular matrices to describe one system. However, the idea is mainly an extension of that of Chapter 19.

Let us continue our analysis by using the matrix $C = [c_{ij}]$.

If the model (B, A) is decomposable, we should be able to rearrange rows and columns of A and B such that

$$(1) \qquad \begin{pmatrix} A_{11} & A_{12} \\ Z & A_{22} \end{pmatrix} + \begin{pmatrix} B_{11} & B_{12} \\ Z & B_{22} \end{pmatrix} = \begin{pmatrix} C_{11} & C_{12} \\ Z & C_{22} \end{pmatrix}.$$

On the other hand, if (B, A) is totally decomposable, we would at least obtain

$$(2) \qquad \begin{pmatrix} A_{11} & Z \\ Z & A_{22} \end{pmatrix} + \begin{pmatrix} B_{11} & Z \\ Z & B_{22} \end{pmatrix} = \begin{pmatrix} C_{11} & Z \\ Z & C_{22} \end{pmatrix}.$$

Of course, decomposition could still be carried further, provided the subeconomies represented by C_{11} and C_{22} are decomposable.

Observe that to perform any of the above rearrangements we need two permutation matrices, of order $m \times m$ and $n \times n$, respectively. The same permutations should be performed on A and B simultaneously.

Let I_1 and J_1 represent the goods and activities, respectively, appearing in C_{11}. It is easy to see that in both Eqs. (1) and (2) these sets form an independent subeconomy. Of course, the same is also true in (2) for the goods and activities appearing in C_{22}.

The economic interest of the indecomposable technology case lies in the fact that (B, A) models, which satisfy this additional assumption, are shown to have a unique balanced growth factor γ^*. This is proved in the following duality theorem.

Theorem 20.5 (Duality)† For a model (B, A) with an indecomposable technology, $\gamma^* = \mu^* = \lambda^*$.

Proof From Theorem 20.3 we have already learned that, in general, $\mu^* \leq \lambda^*$. To prove the assertion of this theorem it only remains to show that, in the indecomposable case, it also holds that $\lambda^* \leq \mu^*$. First, we have $y^*, p^* > \theta$. Thus, $\lambda^* p^* A y^* \leq p^* B y^* \leq \mu^* p^* A y^*$, that is, $(\mu^* - \lambda^*) p^* A y^* \geq 0$.

† Adapted from Gale [3].

Consider next the product p^*Ay^*. Note, first, that for $y^* > \theta$ we must have $By^* \gg \theta$. Should this not hold, that is, should we have, for instance, $By^* > \theta$, then we would have that, for some i, $B^iy^* = 0$. Suppose, for a moment, this is really the case. Since for all i we have $B^iy^* \geq \lambda^* A^iy^*$, this would imply $A^iy^* = 0$. In economic terms, this result would mean that the ith commodity has neither been produced nor used by the intensity vector y^*. The system would clearly be decomposable, since the commodities could be separated into two independent sets, one containing only the ith commodity, and the other containing the remaining commodities. This, of course, violates our assumption. Hence, we must have $By^* \gg \theta$. Now, for $p^* > \theta$, we have $p^*By^* > 0$. On the other hand, $\mu^*p^*Ay^* \geq p^*By^*$ and $\mu^* > 0$. From this, it follows that $p^*Ay^* > 0$. Therefore, $\lambda^* \geq \mu^*$. From the latter result and $\mu^* \leq \lambda^*$, it easily follows that, in the indecomposable case, $\gamma^* = \mu^* = \lambda^*$.

In the indecomposable technology case, we have, according to the preceding theorem, a unique balanced growth factor γ^*. Associated with this factor, we have the unique pair of intensity and price vectors y^*, p^*.

If we remember from Sections 20.2 and 20.3 that the technological and economic questions can be expressed as a maximization and a minimization problem, respectively, then our recent result has a special property. The factor γ^* is at the same time the maximum of the first problem and the minimum of the second one.

EXERCISES

E.20.1 Prove Theorem 20.2.

Hint: The proof follows similar steps as in Theorem 20.1, but consider the greatest lower bound of $\{\mu \mid \mu > 0, p(B - \mu A) \leq \theta\}$.

E.20.2 Observe that if λ is considered the technological expansion factor of a model (B, A), the number $r = \lambda - 1$ may be regarded as its customary "rate of growth." Show how, analogously, $q = \mu - 1$ may be interpreted as the "shadow rate of interest" of the model (B, A).

E.20.3 The maximal rate of growth $\lambda^* - 1$ of a model (B, A) is equal to the rate of growth of the slowest expanding commodity. Explain why.

E.20.4 The minimal rate of interest $\mu^* - 1$ of a model (B, A) is equal to the rate of interest of the most profitable activity. Explain why.

E.20.5 Show that if the technology matrices of a model (B, A) satisfy the condition $A + B \gg Z$, the model is indecomposable.

Hint: This restrictive assumption was originally used by von Neumann; note that $A + B$ is strictly positive.

E.20.6 Give a pair of matrices of order 3×4 and 4×3 which determine decomposable models (B, A).

Hint: Use Definition 20.4 to determine the place of the zeros.

E.20.7 Give pairs of matrices of order 3×4 and 4×3 which, without satisfying the condition $A + B \gg Z$, still determine indecomposable models (B, A).

E.20.8 Let a model (B, A) be defined by the technology matrices

$$A = \begin{pmatrix} 0 & 1 & 0 \\ 1 & 0 & 0 \\ 0 & 0 & 1 \\ 0 & 1 & 0 \end{pmatrix}; \qquad B = \begin{pmatrix} 2 & 2 & 0 \\ 0 & 0 & 3 \\ 0 & 4 & 1 \\ 0 & 0 & 3 \end{pmatrix}.$$

Is the model indecomposable?

Hint: Use Definition 20.4.

E.20.9 Show that $p^* = (2, 1, 2, 1)$ and $y^* = (\frac{3}{4}, \frac{3}{4}, 1)^T$ are optimal price and intensity vectors, respectively, for the model of the preceding exercise. Find λ^* and μ^*, and see how this model satisfies the propositions of Theorems 20.4 and 20.5.

Hint: $\gamma^* = 4$.

E.20.10 Consider an indecomposable model (B, A). Show that (a) in the balanced growth path, the economy grows at a rate equal to the rate of interest $\gamma^* - 1$; (b) activities which at equilibrium prices yield a rate of return less than $\gamma^* - 1$ will not be used.

PROBLEMS

P.20.1 A special case of von Neumann's model occurs when A is $n \times n$ and B the $n \times n$ identity matrix I. Prove that for such a system, if η^* is the dominant characteristic value of A, $\gamma^* = (\eta^*)^{-1}$.

P.20.2 In the framework of a (B, A) model, consider the case when B consists of columns having one and only one nonzero element, the number one. Compare this model with the standard von Neumann model, and with the model referred to in P.20.1. Give the appropriate economic interpretation.

P.20.3 Define and study the function $\phi(p, y) = pBy/pAy$. Show that the relation $\gamma^* = \mu^* = \lambda^*$ occurs at a saddle point of ϕ, where $\phi(p^*, y^*) = \mu^* = \lambda^*$.

Hint: See von Neumann's original article [1], and also Allen's [1].

P.20.4 Consider the $m \times n$ matrix A $(n > m)$ of the model of P.20.2. Taking only one activity to produce each of the goods of the system, we can define a submodel of m goods and m activities with matrix $A^{(k)}$, where $A^{(k)}$ is an $m \times m$ submatrix of A. Note that the set of all such submodels is finite. Prove that the complete model of m goods and n activities has a unique expansion factor $\gamma^* = (\eta^*)^{-1}$, where η^* is the smallest of all the dominant eigenvalues of the matrices $A^{(k)}$. How many matrices $A^{(k)}$ are there possible?

P.20.5 A necessary and sufficient condition for $\gamma^* > 1$ in an indecomposable model of the type referred to in P.20.1 is that $\eta^* < 1$. What could you establish, analogously, for a von Neumann model of the general type?

NOTES

The main subject of this chapter was first discussed by von Neumann [1] and later generalized by Kemeny, Morgenstern, and Thompson [1], and Gale [4] (see also Howe [1]). Here we have essentially followed Gale [3]. Different treatments of the topic are presented by Karlin [1], Nikaido [4], and Morishima [1]. For a more general model of balanced growth see Gale [4] and Karlin [1]. The definition of an indecomposable rectangular system which we give here is based on the work of Weil [1, 2]. For a different definition of indecomposability see Gale [3].

REFERENCES†

ALLAIS, M.
* "Traité d'économie pure." Imprimerie Nationale, Paris, 1952.

ALLEN, R. G. D.
[1] "Mathematical Economics." Macmillan, New York, 1960.

ALEXANDROFF, P.
* "Einfachste Grundbegriffe der Topologie." Springer-Verlag, Berlin and New York, 1932
[2] "An Introduction to the Theory of Groups." Hafner, New York, 1959.

ALEXANDROFF, P., and HOPF, H.
[1] "Topologie," Vol. I. Chelsea, Bronx, New York, 1965.

APOSTOL, T. M.
[1] "Mathematical Analysis." Addison-Wesley, Reading, Massachusetts, 1963.

ARROW, K. J., and DEBREU, G.
[1] Existence of an equilibrium for a competitive economy. *Econometrica* **22**, 265–290 (1954).

† References preceded by an asterisk have not been cited in text and are included here either for their historical importance or as suggestions for further reading.

Arrow, K. J., Block, H. D., and Hurwicz, L.

[1] On the stability of the competitive equilibrium II. *Econometrica* **27**, 82–109 (1959).

Arrow, K. J., and Hahn, F.

[1] "General Competitive Analysis." Holden-Day, San Francisco, California, 1971.

Arrow, K. J., and Hurwicz, L.

[1] On the stability of the competitive equilibrium I. *Econometrica* **26**, 522–552 (1958).

Aumann, R.

[1] Markets with a continuum of traders. *Econometrica* **32**, 39–50 (1964).

[2] Existence of competitive equilibria in markets with a continuum of traders. *Econometrica* **34**, 1–17 (1966).

Begle, E. G.

[1] A fixed point theorem. *Ann. of Math.* **51**, 544–550 (1950).

Bellman, R.

[1] "Introduction to Matrix Analysis." McGraw-Hill, New York, 1960.

Berge, C.

* "The Theory of Graphs and Its Applications." Wiley, New York, 1962.

[2] "Topological Spaces, Including a Treatment of Multi-Valued Functions, Vector Spaces, and Convexity." Oliver & Boyd, Edinburgh, 1963.

Birkhoff, G.

[1] "Lattice Theory" (Amer. Math. Soc. Colloq. Publ., Vol. 25). Amer. Math. Soc. Providence, Rhode Island, 1940.

Birkhoff, G., and MacLane, S.

[1] "A Survey of Modern Algebra." McGraw-Hill, New York, 1957.

Bourbaki, N.

[1] "Eléments de mathématiques." Hermann, Paris, 1951.

Brouwer, L. E. J.

[1] On continuous vector distributions on surfaces. *In* "Proceedings, Akademie van Wetenschappen," Vols, 11–13, Amsterdam, 1909 and 1910.

Cairns, S. S.

[1] "Introductory Topology." Ronald Press, New York, 1961.

Champernowne, D. G.

* A note on J. v. Neumann's article on "A Model of Economic Equilibrium." *Rev. Econ. Studies* **13**, 10–18 (1945–1946).

Charnes, A., Cooper, W. W., and Henderson, A.

* "An Introduction to Linear Programming." Wiley, New York, 1953.

Chipman, J. S.

* The multi-sector multiplier. *Econometrica* **18**, 355–374 (1950).

Debreu, G.

[1] See Arrow and Debreu [1].

* Representation of a preference ordering by a numerical function. *In* "Decision Processes" (R. M. Thrall, C. H. Coombs, and R. L. Davis, eds.). Wiley, New York, 1954.

* Valuation equilibrium and Pareto optimum. *Proc. Nat. Acad. Sci. U.S.A.* **40,** 588–592 (1954).

[4] "Theory of Value: An Axiomatic Analysis of Economic Equilibrium" (Cowles Monograph 17). Wiley, New York, 1959.

[5] New concepts and techniques for equilibrium analysis. *Internat. Econ. Rev.* **3**, 257–273 (1962).

[6] On a theorem of Scarf. *Rev. Econ. Studies* **30**, No. 3 (1963).

DEBREU, G., and HERSTEIN, I. N.

[1] Nonnegative square matrices. *Econometrica* **21**, 597–607 (1953).

DEBREU, G., and SCARF, H.

[1] A limit theorem on the core of an economy. *Internat. Econ. Rev.* **4,** 235–246 (1963).

DIEUDONNÉ, J.

* "Foundations of Modern Analysis" (enlarged and corrected printing). Academic Press, New York, 1969.

DORFMAN, R., SAMUELSON, P. A., and SOLOW, R. M.

[1] "Linear Programming and Economic Analysis." McGraw-Hill, New York, 1958.

DUNFORD, N., and SCHWARTZ, J. T.

[1] "Linear Operators." Pt. I; "General Theory." Wiley (Interscience), New York, 1958.

EDGEWORTH, F.

[1] "Mathematical Psychics." Kegan Paul, London, 1881.

EILENBERG, S., and MONTGOMERY, D.

[1] Fixed point theorems for multi-valued transformations. *Amer. J. Math.* **68,** 214–222 (1946).

FADDEEVA, V.

[1] "Computational Methods of Linear Algebra." Dover, New York, 1959.

FAN, K.

* On systems of linear inequalities. *In* "Linear Inequalities and Related Systems" (H. W. Kuhn and A. W. Tucker, eds). (Ann. Math. Studies No. 38). Princeton Univ. Press, Princeton, New Jersey, 1956.

FANG, J.

[1] "Abstract Algebra." Shaum, New York, 1963.

FINKBEINER, D. T.

[1] "Introduction to Matrices and Linear Transformations." Freeman, San Francisco, California, 1966.

FRAENKEL, A.

[1] "Abstract Set Theory." North-Holland Publ., Amsterdam, 1953.

FRANZ, W.

[1] "General Topology." Ungar, New York, 1965.

[2] "Algebraic Topology." Ungar, New York, 1968.

FRECHET, M.

* "Les Espaces Abstraits." Gauthier-Villars, Paris, 1951.

FROBENIUS, G.

[1] Über Matrizen aus nicht-negativen Elementen. *S.-B. Kgl. Preuss. Akad. Wiss.* 456–477 (1912).

GALE, D.

[1] Convex polyhedral cones and linear inequalities. *In* "Activity Analysis of Production and Allocation" (T. C. Koopmans, ed.). Wiley, New York, 1951.

[2] The law of supply and demand. *Math. Scand.* **3**, 155–169 (1955).

[3] "The Theory of Linear Economic Models." McGraw-Hill, New York, 1960.

[4] The closed linear model of production. *In* "Linear Inequalities and Related Systems" (H. W. Kuhn and A. W. Tucker, eds.). Princeton Univ. Press, Princeton, New Jersey, 1956.

GANTMACHER, F. R.
[1] "The Theory of Matrices," Vols. I and II. Chelsea, Bronx, New York, 1960.
GOLDBERGER, A.
[1] "Econometric Theory." Wiley, New York, 1964.
HADLEY, G.
[1] "Linear Algebra." Addison-Wesley, Reading, Massachusetts, 1961.
HAHN, W.
[1] "Theorie und Anwendung der direkten Methoden von Lyapunov." Springer-Verlag, Berlin and New York, 1959.
HALANAY, A.
[1] "Differential Equations: Stability, Oscillations, Time Lags." Academic Press, New York, 1965.
HALL, D. W., and SPENCER, G. L.
* "Elementary Topology." Wiley, New York, 1955.
HALMOS, P.
[1] "Naive Set Theory." Princeton Univ. Press, Princeton, New Jersey, 1960.
[2] "Finite Dimensional Vector Spaces." Van Nostrand-Reinhold, Princeton, New Jersey, 1958.
HAUSDORFF, F.
[1] "Set Theory." Chelsea, Bronx, New York, 1957.
HAWKINS, D., and SIMON, H. A.
[1] Note: Some conditions of macroeconomic stability. *Econometrica* **17**, 245–248 (1949).
HERSTEIN, I. N., and MILNOR, J.
[1] An axiomatic approach to measurable utility. *Econometrica* **21**, 291–97 (1953).
HICKS, J. R.
* "Value and Capital." Oxford Univ. Press (Clarendon), London and New York, 1939.
HILDENBRAND, W.
[1] Existence of equilibria for economies with production and a measure space of consumers. *Econometrica* **38**, 608–623 (1970).
HOWE, C. W.
[1] An alternative proof of the existence of general equilibrium in a von Neumann model. *Econometrica* **28**, 635–639 (1960.)
HUREWICZ, W., and WALLMAN, H.
* "Dimension Theory." Princeton Univ. Press, Princeton, New Jersey, 1941.
INTRILIGATOR, M. D.
[1] "Mathematical Optimization and Economic Theory." Prentice-Hall, Englewood Cliffs, New Jersey, 1971.
KAKUTANI, S.
[1] A generalization of Brouwer's fixed point theorem. *Duke Math. J.* **8**, 457–459 (1941).
KARLIN, S.
[1] "Mathematical Methods and Theory in Games, Programming and Economics," Vol. I. Addison-Wesley, Reading, Massachusetts, 1959.
[2] Positive operators. *J. Math. Mech.* **8**, 907–937 (1959).
KELLEY, J. L.
[1] "General Topology." Van Nostrand-Reinhold, Princeton, New Jersey, 1955.
KELLEY, J. L., *et al.*
* "Linear Topological Spaces." Van Nostrand-Reinhold, Princeton, New Jersey, 1963.

KEMENY, J. G., MORGENSTERN, O., and THOMPSON, G. L.
[1] A generalization of the von Neumann model of an expanding economy. *Econometrica* **24**, 115–135 (1956).

KEMENY, J. G., MIRKIL, H., SNELL, J. L., and THOMPSON, G. L.
[1] "Finite Mathematical Structures." Prentice-Hall, Englewood Cliffs, New Jersey, 1958.

KNASTER, B., KURATOWSKI, K., and MAZURKIEWICZ, S.
[1] Ein Beweis des Fixpunktsatzes fur *n*-dimensionale Simplexe. *Fund. Math.* **14**, 132–137 (1929).

KOOPMANS, T. C.
[1] "Three Essays on the State of Economic Sciences." McGraw-Hill, New York, 1957.
[2] Analysis of production as an efficient combination of activities. *In* "Activity Analysis of Production and Allocation" (T. C. Koopmans, ed.). Wiley, New York, 1951.

KUHN, H. W., and TUCKER, A. W.
* *In* "Contributions to the Theory of Games I" (H. W. Kuhn and A. W. Tucker, eds.). Princeton Univ. Press, Princeton, New Jersey, 1953.

KURATOWSKI, K.
* "Topologie," Vols. I and II. Pańštwowe Wydawnictwo Naukowe, Warsaw, 1958 and 1961.
[2] "Introduction to Set Theory and Topology." Pergamon, Oxford, 1961.

LANCASTER, K.
[1] "Mathematical Economics." Macmillan, New York, 1968.

LANG, S.
[1] "Linear Algebra." Addison-Wesley, Reading, Massachusetts, 1966.

LANGE, O.
[1] "Introduction to Econometrics." Pergamon, Oxford, 1962.

LEFSCHETZ, S.
* "Introduction to Topology." Princeton Univ. Press, Princeton, New Jersey, 1942.

LEONTIEF, W. W.
[1] "The Structure of the American Economy." Harvard Univ. Press, Cambridge, Massachusetts, 1941.

LIPSCHUTZ, S.
[1] "General Topology." Schaum, New York, 1965.

MCKENZIE, L.
[1] On the existence of general equilibrium for a competitive market. *Econometrica* **27**, 54–71 (1959).

MENDELSON, B.
[1] "Introduction to Topology." Allyn & Bacon, Rockleigh, New Jersey, 1962.

MORGENSTERN, O.
[1] "Economic Activity Analysis" (O. Morgenstern, ed.). Wiley, New York, 1954.

MORISHIMA, M.
[1] "Equilibrium, Stability and Growth." Oxford Univ. Press, London and New York, 1964.
[2] Existence of solution to the Walrasian system of capital formation and credit. *Z. Nationalökomomie* **20**, 238–243 Nos. 1–2 (1960).

MURATA, Y.
[1] An alternative proof of the Frobenius Theorem. Tech. Rep. No. 2. Workshop in

Math. Econ., Econometrics and Optimization, Dalhousie Univ., Halifax, Nova Scotia, 1971. (A new version will be forthcoming in the *Journal of Economic Theory*, **5**, No. 2).

NIKAIDO, H.

[1] On the classical multilateral exchange problem. *Metroecon.* **8**, 135–145 (1956).

[2] A supplementary note to "On the Classical Multilateral Exchange Problem." *Metroecon.* **9**, 209–210 (1957).

[3] "Introduction to Sets and Mappings in Modern Economics." North-Holland Publ., Amsterdam, and Amer. Elsevier, New York, 1970.

[4] "Convex Structures and Economic Theory." Academic Press, New York, 1968.

OWEN, G.

[1] "Game Theory." Saunders, Philadelphia, Pennsylvania, 1968.

PARETO, V.

* "Cours d'economie politique." Rouge, Lausanne, 1896–1897.

* "Manuale di economia politica." Soc. Editrice Libraria, Milano, 1906.

PATTERSON, E. M.

* "Topology." Oliver & Boyd, Edinburgh, 1956.

PEASE, M. C.

[1] "Methods of Matrix Algebra." Academic Press, New York, 1965.

PERRON, O.

[1] Zur Theorie der Matrizen. *Math. Ann.* **64**, 248–263 (1907).

QUIRK, J., and SAPOSNIK, R.

[1] "Introduction to General Equilibrium Theory and Welfare Economics." McGraw-Hill, New York, 1968.

RADER, J. T.

[1] Edgeworth exchange and general economic equilibrium. *Yale Econ. Essays* **4**, No. 1 (1964).

[2] "Theory of General Economic Equilibrium," Academic Press, New York, 1972.

SAMUELSON, P. A.

* "Foundations of Economic Analysis." Harvard Univ. Press, Cambridge, Massachusetts, 1947.

* *In* "Collected Scientific Papers" (J. Stiglitz, ed.). MIT Press, Cambridge, Massachusetts, 1966.

SCARF, H.

[1] An analysis of markets with a large number of participants. *In* "Recent Advances in Game Theory." Princeton Univ. Press, Princeton, New Jersey, 1962.

SHUBIK, M.

[1] Edgeworth market games. *In* "Contributions to the Theory of Games" (A. W. Tucker and R. D. Luce, eds.), Vol. IV. Princeton Univ. Press, Princeton, New Jersey, 1959.

SPANIER, E. H.

* "Algebraic Topology." McGraw-Hill, New York, 1966.

SPERNER, E.

[1] Invarianz der Dimensionszahl und des Gebietes, *Abh. Math. Sem. Univ.* Hamburg **6,** 265–272 (1928).

SUPPES, P.

[1] "Axiomatic Set Theory." Van Nostrand-Reinhold, Princeton, New Jersey, 1960.

THRALL, R., and TORNHEIM, L.

[1] "Vector Spaces and Matrices." Wiley, New York, 1957.

TITCHMARSH, E. C.
[1] "The Theory of Functions." Oxford Univ. Press, London and New York, 1939.

TOMPKINS, C. B.
[1] Sperner's lemma and some extensions. *In* "Applied Combinatorial Mathematics" (E. F. Beckenbach, ed.). Wiley, New York, 1964.

UZAWA, H.
[1] Note on the existence of an equilibrium for a competitive economy. Mimeographed. Stanford Univ., Stanford, California, 1956.
* Preference and rational choice in the theory of consumption. *In* "Mathematical Methods in the Social Sciences" (K. J. Arrow, S. Karlin, and P. Suppes, eds.). Stanford Univ. Press, Stanford, California, 1960.

VAIDYANATHASWAMY, R.
[1] "Set Topology." Chelsea, Bronx, New York, 1960.

VAN DER WAERDEN, B.
[1] "Modern Algebra," Vols. I and II. Ungar, New York, 1950.

VIND, K.
[1] Edgeworth allocation in an exchange economy with many traders. *Internat. Econ. Rev.* **5**, No. 2 (1964).

VON NEUMANN, J.
[1] A model of general economic equilibrium. *Rev. Econ. Studies* **13**, 1–9 (1945).

WALD, A.
[1] On some systems of equations of mathematical economics. *Econometrica* **19**, 368–403 (1951).

WALLACE, A. H.
* "An Introduction to Algebraic Topology." Pergamon, Oxford, 1957.

WALRAS, L.
[1] "Elements of Pure Economics." Irwin, Homewood, Illinois, 1954.

WEIL, R. L.
[1] The decomposition of economic production systems. *Econometrica* **36**, 260–278 (1968).
[2] Solutions to the decomposable von Neumann model. *Econometrica* **38**, 276–280 (1970).

INDEX

Asterisk preceding page number indicates that reference is to be found in footnote.

D

E